Energy Law and Policy

Edited by
Usha Tandon

OXFORD
UNIVERSITY PRESS

Oxford University Press is a department of the University of Oxford.
It furthers the University's objective of excellence in research, scholarship,
and education by publishing worldwide. Oxford is a registered trademark of
Oxford University Press in the UK and in certain other countries.

Published in India by
Oxford University Press
2/11 Ground Floor, Ansari Road, Daryaganj, New Delhi 110 002, India

ISBN-13: 978-0-19-948297-9
ISBN-10: 0-19-948297-7

Typeset in Dante MT Std 10.5/13
by The Graphics Solution, New Delhi 110 092
Printed in India by Replika Press Pvt. Ltd

To
my beloved Anil
for
regulating my energy for a sustainable relationship

Contents

Part II   National Perspectives

Foreword

The undertaking of this comprehensive study of the challenges of establishing laws that are globally sustainable, just, and effective to provide for the expanding energy needs for economic development, while preserving the environment for present and future generations, makes an important contribution to this most important of global issues.

Usha Tandon has brought together essays of various international energy experts to illuminate these international and national energy law challenges. These experts include the renowned Klaus Bosselmann and several experts from India and some developing countries, to explore the difficulties and possible solutions for their countries' domestic energy laws.

The study posits fundamentally that continued predominant reliance on fossil fuels to meet increasing energy for development demands threatens the future of the planet from the consequences of climate change and the health of human populations from pollution. It points to renewable energy and energy efficiency as the most promising solutions. It discusses the need for nuclear energy to achieve sustainability goals, but faces the accident dangers, the inadequacy of International Atomic Energy Agency (IAEA) regulations for inspections and safeguards, and the possibilities of uranium exhaustion. For all considered solutions, the obstacles are discussed as well as the opportunities; no easy magic bullets are proposed, and the compromises required for adoption are addressed with candour. There is a lot of food for thought.

The study proceeds to consideration of international laws to cope with the legal requirements to resolve these challenges, dealing with the inadequacies of international law enforcement powers, the tensions between them and national sovereignty demands, the elevation of the private marketplace and eco-taxes ostensibly to handle all social and economic problems, and the enormous powers of giant international companies to preserve their industrial and profit ambitions at the expense of the environment; but the study points to ever more evident and costly environmental crises that must educate all these competing forces to recognize that none of their ambitions will be possible without making compromises to preserve the atmospheric commons.

It points to the desirability of utilizing the WTO a policy enforcement mechanisms with respect to trade in energy, and its possible role in resolving the conflict between the need to protect intellectual property to promote energy technology innovation and the need to make advanced technologies available to developing countries.

It goes on to discuss the great importance of subnational laws and the private sector to deal with these challenges, and the fact that often state, provincial, city, and local governments can do, and are doing, more than their national governments to address these issues. But it points out that national and state government support is often required (or lack thereof harmful) in this regard.

The study observes with beautiful eloquence that the atmosphere is a global commons for the benefit of all mankind that must not be privatized or commercialized, and that economic growth is not achievable without energy security and justice.

Lastly, the study examines in depth the laws of India, pointing to the salutary but perhaps overambitious goals of the newly adopted laws to provide a transition from fossil fuels to efficiency and renewable energy, and the need for more effective implementation provisions. It then demonstrates with the laws of Mauritius and Nigeria the great problems experienced by smaller developing countries in making this transition.

Tandon has done a superb job of editing this opus, not surprising considering her membership on the editorial board of the *US-China Law Review* and position as editor-in-chief of the *Journal of the Campus Law Centre*; also, given her extensive background and writings in environmental law and her receipt of several awards including the Max Plank Fellowship, Germany in 2011. She is superbly well-qualified to undertake this challenging undertaking and she does it with style.

Richard L. Ottinger
Dean Emeritus, Elizabeth Haub Law School, Pace University,
New York, USA

Acknowledgements

First and foremost, I am grateful to all the authors who, supporting my idea to bring out a book on this upcoming area of energy law, warmly and enthusiastically contributed for this volume. Having worked on climate change and biodiversity law, I realized that the broader issues of mitigation of climate change or conservation of biodiversity cannot be addressed without understanding the legal issues involved in the use of various forms of energy. I discussed the viability of this project with Professor Jariwala from whom I got electrifying positive response, which was a wonderful beginning for me. When I conversed with Professor Baxi, his amazingly encouraging gesture boosted my energy to move faster. Professor Klaus could not turn down my request this time and sent me his keynote address on atmosphere, which was then further developed by him along with Dr Lakshmanan, connecting it with energy for cleansing the atmosphere. Professor Rosencranz readily accepted my proposal for his contribution to this volume. Professor Wani, Professor Ansari, Professor Kumar, and Mr Upadhyay, also happily agreed to write for this volume. Dr Erimma and Dr Mahadew have been writers for me in my work for climate change. Both of them spontaneously consented to contribute, presenting the energy law in Nigeria and Mauritius. In the month of February 2016, as a director and then the professor-in-charge of Campus Law Centre, University of Delhi, India, I organized an international conference, in which one of the sessions was on energy law. The papers presented by Ms Chandralekha, Ms Pattajoshi, and Ms Verma at this session were selected for this volume and later got revised and updated from the authors. Mr Gupta and Ms Tandon immediately grabbed the opportunity when I opened it for them. I lovingly and gratefully acknowledge the patience and understanding of all authors for seriously attending the clarifications, and editorial comments during various rounds of editing of this volume. The grace with which the senior authors honoured the timeline was blissful for me for which I extend my special thanks to them.

I am thankful to my colleagues Ms Sinha, Ms Kaur, and Mr Anand for checking the manuscript from various angles of editing. The checking and cross-checking of endnotes, complete information in endnotes, uniformity of citations, and proof reading have been done by Neeraj Kumar Gupta, Akash Anand, Priti Rana, Nancy Dhuma, and Pratibha Tandon. I cordially appreciate their efforts and extend my thanks to them. The revision and final setting of the manuscript has been wonderfully and meticulously done by Mr Gupta who worked very systematically and methodologically. He has been associated with this project right from its beginning. I hugely appreciate his hard work and sincerity exhibited throughout the whole process of the editing of this book.

I am thankful to Dr Pushpa for giving his inputs on the title of the book and arrangement of various chapters under different parts. I also thank Professor Klaus for giving green signal to go ahead with the title and plan of contents.

I sincerely thank Professor Richard L. Ottinger for graciously writing the foreword to the book, highlighting the importance of the volume in dealing with the global issues of energy needs.

Last, but not the least, I am thankful to Oxford University Press, India for undertaking the task of publishing this book.

Introduction

This volume argues that the discourse on various issues relating to energy law has yet to pick up serious legal academic exercise, especially in India. Based on this argument, it seeks to contribute, albeit little, to the understanding of some aspects of energy law—both in international and national settings—such as global climate governance, energy trade, energy justice, energy tax, IPR in clean energy, the linkages between energy and sustainable development, energy and natural resources. It also discusses domestic laws of India, Mauritius, Nigeria, and China on energy conservation, clean renewable energy, and nuclear energy.

The volume explains that the need for energy would increase in future and dependence on fossil fuels alone can be disastrous in the long run. The fast-depleting conventional energy sources, the concerns on clean technology, and most recently the visible impacts of climate change make a strong case for the use of renewable energy as a mitigating tool to combat climate change. The book argues that it is time to look into the existing legal imperatives with more seriousness to frame a robust comprehensive legal framework on renewable energy at the domestic levels. At the same time, it warns that the renewable energy technology also has negative impacts on environment, especially on habitat loss of biodiversity—flora and fauna—and noise nuisance issues which can adversely impact the human health and well-being of other life forms. Therefore, before establishing mega renewable energy plants, a detailed environmental impact study must be undertaken to assess their impact on local environment. The volume emphasizes the need for research and development in this context to reduce their environmental impacts.

Though the source of clean energy for sustainable development, indeed, the renewable energy source, it is nonetheless supplemented with alternative energy sources where instead of fossil fuels some other substances are used. The chapters in the volume examine that among the alternatives, nuclear energy is a prominent choice as it has the potential to address energy needs of fast-growing economies and burgeoning population. In this context, they provide a critical overview of international regulation of nuclear energy focusing on the specific role of International Atomic Energy Agency (IAEA) and investigate, at the domestic level, into certain pertinent questions, including legal control and monitoring of nuclear power generation in India and China. The various forms of energy, whether derived from fossil fuel, nuclear substances, or renewable sources of energy, need conservation and its efficient use in various sectors to meet the increasing demand of energy. The book contains a comprehensive approach to energy conservation which refers to efficient management in the use of energy. Energy conservation requires that in order to use energy more efficiently and rationally, those measures should be adopted which are technologically feasible, economically rational, and environmentally and socially acceptable.

The volume discovers that there is no specific international energy law to regulate energy trade. It urges that increasing needs of energy and privatization of state-owned entities, the relationship of energy trade and environmental pollution, debate on sustainable development, and so on, are strong factors to compel World Trade Organization (WTO) to adopt some policy framework towards the regulation of trade in energy. The tautness between strong Intellectual Property (IP) rights to promote clean technology innovations and enabling global access to these technologies has been examined in the book as analogous to the scuffle between IP rights to protect pharmaceutical inventions and the claim for access to life-saving drugs in developing countries.

It vehemently argues that atmosphere is a global commons for the benefit of everyone on this planet and cannot be privatized or commercialized. It strongly argues, further, that the shift, from human rights of individuals, which are universal to the human rights of multinationals, which are trade-related and market-friendly, is now being severely challenged. It raises serious concerns over the 'creative destruction' that privatize commons in the name of development in this Anthropocene era. It argues, furthermore, that it is indeed impossible to achieve economic growth unless the energy security and justice is made the goal of all energy-related policies.

Having briefly summarized the core contents of the book, I now introduce to the readers, in a chronologically manner, the key issues and arguments put forward by authors in their chapters.

Klaus Bosselmann and Pushpa Kumar Lakshmanan in their chapter, 'The Atmosphere as a Global Commons and Cleansing It with New Energy Options', examine atmosphere as a global commons from the perspective of energy to emphasize that atmosphere, an important global commons, faces critical problem by human intervention. Global climate change is the classic example of indiscriminate air pollution caused to the atmosphere due to faulty energy choices that emitted greenhouse gases and ozone depleting substances. The authors establish the normative basis for declaring atmosphere as a global commons for the use of all on this earth that cannot be nationalized, or otherwise appropriated by some at the expense of others. It offers a new thinking on governing atmosphere as a global commons in the era of global warming and climate change due to swelling atmospheric pollution. The chapter probes ideas for improved climate governance through incremental reforms with the help of climate rent and carbon tax, the necessity of global cooperation, regional and national level actions, and the possibility of trusteeship governance through a transformational approach from state-centred governance to Earth-centred governance. As energy choices hold the key to minimize atmospheric pollution, this chapter provides new energy options for cleansing the atmosphere in the context of global climate change.

Upendra Baxi in his chapter titled 'Making Sense of Energy and Natural Resource Law in the Anthropocene Epoch' raises certain important questions on Energy and Natural Resource Law, such as, what sense do we make of 'natural resources' and permanent sovereignty over natural resources? What does it mean to say juristically and politically that the 'people' or the 'state' own these in these days of neo-liberal secular theology of free markets and authoritarian global governance? What may be the ways forward for efficient usage of natural resources in human rights and justice-friendly ways? What sense do we make of the notions of sustainable 'development' in the Anthropocene? The author maintains that the existing international law allows immunity of multinational corporations and other entities from human rights obligations and responsibilities and allows even green washing in geoengineering, or climate engineering, called by common consent as the field of deliberately manipulating the Earth's climate to alleviate climate change. While discussing in detail the issue of sovereignty over natural resources, the author claims that the transition from universal human rights of all individual human beings to TRMF (trade-related, market-friendly) human rights of multinationals and their cohorts is now being severely challenged. He argues that we clearly need to rethink law and justice from the Anthropocene standpoint and concludes, inter alia, that in the Anthropocene era, we need to consider how to make impossible the illegal 'contemporary enclosures' that privatize the erstwhile commons in the name of development, alias 'creative destruction'.

Usha Tandon discusses international regulation of nuclear energy with special reference to International Atomic Energy Agency (IAEA) in 'Regulation of Nuclear Energy for Sustainable Development: A Critical Overview of International Regime with Special Reference to IAEA'. The author explains that the focus of the international community has always been to ensure that nuclear energy is used peacefully and safely. Though the primary responsibility for the regulation of the use of nuclear energy rests with individual nations, however as other countries may be affected as well, it has been recognized that the regulation of nuclear energy necessitates the international regulation to ensure, inter alia, uniformity of standards, coordination, pooling of resources and services, as well as compliance. In this respect, IAEA, has played a pivotal role. She outlines global legal framework for nuclear energy's safe and peaceful development; provides an overview of the Statute of International Atomic, discussing various conventions for the safe use and handling of nuclear materials, and critically analyses IAEA in terms of its leadership and absolute efficiency. She argues that nuclear safety measures must be carried out with utmost transparency and coordination among the countries and international organization and all states using nuclear energy must give due regard to the provisions contained in various conventions.

The international energy law to regulate energy trade is explored by V. Chandralekha in her chapter, 'International Energy Law and WTO: Issues and Challenges'. The author describes as to how the production and transmission of energy at the international level is a complex operation which often involves both goods and services. All countries are interdependent, yet the energy trade at the international level is mainly regulated by the domestic law. There is no international energy law as such to regulate energy trade, though there are many Conventions and Treaties at the international level to protect environment and prevent climate change, through energy issues. From the point of international law, all these agreements regulate the international energy market, but the establishment of WTO has created a lot of confusion by ignoring the energy trade totally. Importantly, some of the energy exporting countries are not the members of WTO. She highlights the position of energy market before and after the establishment of WTO, the impact of WTO upon the international energy trade, and the challenges faced by the WTO in establishing policy framework for energy trade. She discusses the role played by the WTO in regulating the energy market and argues for the adoption of some policy framework for the regulation of trade in energy.

In 'Intellectual Property in the Way of a Clean and Green Environment: Is Licensing the Solution?', intellectual property issues for clean energy are analysed by Nikita Pattajoshi and Akash Kumar. There are myriad views about whether IP presents a barrier to technology transfer and diffusion in the field of climate change mitigation by use of clean and energy-efficient technology. While on the one hand, it is believed that strong IP rights and protection urges continued growth of a clean technology industry, on the other hand, such strong rights would impede developing countries' access to the most efficient available technologies. In this context, the authors analyse the existing standards and flexibilities under various legal frameworks in general and TRIPS in particular. They also study the current patterns of technology transfer in clean energy technologies. They argue that in this context, the perceived tension between relying on strong IP rights to promote innovation in clean technology and enabling global access to these technologies can be seen analogous to the tussle between strong IP rights to protect pharmaceutical inventions and fight for access to life saving drugs in developing countries.

Armin Rosencranz along with Rajnish Wadehra, Neelakshi Bhadauria, and Pranay Chitale critically analyses India's renewable energy targets for 2022 in their chapter, 'Clean Energy in India: Supply and Prospects'. The authors explain that energy supply has always been a major concern for India. The need to shift to renewable energy is greater now than ever. This need led Prime Minister Modi and his cabinet to set an ambitious goal at the Global Climate Conference in Paris, in December 2015, of 175 gigawatt (GW) of renewable energy to be achieved by 2022. The major part of this target is to be achieved through solar energy (100 GW). The authors address the policy changes taken up by the Government of India since the declaration of the targets and argue as to how unrealistic these targets are as measured by the progress made to date. The chapter starts by giving an overview of India's energy policy, partly under the Electricity Act of 2003 and addresses the resistance against renewable energy, especially solar energy, by people in India. The authors further point out the problems with setting unrealistic goals and suggest certain policy changes to achieve sustainable development through the use of renewable energy and conclude by addressing the alternatives to solar energy to achieve the nation's renewable energy goals, which may not require high investments and costs.

Sanjay Upadhyay in 'Renewable Energy Development in India: The Need for a Robust Legal Framework' emphasizes the need for developing renewable energy to compliment conventional energy sources in India. The author claims that the rapid industrialization and urbanization, the fast-depleting conventional energy sources, the growing need for green and clean technology, ill effects of climate change demand the enactment of a comprehensive legislation for renewable energy. The advantages of RE sources as indigenous non-polluting and virtually inexhaustible resources, especially in this uncertainty of global climate change make a fit case for promoting renewable energy. It also provides national energy security at a time when decreasing global reserves of fossil fuels threatens the long-term sustainability of the Indian economy. India has a vast supply of renewable energy resources, and thus is also a major energy producer and consumer. It is the eleventh largest economy in the world and is poised to make tremendous economic strides over the next decade or so, with significant development already in the planning stages. The author argues that it is time to look into the existing legal imperatives with more rigour to frame a robust comprehensive legal framework on renewable energy in India.

Erimma Gloria Orie highlights the legal impediments in the development of renewable energy in Nigeria in her chapter, 'Examining the Legal Impediments in the Development of Renewable Energy in Nigeria'. The author explains that presently, energy production from the sources such as, solar, hydro, biomass, biogas, and wind energy is abysmally low despite efforts of government to increase renewable energy generation. She finds that there are several legal issues and challenges that tend to thwart the Nigerian government's efforts. Some of the obstacles include policy and regulatory barriers, non-existing structure for power purchase agreements and institutional framework. She takes up the comparison of China that has an impressive renewable energy credentials. Its newly installed capacity of renewable energy RE power surpassed its combined newly installed capacity of fossil fuel and nuclear power for the first time. China accounts for 61 per cent of renewable energy in developing countries and its total investment is higher than that of all the European countries. Therefore, the author argues that for Nigeria to attain optimal development in line with the global search for green energy, the Chinese experience presents credible lessons for emulation.

Abdul Haseeb Ansari, in his chapter, 'Sustainable Energy for Sustainable Development: World View on Nuclear Energy with Special Reference to India', discusses nuclear energy in the context of sustainable development. He clarifies that it is a clean source of energy but may not be sustainable as uranium is an exhaustible natural resource; nonetheless, it is persuasive because it has potential to fulfil energy needs of fast growing economies. Furthermore, this energy source suffers from a lot of inhibitions which may or may not be true and some of which have been compounded by the recent Fukushima Daiichi nuclear disaster. On the other hand, the author explains, that countries like India and China have been compelled to agree to alleviate carbon emissions and like other countries, they also have submitted their emission reduction targets to the United Nations and are already working on it. But they are suffering from acute power shortage causing sufferance to their people and impeding their economies. In view of this, the author argues that these countries and some others countries in similar situations have no choice but to resort to nuclear power generation. In view of all these, the author discusses certain pertinent multi-disciplinary issues, including legal control and monitoring of nuclear power generation.

C.M. Jariwala presents a critical overview of Indian Law on Energy Conservation. The author, in his chapter titled 'The Indian Energy

Conservation Law: A Critical Overview', first demonstrates that as the development process starts, it requires more and more consumption of energy resulting in its scarcity, attracting its inefficient and unsustainable use. Such an approach further poses the problem for the environment and in turn challenges for the climate change, resulting in a question mark on the very existence of life on this planet. And therefore, conservation of energy becomes the need of the time. The author first, analyses the Constitution of India, which has yet to come out, according to him, of the traditional contour and change with the changing time, and then provides a critical review of Indian legislation of 2001 on conservation of energy. While examining the powers and functions of regulatory body under the Act, that is, Bureau of Energy Efficiency (BEE), he emphasizes that BEE must effectively coordinate with the activities of different role players for the efficient use of energy. The author also discusses the comparative energy conservation law in other countries to find out where India stands and argues that all those dealing, handling, or consuming energy must be educated about the energy and energy law, so that a green energy discipline and conservation culture are internalized in their behaviour and conduct.

Shivananda Shetty and Surender Kumar explain in 'Information Disclosure: A Policy Tool for Managing Environmental and Energy Challenges' as to how the information disclosure has been effectively used as a policy mechanism to improve the corporate environmental and energy compliance behaviour across countries. Market forces influence the firms' behaviour and motivate them to produce energy and environmentally efficient products. The authors look at the status of current policies which aim at influencing firms' behaviour and the effectiveness of the policies in achieving the desired objective and further identify the gap in the current information gathering process and suggest options to improve the same to enable the regulator to effectively implement information disclosure policy.

Amar Roopanand Mahadew assesses the legal framework of Mauritius on sustainable energy and the environment. The chapter titled, 'Assessing the Legal Framework of Mauritius on Sustainable Energy: Is It Robust Enough to Achieve the Dream of "Mauritius: A Sustainable Island"?' provides an overview of energy sector in Mauritius to understand the energy demands and requirements of the country. The diverse laws and regulations governing energy in Mauritius have been thoroughly gauged. The author argues that the issue of sustainable development through efficient energy use is so pressing, principally, with the effects of climate change that a document with a binding force than simple policy papers from the relevant ministries are the need of the hour and concludes with making some recommendations for the improvement and amendment to legislation that can be brought to achieve the dream of 'Mauritius: A Sustainable Island' through sustainable energy.

In the chapter, 'Eco-Tax on Energy Resources: A Critical Appraisal with Special Reference to India', Neeraj Kumar Gupta and Pratibha Tandon analyse the concept of eco-tax, which is essentially an economic concept and revolves around the theory of externality to argue for the levy of eco-tax as a measure to address the issue of externality to protect the environment. They further discuss international environmental law which provides that financial instruments including eco-tax should be utilized for the purposes of implementing the 'polluter pays principle'. The authors describe various types of environment taxes and the difference between eco-tax and environment related taxes. After this, they analyse the Indian environmental jurisprudence, arguing that it is growing into a mature legal regime experimenting various methodss for environmental protection, including financial measures.

M. Afzal Wani explores the role of law in the prevention of excessive production, excessive use and misuse of energy as given the emerging legal culture at the national and global levels the expectations from law are momentous. Building the correlation among humanity, energy, and law, the author in his chapter, 'Humanity, Energy, and Law: Urgencies and Challenges', argues that cultivation of ethics and proper ethical standards would be greatly supportive to the cause. He points out that a kind of dialogue among common people, academics, and the policy makers is informally and formally going on in the whole world discussing issues of energy security, impact of unguided production and use of energy and failure of nations to forge consensus on pressing issues. He gives a detailed account of Indian legislation to reflect the trends so far in the area of energy laws and concludes that legislations so far are more are less skeletal, symbolic, and inadequate. He argues that regulation has to be made largely by law to ensure saner and safer use of resources and careful production of energy and emphasises the need of reliable research in this regard.

Maansi Verma's chapter, 'Achieving Redistributive Energy Justice: A Critical Analysis of Energy Policies of India', discusses various facets of energy justice and critically analyses the Energy Policies of India for achieving Redistributive Energy Justice. She starts with identifying the magnitude of the problem where facts and narratives are given to show the various dimensions of injustice. Thereafter, the author delves deeper into the inter-connection between energy justice and other essential variables namely economic development, energy security, and environmental concerns and then deals with the current policy scenario to evaluate it on the energy justice parameter. She opines that it is indeed impossible to achieve economic growth unless issues of energy poverty are addressed and argues for making the energy justice the goal of all energy related policies.

Abbreviations

AEC	Atomic Energy Commission
AERB	Atomic Energy Regulatory Board
ALARA	As Low As Reasonably Achievable
ASADI	African Science Academy Development Initiative
AT&C	Aggregate Technical and Commercial Losses
BARC	Bhabha Atomic Research Centre
BAU	Business As Usual
BCAC	Building Control Advisory Council
BEE	Bureau of Energy Efficiency
BITs	Bilateral Investment Treaties
BOS	Balance of System
BPO	Business Process Outsourcing
BSE	Bombay Stock Exchange
CA	Chartered Accountant
CAC	Command and Control
CAGR	Compounded Annual Growth Rate
CDP	Carbon Disclosure Project
CEA	Central Electricity Authority
CEB	Central Electricity Board
CERC	Central Electricity Regulatory Commission
CFI	Capital Finance International
CHP	Combined Heat and Power
CLND Act	Civil Liability for Nuclear Damage Act
CMIE	Centre for Monitoring Indian Economy
CNCPC	China National Cleaner Production Centre

CNG	Compressed Natural Gas
CoP	Conference of Parties
CP	Cleaner Production
CPCB	Central Pollution Control Board
CREN	Council for Renewable Energy, Nigeria
CRZ	Coastal Regulatory Zones
CSC	Convention on Supplementary Compensation for Nuclear Damage
CTE	Consent to Establish
CTO	Consent to Operate
DAE	Department of Atomic Energy
DDG	Decentralized Distributed Generation
DISCOs	Electricity Distribution Companies
DPR	Department for Petroleum Resources
EA	Energy Auditor
ECN	Energy Commission of Nigeria
ECT	Energy Charter Treaty
EEA	Energy Efficiency Act
EEMO	Energy Efficiency Management Office
EEZ	Exclusive Economic Zone
EIA	Environment Impact Assessment
EIA/IEO	Energy Information Administration/ International Energy Outlook
EM	Energy Manager
EPA	Environment (Protection) Act
EPO	European Patent Office
EPSRA	Electric Power Sector Reform Act
EPZ	Emergency Planning Zone
ESAs	Ecologically Sensitive Areas
ESLs	Environmental Survey Laboratories
ESMAP	Energy Sector Management Assistance Program
ESTP	Economic and Social Transformation Plan
ESTs	Environmentally Sound Technologies
EU	European Union
FAA	Finance and Audit Act
FAME	Faster Adoption and Manufacturing of Hybrid and Electric Vehicles in India

FAO	Food and Agricultural Organization
FCEVs	Fuel Cell Electric Vehicles
FEC	Federal Executive Council
FIT	Feed-in-Tariff
FMP	Federal Ministry of Power
FMST	Federal Ministry of Science and Technology
FT	Fischer-Tropsch
FYP	Five-Year Plan
GATS	General Agreement on Trade in Services
GATT	General Agreement on Tariffs and Trade
GDP	Gross Domestic Product
GEC	Green Energy Corridor
GHG	Greenhouse Gas
GNP	Gross National Product
GPCB	Gujarat Pollution Control Board
HDI	Human Development Index
HPU	Health Physics Unit
HS Nomenclature	The International Convention on the Harmonized Commodity Description and Coding System
IAEA	International Atomic Energy Agency
IAEA GSR Part-1	Governmental Legal and Regulatory Framework for Safety
IAEA-SF-1	Fundamental Safety Principles
IBRD	International Bank for Reconstruction and Development
ICLEI	International Council for Local Environment Initiatives
ICRP	International Commission on Radiological Protection
ICTSD	International Centre for Trade and Sustainable Development
IEA	International Energy Agency
IFC	International Finance Corporation
IHDS	India Human Development Survey
IIP	Indian Institute of Petroleum
IICT	Indian Institute of Chemical Technology
ILI	Indian Law Institute

ILO	International Labour Office
IMF	International Monetary Fund
INDCs	Intended Nationally Determined Contributions
IOCL	Indian Oil Corporation Limited
IPCC	Intergovernmental Panel on Climate Change
IREDA	Indian Renewable Energy Development Agency
IRRS	Integrated Regulatory Review Services
ISO	International Organization for Standardization
ITO	International Trade Organisation
IUCN	International Union for Conservation of Nature
LAA	Land Acquisition, Rehabilitation, and Resettlement Act, 2013
LFN	Laws of the Federation of Nigeria
MAB	Marginal Abatement Benefit
MAC	Marginal Abatement Cost
MCM	Ministerial Council Meeting
MDGs	Millennium Development Goals
MEPU	Ministry of Energy and Public Utilities
MFN	Most Favoured Nations
MNRE	Ministry of New and Renewable Energy
MOEFCC	Ministry of Environment, Forests and Climate Change
MOSPI	Ministry of Statistics and Programme Implementation
MRET	Mandatory Renewable Energy Target Scheme
MRT	Mass Rapid Transit
MRV	Monitoring, Reviewing, and Verification
MSI	Maurice Ile Durable / Mauritius—Sustainable Island
MSIPSAP	Maurice Ile Durable Policy, Strategy, and Action Plan
MTs	Climate Change Mitigation Technologies

MYTO	Multi-year Tariff Order
NAFTA	North American Free Trade Agreement
NAMA	Nationally Appropriate Mitigation Action
NAPCC	National Action Plan on Climate Change
NASPA-CCN	National Adaptation Strategy and Plan of Action for Climate Change Nigeria
NBPI	Nigerian Biofuel Policy and Incentives
NCCPRS	National Climate Change Policy and Response Strategy
NCEF	National Clean Energy Fund
NDMA	National Disaster Management Authority
NDRC	National Development and Reform Commission
NEA	National Energy Administration
NEC	National Energy Commission
NEEDS	National Economic Empowerment and Development Strategy
NEP	National Energy Policy
NEPA	National Electric Power Authority
NERC	Nigerian Electricity Regulatory Commission
NESCO	Nigerian Electricity Supply Company
NGO	Non-governmental Organization
NITI AAYOG	National Institute for Transformation of India Aayog
NIWE	National Institute of Wind Energy
NMEEE	National Mission on Enhanced Energy Efficiency
NNPC	Nigerian National Petroleum Corporation
NPA	Non-performing Asset
NPC	National People's Congress
NPCIL	Nuclear Power Corporation of India Limited
NPPs	Nuclear Power Plants
NPT	Nuclear Non-Proliferation Treaty
NREAG	National Renewable Energy Advisory Group
NREC	National Renewable Energy Committee
NREDA	National Renewable Energy Development Agency

NREEEP	National Renewable Energy and Energy Efficiency Policy
NSG	Nuclear Suppliers Group
NSS	National Sample Survey
OECD	Organization for Economic Co-operation and Development
OPEC	Organization of the Petroleum Exporting Countries
PACE	Partnership to Advance Clean Energy
PACP	Presidential Action Committee on Power
PAT	Perform Achieve and Trade
PAU	Punjab Agricultural University
PHCN	Power Holding Company of Nigeria
PPA	Power Purchase Agreement
PPP	Public-Private Partnership
PSNR	UN Declaration of Permanent Sovereignty over Natural Resources
PSU	Public Sector Undertakings
PTFP	Presidential Task Force on Power
R&D	Research and Development
RCC	Reinforced Cement Concrete
RE Committee	National Renewable Energy Committee
REA	Rural Electrification Agency
REAP	Renewable Electricity Action Programme
REC	Renewable Energy Certificate
REL	Renewable Energy Law
REMP	Renewable Energy Master Plan
REPF	Renewable Energy Procurement Framework
REPG	Renewable Electricity Policy Guidelines
RESIP	Draft Rural Electrification Strategy and Implementation Plan
RGGVY	Rajiv Gandhi Grameen Vidyutikaran Yojana (Rajiv Gandhi Rural Electrification Scheme)
RPO	Renewable Purchase Obligation
RTI	Right to Information
S&L	Standards and Labelling Program
SC	Scheduled Caste

SCM Agreement	Agreement on Subsidies and Countervailing Measures
SCP	Sustainable Consumption and Production
SDA	Sustainable Development Act
SDGs	Sustainable Development Goals
SECC	Socio-economic and Caste Census
SECI	Solar Energy Corporation of India
SERC	State Electricity Regulatory Commission
SHP	Small Hydro Power
SIDS	Small Island Developing States
SNAs	State Nodal Agencies
SPCB	State Pollution Control Board
SRES	Special Report on Emissions Scenarios
ST	Scheduled Tribe
STCA	State Trading Corporation Act
SYNGAS	Synthesis Gas
TEPCO	Tokyo Electric Power Company
TERI	The Energy and Resources Institute
TPES	Total Primary Energy Supply
TRI	Toxic Release Inventory
TRIPS	Trade Related Intellectual Property Rights
UDAY	Ujwal DISCOM Assurance Yojana
UNAEC	United Nations Atomic Energy Commission
UNCED	United Nations Conference on Environment and Development
UNDP	United Nations Development Programme
UNEP	United Nations Environment Programme
UNESCO	United Nations Educational, Scientific and Cultural Organization
UNFCCC	The United Nations Framework Convention on Climate Change
UNGA	United Nations General Assembly
UNSC	United Nations Security Council
UNSCEAR	United Nations Scientific Committee on the Effects of Atomic Radiation
URA	Utility Regulatory Authority
URAA	Utility Regulatory Authority Act

USEPA	United States Environmental Protection Agency
VGF	Viability Gap Funding
WCED	World Commission on Environment and Development
WHO	World Health Organization
WIPO	World Intellectual Property Organization
WNA	World Nuclear Association
WSSD	World Summit on Sustainable Development
WTO	World Trade Organisation

INTERNATIONAL PERSPECTIVES

PART ONE

INTERNATIONAL
PERSPECTIVES

KLAUS BOSSELMANN
PUSHPA KUMAR LAKSHMANAN

The Atmosphere as a Global Commons and Cleansing It with New Energy Options

The four elements of life—earth, water, air, energy—are common property of the human race. The management and use of such portions thereof as are vested in or assigned to particular ownership, private or corporate or national or regional, of definite or indefinite tenure, of individualist or collectivist economy, shall be subordinated in each and all cases to the inherent interest of the common good.

—Preliminary Draft of a World Constitution, 1947–8.

The Global Commons and the Law

Atmosphere is the protective shield and primary source of life of the planet Earth. The following pertinent observation of Susan Buck on atmosphere needs introspection:

> The atmosphere is the envelope of gases that surrounds our planet. Held in by gravity, propelled outward by centrifugal force, it shields us from the vacuum of space and the sun's radiation. It provides water and oxygen to sustain life on Earth. It is a highway for commerce and recreation

and a canvas for heart-stopping displays of beauty. *Our contribution to this wonder, when we think about it at all, is pollution.*[1] (emphasis added)

Atmosphere is an evolutionary creation of nature for over 4.5 billion years.[2] It has been shared and used in common by all living creatures across the globe and states without any restrictions or discrimination. However, intensive pollution of atmosphere for the past few centuries played havoc to this vital source of life.

Intuitively, the atmosphere is a global commons. Somehow we all know that the atmosphere belongs to no one, cannot be owned by anyone in a legal sense and must be available to every person, each country, and at all the time. In fact, before global climate change became an issue, in the 1980s, the environmental literature typically described areas outside national jurisdiction (oceans, atmosphere, outer space, and Polar regions) as 'shared natural resources' or 'global commons'.[3] There was a general assumption that these areas cannot in any manner be nationalized, privatized, commercialized, or otherwise be appropriated by some at the expense of the others.

This assumption rested in the long-standing tradition of all cultures that one cannot own, what essentially belongs to all. In Western legal history, the concept of 'the commons' dates back to the Roman law distinguishing things that can be owned individually (*res privatum*) from things which are common to all (*res publicum*). By definition, the commons required a form of governance that ensured their sustained availability for all. Governance of this kind was inevitably grounded in a sense of stewardship or trusteeship.[4] It can't be any other way, if you are asked to manage what belongs to someone else, you are essentially acting as a trustee for the benefit of someone else.

The relationship between *res publicum* and the governance of it based on trusteeship is an essential part of public law doctrines (especially in civil law traditions, and less so in common law traditions such as New Zealand or India). This understanding has never changed and is probably still being taught in many public law courses.

What has changed is the law itself. In most jurisdictions, the thinking of *res privatum* dominates what essentially belongs to *res publicum*. We need to think of the massive wealth accumulated by privately owned assets over a period of more than the last 25 years, to be able to appreciate the significance of private properties and how effectively it isolates itself from the responsibilities towards the commons. The big divide between the increasing wealth of a few and the decreasing wealth of many (including the states) is, of course, the result of economic liberalization and free market ideology. But lesser-known, is perhaps, the specific role that the law plays in this process.

Legal theorists widely accept that the law reflects the dominant values of any society, but also entrenches them. Moral values and legal norms of a society cannot be entirely separated from each other, as political liberalism would have it, they are in a dialectic relationship. This means that a hard law not only trumps soft morality, but it changes the morality itself and in doing so, it shapes new laws. What has happened over the last few decades is an incremental change of public morality that favours acceptance of market demands and private goods (*res privatum*)–over the governance of common goods (*res publicum*). With respect to the global commons such as the atmosphere, this shift of morality has meant that the atmosphere has been treated as a res nullius, a legal no-man's sphere, a void that has been filled with open access to anyone wishing to deposit carbon and other greenhouse gases (GHG) into.

The point we are concerned with in this chapter is not how badly things went wrong or how the states failed to negotiate an effective climate regime—the recently adopted Paris Agreement on Climate Change too has not changed that attitude much[5]—but how rational and effective the global climate governance architecture could look like in future.

The ideas of 'historical contributions' and 'common but differentiated responsibilities and respective capabilities (CBDR)' still steer the negotiation strategies of many states. They fail to look beyond these contested notions. Even though the principle of CBDR strives to fix responsibilities or duties on the industrialized states, it primarily promotes a rights based approach of respective states. In all practical terms, the developing and least developed states promote an argument that reserves a right for them to pollute the atmosphere in the name of their development. This argument demonstrates that, 'earlier you holed our ship that is drowning presently, and we should have legitimate right to hole it now'. This argument is absurd and unacceptable when all of us jointly travel in the same ship and the same is already sinking. States do not realize the fact that putting more holes in the ship will only take them to inescapable abyss. All states should come forward to plug the holes of our sinking planet instead of introducing more holes that will destroy their 'common future'. Of course, development of nations is necessary, but it should not be at the cost of destruction. Alternative means of development have to be seriously pursued to have meaningful development.

A new line of thinking is needed now to save the atmosphere from its destruction. In this time of crisis where climate change is taking a heavy toll on all living organisms, and drastically changing the nature of the global atmosphere, 'a duty based approach' on the part of all states is necessary, instead of justifying their historical or future rights to pollute. It is time that our notions of CBDR and sustainable development are overhauled with open mind. Though the fossil fuel based energy sources could serve as cheap propellants of development, all countries should take advantage of the advancements in renewable energy technologies available now and apply them in their developmental plans. Rather than insisting on rights based approach, states must think in terms of their duty to protect the atmosphere collectively that will indeed provide larger avenues of development in the long run.

The disregard for global commons is not just a physical, but legal phenomenon, so chances are that the law—as a cultural and social construct—can make a difference to the way we govern ourselves. Considering that public morality can change, and given that we can build on centuries-old traditions of effective governance of the commons, it is at least conceivable that an international climate regime can be based on the notion of the global commons. It should be built on effective atmosphere-friendly–sustainable energy options.

New Legal Approaches to Govern the Global Commons

In view of the drastic adverse impacts occurring in the global atmosphere, there is a need to evolve new legal norms and adopt regulatory approaches to save the future of all life forms on this planet.

Common Concern of Mankind

The United Nations General Assembly Resolutions on Protection of Global Climate for present and future generations of mankind[6] welcomed the proposal of Malta for conservation of climate as a part of the common heritage of mankind in 1988 and recognized climate change as a common concern of mankind. The General Assembly called on the member states to collaborate with each other and inter-governmental organizations to take timely actions to deal with climate change issues. This approach has to be harnessed well so that the atmosphere can be safeguarded in a holistic way. International community should also understand that besides climate change, the global atmosphere itself becomes a common concern of mankind. This is not only for the atmosphere, many other natural resources also need to be looked at through the lens of common concern of mankind. There is a growing awareness to protect different kinds of natural domains such as, land, water, and so on, (not to be called merely resources) which are on the verge of complete destruction. The movement to protect freshwater resources is a case in point.[7] Maude Barlow strongly argues that:

> *No one owns water.* Social justice and human rights groups, local com-
> munities, small farmers, peasants, and indigenous peoples fighting to
> maintain traditional control over their *lands*, and a growing movement
> around the world to stop the privatization of areas *once considered the*
> *common heritage of humanity* … assert that *water is a commons that belongs*
> *to the earth, other species, and future generations as well as our own.* Because
> it is a flow source necessary for life and the health of the ecosystem, and
> because *there is no substitute for it, water must be regarded as a public good* to
> be preserved as such for all time in both law and practice.[8]

This same argument and analogy applies to the global atmosphere. In this knowledge era, it is imperative for states to share the common concerns and to find out new ways of protecting the global commons in order to ensure a meaningful living for everyone that respects quality of each other's life.

Erga omnes Obligations and Jus cogens

One of the possible ways to save the planet and the global commons is to clearly articulate the nuances of erga omnes obligations and jus cogens to include protection of unique natural sources like atmosphere within their ambit.

The celebrated contribution of the International Court of Justice (ICJ) in the *Barcelona Traction* case is worth mentioning here. The court in that case observed that 'an essential distinction should be drawn

between the obligations of a state towards the international community as a whole, and those arising vis-à-vis another state'.[9] According to
the ICJ, the obligations of a state towards the international community
are the concerns of all states and they all have obligations erga omnes
to protect them.[10] At that time when the *Barcelona Traction* case was
decided, the prominent issues were outlawing of acts of aggression,
protection of basic human rights, genocide, racial discrimination, and
slavery. Hence, the court applied the erga omnes obligations on the
states to deal with these issues.

Even though the facts and circumstances of the *Barcelona Traction*
case were different, the principle laid down by the ICJ in this case is
timeless and very relevant in the present context. In the contemporary
era, the obligations of a state towards the international community are
widening that calls for broadening the scope of erga omnes obligations.
Global climate change due to the excessive atmospheric pollution,
large scale biodiversity destruction, and desertification are some of the
pressing issues plaguing the world today in addition to those already
highlighted in *Barcelona Traction* case. It is time that states consider these
issues under the umbrella of erga omnes obligations and save the planet
from its further destruction.

International peace and security cannot be achieved only by avoiding wars, but by providing happy and healthy living to all in a liveable
atmosphere. Protection of the atmosphere as a global commons should
therefore be recognized by the states as one of the erga omnes obligations in a time where emissions of GHG and global warming are at
their peak. By doing this, no state would surrender its sovereignty in any
manner, rather it will consolidate and strengthen its sovereignty to live
in a healthy planet.

Justice Weeramantry's separate opinion in the *Gabčíkovo* case is pertinent to note in this context as it suggested inclusion of environmental obligations within the ambit of erga omnes. Justice Weeramantry
opined that the 'great ecological questions will call for a thought' within
the realm of erga omnes and international environmental law should
move beyond the individual rights and obligations of state parties and
concentrate on 'the global concerns of humanity as a whole'.[11] This
observation deserves a serious thought by all concerned at this juncture.

Another important development worth taking into consideration
is the Resolution on Obligations erga omnes in International Law

('Obligations erga omnes Resolution') adopted by the Institut de Droit International (Institute of International Law), based on a report by Giorgio Gaja in 2005. Article 1 (b) of the Resolution specifically includes 'the obligations relating to the environment of common spaces'.[12]

The German Constitutional Court has gone a step further and viewed the 'basic rules for the protection of the environment' as part of jus cogens.[13] This is a very encouraging trend. It is time for climate change negotiators and the international community in general to recognize the fact that scarce and unique natural sources like atmosphere are fast losing their character and composition. Very soon the atmosphere may lose its nature as a life support system or will by itself emerge as disastrous climate change that will make the earth inhospitable for any kind of life forms. Again, it must be noted that the states must move ahead from general protection of environment to the specific approaches to protect the global commons, that is, the earth, water, air, energy, and so on, with new legal approaches and norms.

Public Trust Doctrine

The public trust doctrine was the first principle of legal control that recognized 'the universal notion of commons' and strived to protect certain natural resources such as air, water, and so on, which must be protected for common good. In this regard, the Codex Justinianus (AD 529) stated that: 'By the laws of nature, these things are common to all mankind: the air, running water, the sea and consequently the shores of the sea.'[14]

The public trust doctrine imposes three types of restrictions on the governmental authorities. The public property subject to the trust should be used only for public purpose and it should be available for the use of general public; such property cannot be sold, even for a fair cash equivalent; and, it should be maintained for particular types of uses.[15]

Though the public trust doctrine serves the common good to some extent, there are three problematic notions in public trust doctrine: first, the *state* is at the centre of regulation as a trustee and the public have no role in it; second, the vital things in common such as, air, running water, and sea are reduced to the level of *property*; third, the public are conceived as the *owners* of the so-called property. There are serious flaws in understanding the notion of public trust as 'property'. However, the doctrine has been applied and adjudicated in many cases in different jurisdictions differently.

Even if we assume that the state has the authority to regulate the global commons by virtue of its sovereign power, the state-citizen relationship under the public trust doctrine would still need to be recognized in order to include the responsibility of citizens to protect the global commons.

There is a need to have a serious relook at the public trust doctrine vis-à-vis the global commons from a different legal perspective. So far we perceive that states are at the centre of regulating the global trust (natural resources) for the benefit of their citizens. However, under this analogy, the state endeavours to prevent the usurpation or the misuse of the public resources by a select few. This argument is premised on the fact that the state as a sovereign power, can interfere and ensure that the global commons are for the use of public at large and not for the usurpation by anyone for private gains.

The same public trust doctrine could also be looked at from a different angle. In addition to the citizen-state relationship in maintaining and regulating the trust, the citizen-to-citizen responsibilities should be thought of. Public trust should also include responsible behaviour of citizens in maintaining the sanctity of common resources.

One may argue that energy may not qualify here as a global commons, as it is produced by private players and marketed as a commodity. Nevertheless, one needs to understand that the vast amount of energy that pervades the globe and the universe is a form of the global commons that needs to be respected and not restricted by anyone for any reason. In the liberalized market era, when we gear up to tap the solar energy and wind energy, a circumstance may emerge where access to sunlight and wind power could surface as justiciable issues. Though there will be no difficulty in recognizing the private ownership of 'generated energy', the states must ensure that access to energy sources available in nature shall not be subject to monopoly, discrimination, or deprivation.

Precautionary Principle

As far as emission of GHG and the resulting global climate change are concerned, the scientific uncertainty no longer remains an issue.[16] When certainty about the cause of climate change is clearly established, the applicability of precautionary principle or approach becomes defunct.[17] It is time that the states moved forward from precautionary principle and applied the principle of prevention and the polluter pays principle in dealing with atmospheric pollution. A strong liability regime for air pollution and climate change is the need of the hour to safeguard the atmosphere.

In addition to these principles, the states need to seriously reformulate their policies incorporating inter-generational equity and intra-generational equity in the wake of increasing adverse effects of climate change and the atmospheric pollution.

Need for New Renewable Energy Options to Cleanse the Atmosphere

Energy development is considered to be the barometer of economic growth.[18] But the question is that what kind of energy are we talking about? Is it a renewable form of energy that promises healthy and sustainable future? Or, are we talking about a form of energy that affects the health of everyone through the noxious air pollution and destroys the global atmosphere and the climate itself?

With the advent of industrial revolution, development was primarily propelled by coal and other petrochemical fuels. Since the invention of steam engine technology to the present day ultra-modern automobile technology, fossil fuels played a major role in transforming the developmental scenario of many states. Besides the transport sector, most of the industrial activities, power generation, and other mechanized operations depended on coal based energy or other forms of fossil fuels. One of the reasons for the widespread applications of coal based energy was its easy availability and affordability. Due to this very reason, unfortunately, all the technologies that were developed since sixteenth century onwards were mostly aimed at using coal or other forms of fossil fuels. The world would have witnessed a different kind of development trajectory if coal and other carbon based energy sources were not discovered by the human race in the first place. In such case, the inventions might have been based on eco-friendly renewable energy sources. It is also unfortunate that some of the early inventions of renewable energy sources were not encouraged properly, but rather suppressed. This scenario seems to continue globally even today.

Vigorous coal based inventions and developmental thrust of states led to enormous release of GHG and other noxious gases into the atmosphere. The Intergovernmental Panel on Climate Change states that since pre-industrial times, the level of carbon dioxide concentrations have increased by 40 per cent globally, mainly due to fossil fuel emissions.[19] It is alarming to know that, presently the atmospheric concentrations of carbon dioxide, methane, and nitrous oxide have increased at an extraordinary rate compared to the levels in the past 800,000 years.[20] Indeed, one of the prime reasons for this deplorable state of atmosphere today can be attributed to the wrong choice of energy for hundreds of years. The coal-based developmental process has taken heavy toll on atmosphere for over three centuries.[21]

The cost of mitigation and adaptation due to climate change appears to be too high for the states. Global efforts to share and build adequate renewable technologies and financial resources are still lacking in this direction. When the human genome project was visualized in 1990 to sequence the chemical base pairs of human DNA, soon it emerged as the world's largest collaborative biological project with the involvement of many countries such as the US, the UK, Japan, France, Germany, and China.[22] The direct impacts of this project included employment generation of about 711,000 direct job-years (about 3,825,500 job years including indirect and induced impacts) and a direct impact of combined personal income was estimated more than US$ 71 billion (about US$ 240 billion including indirect and induced impacts).[23] Such collaborative efforts amongst the states should be rolled out to invent cost effective renewable energy production systems that will improve the global energy scenario and cleanse the atmosphere.

Atmospheric global warming has become a reality and the adverse effects of climate change is going beyond the capacity of states to deal with. The costs of adaptation and mitigation are spiralling up. The United Nations Environment Programme's (UNEP) Adaptation Finance Gap Report 2016 estimates that 'the costs of adaptation could range from US$ 140 billion to US$ 300 billion by 2030, and between US$ 280 billion and US$ 500 billion by 2050'. This, in real terms, means 'two-to-three times higher than current global estimates by 2030, and potentially four-to-five times higher by 2050'.[24] In this situation, the developing countries would be the biggest sufferers in terms of health, quality of living, and sustainability.

The recently concluded Paris Agreement looked at this scenario and called for the cooperation of both developed and developing countries to shoulder their responsibilities. In order to bear the financial burden, the developed countries reiterated their 2020 commitment to mobilize US$ 100 billion per year for adaptation and mitigation until 2025.

As the experts say, switching to renewable energy and other alternative forms of clean energy is not just the best choice under the present circumstances, but it is our 'only option'.[25] This is for two reasons—one is that the reserves of conventional oil and gas resources are constantly depleting, and another reason is that the cost of meeting the damage caused by the conventional form of energy is becoming unmanageable for the states. The International Energy Agency (IEA) cautions that the world will witness the fall of 40–60 per cent of conventional oil and gas reserves by 2030.[26]

This calls for serious efforts to pursue new energy options in order to meet the growing energy demand. This will eventually contribute to cleanse the atmosphere and to avert drastic climatic changes worldwide. New energy alternatives will reduce the pollution pressure on the atmosphere and will provide a healthy and sustainable development for the states.

The Sustainable Development Goals (SDGs) committed to provide: (a) universal access to affordable, reliable and modern energy services; (b) improving the share of renewable energy; and, (c) doubling of energy efficiency by the year 2030.[27] The clarion call of the global community is—be it through the Paris Agreement on Climate Change or through the SDGs—to work towards reducing the dependence of states on fossil fuels and to adopt renewable forms of energy that will minimize GHG emissions and other forms of pollution.

Energy efficiency and energy saving are important at this juncture. Energy production loss, transmission loss, and injudicious means of energy utilization should be avoided with proper understanding because one unit of energy saved is much bigger than one unit of energy produced. One unit of energy saved not only signifies saving of one unit cost, but also the production cost, transmission loss, prevention of environmental pollution in generating that one unit, and prevention of pollution while using it. In addition, it also saves that one unit of energy for future use. This stresses judicious usage of energy by all concerned.

The solar energy, wind energy, hydroelectric power, and nuclear energy are seen as viable alternatives to the fossil fuels. Every geographical location may not ensure solar or wind energy. Wherever these forms of energy are available, they should be optimally tapped. However, the solar energy, bioenergy, and wind energy could not take off as expected due to their low efficiency rate. Researchers all over the world are trying to perfect solar and wind energy technologies for future utilization. The technology to develop other renewable forms of energy such as geothermal energy, ocean energy, and use of hydrogen to generate electricity are still in experimental stages. The other forms of energy production from waste, biomass, and tidal sources are in the pipeline. Energy from exothermic chemical reactions and from the actions of bacteria are some other probable options the scientists are exploring. While we think of new energy options, rather than debating only on renewable and non-renewable sources of energy, our focus should be on developing clean and green energy sources.

Even though nuclear energy is mooted as a source of clean energy by some quarters, the safety concerns associated with it make it unacceptable in many countries. After all, life of citizens is more precious than the amount of energy generated. Unless the nuclear science and the states reach that level of maturity where safety of nuclear operation and disposal of its wastes are convincingly assured as safe and the states ensure a credible system for not using the nuclear energy for military purposes, the use of nuclear technology shall always remain controversial.

The United Nations' (UN) efforts to kindle renewable energy development, global energy access, and energy efficiency are showing new direction in securing a healthy future.[28] The UN-Energy was established as an inter-agency collaborative knowledge network for energy in 2004. The Food and Agriculture Organization (FAO), UNEP, and United Nations Educational, Scientific and Cultural Organization (UNESCO) have collaborated with the UN-Energy in its mission to develop renewable energy. International efforts such as this will certainly go a long way in ensuring better energy and cleaner atmosphere soon. Studies also show that the World Trade Organization (WTO)'s law can also be used to promote clean energy policies in the WTO member countries. Experts claim that national policies aiming to reduce GHG can be brought into compliance with international trade law. This can be achieved through labelling electricity from renewable sources, trading systems for green certificates, and energy taxes.[29]

Paris Agreement on Climate Change

The success of Paris Agreement on Climate Change is its change in approach between Annex and non-Annex countries. With the voluntarily submission of Intended Nationally Determined Contributions (INDCs) by the states, for the first time, the world is united to work towards climate change mitigation and adaptation. With the adoption of the Paris Agreement, the gap between the developing and the developed countries is evaporating in mitigation and adaptation measures in climate change.

It is very encouraging to note that many states have committed to switch over to renewable energy options. This includes specific measures to improve access to clean energy within their states, encouraging investment schemes for renewable energy generation, and improvement of the grid infrastructure. Some states have also committed to achieve 100 per cent renewable energy supply for the electricity sector.[30]

One of the major challenges for global climate governance is the fragmented structure of climate, air pollution, and energy policies. Fragmentation of pollution control strategies and faulty energy choices are the primary reasons for unabated levels of atmospheric pollution. Energy, air pollution, and climate are intricately linked and they create a combined effect on atmosphere. In order to protect atmosphere, coordinated efforts are required to choose healthy energy options. Integrated policy solutions are crucial for maintaining atmospheric health and climatic stability.[31]

In the following sections, two different avenues based on global commons governance are explored. One is the avenue of incremental reforms usually advocated by climate economists and more in line with exiting practices (carbon pricing, state negotiations, market conform instruments) and the other avenue is a transformation from state-centred governance to Earth-centred governance. Both avenues are not mutually exclusive, provided that the limitations of state-centredness will eventually be overcome.

Improving Climate Governance

Ideas around improved governance of the atmosphere focus around three areas, carbon pricing and taxation, global cooperation, and national and sub-national action.[32]

Climate Rent and Carbon Tax

To an extent, we can judge emissions trading schemes–no matter how inefficient they have been so far–as an acknowledgement of the polluter pays principle. Underpinning the polluter pays principle is the concern for a commons to keep it protected from individual overuse. In reality, however, emission trading has been guided by the opposite concern. Many of its flaws around offsetting, permits, perverse incentives, and general inefficiency are rooted in concerns for the protection of property rights rather than the global commons.

Crucial here would be a global carbon budget that is tightly monitored and fairly allocated. If that happens, the character of governance could change from a state-negotiated minimal compromise ('business-as-usual') to governance of a global commons assuming some kind of common property of humanity. The 'owner' is all humanity and charges the user of its property in the same way as landlords charge their tenants. We can then meaningfully speak of a 'climate rent'.[33] The concept of climate rent reflects common ownership and common responsibility. If used intelligently, a climate rent–like a global carbon tax scheme–would not only generate huge revenue, but restrict the use of the atmosphere as a carbon sink. As a consequence, property titles of the owners of coal, oil, and gas would be devalued. Eventually, fossil fuels would become uneconomical.

The counter argument against such a global rent or tax scheme is the usual concern for property rights–which from a commons perspective—should not be an argument at all. By the very definition, private property interests are subordinate to the inherent interests of the common good. It is only from a libertarian perspective, this smacks of unlawful expropriation and communism. From any other perspective including legal history and contemporary constitutionalism, it is clear that private property is never absolute, but embedded in social relationships. The idea that private property in natural resources is ethically justified only if it serves the common good is as old as the very institution of private property itself. It goes back to Justinian's (AD 483–565) concept of property, sits among Thomas Aquinas' (1225–1274) 'first principles of action', and even conforms with John Locke's (1632–1704) so called 'enough-and-as-good' proviso which aims for an equal share of each person in the Earth's material resources.[34]

So there is nothing revolutionary in the idea to restrict private property for the sake of the common good. For example, Article 14 (2) of the German Basic Law (that is, the Constitution) says, 'Property entails obligations. Its use shall also serve the public good'. Article 14 (3) allows for 'expropriation ... for the public good', and Article 15 specifically provides that 'land [and] natural resources ... may for the purpose of socialization be transferred to public ownership'. Similarly, the concepts of 'eminent domain' and 'public interest' in the common law jurisprudence justify the state to acquire the private property for public good. The land acquisition laws are primarily devised to safeguard the public interest. There are also instances in some countries wherein the fundamental right to property under the Constitution was amended to make it a constitutional right subject to modification by a statute to enable the state to deprive of anyone's property in public interest. For example, Article 19 (1) (f) of the Constitution of India, 1950 that recognized the right to property as a fundamental right was amended by the 44th Constitution Amendment Act, 1978 to make it a constitutional right under Article 300-A.[35]

The ethical and legal argument for the priority of the common good over private property protection is indeed compelling.[36] What is lacking, of course, is the political support for it. And this is not just a matter of good political argument. We are talking about nothing less than trying to tame capitalism and reconcile it with an overarching concern for the atmosphere, the global commons, social justice, and ultimately, human survival. Currently, the dynamics of global capitalism and prospects for sustainability are at odds with each other, to say the least, so everything depends on a strategy that is both, ambitious and feasible. In this pursuit, we can now look at the challenge of global cooperation.

Global Cooperation

The slow and cumbersome process of international climate change negotiations is well known and subject to much criticism. At the core of the critique is the 'dilemma of international environmental law'.[37] On the one hand, it might be expected that the natural world forms a non-negotiable absolute for any agreement; on the other hand, nation states see the natural world merely as relative, competing with other concerns such as economic prosperity, political strategizing or downright power interests. This creates the dilemma of wanting to be seen as ambitious without actually being committed. Inge Kaul describes this phenomenon as a 'sovereignty paradox' explaining it in the following way, '[T]he paradox is that states, notably their governments, are losing policy-making sovereignty precisely because they hold on to conventional strategies of realising sovereignty, which make them shy away from international cooperation. If nations want to win back their policy-making capacity, only one path remains: cooperation.'[38]

It is indeed fascinating to see how, for example, New Zealand is losing its policy-making sovereignty, by eagerly pursuing free trade agreements[39] that could ultimately undermine social and environmental security, that is, the essence of state sovereignty. In stark contrast, New Zealand vehemently defends its climate sovereignty against global climate interests. Such a mismatch is, of course, absurd by any standards of logic or morality, but consistent with 'conventional strategies'[40] of putting the national interests over global interests.

Much has been written about the free rider problem,[41] about the prisoner's dilemma,[42] the chicken game,[43] and other obstacles[44] in international negotiations. The game theory literature is full of descriptions why consensus building among states is so difficult. Game theorists, however, also offer strategies for improvement. One popular idea is that the very existence of international environmental agreements may change the rules of the game. With new treaties and their associated bureaucracies, cooperation becomes more likely than non-cooperation.[45] Moreover, recent studies have shown that catastrophic events such as tornados, hurricanes, tsunamis, severe draughts, and floods can substantially increase the desire for cooperation.[46] In time, Mother Nature may well be the most powerful driver for cooperation.

Regional and National Actions

A promising upside of the failure of global climate change policy is that a number of regions adopt their very own climate policies. Frustrations with international passivity, especially of the United States (US), are endemic, perhaps most so in Europe, and quite possibly, unilateral actions are becoming more important than the negotiation process. The European Union (EU) has adopted a far-reaching package of climate policies and aims at reducing GHG emissions by 20 per cent in 2020 (relative to 1990) and an increase of renewable energies by 20 per cent in 2020.[47]

Germany, in particular, has adopted the famous 'energy U-turn', now commonly known as *Energiewende*. The goal is to simultaneously phase out nuclear energy and coal, and reduce GHG emissions by 40 per cent in 2020 and up to 95 per cent by 2050, relative to 1990. The *Energiewende* may be new, but its driving-forces go back to the anti-nuclear and Green movement since the 1970s. This eventually leads to a cultural shift and a firm commitment to a low-carbon economy that no government—no matter how conservative—could ever ignore. Rather independently from political parties and ruling governments, this cultural shift is real and arguably crucial for future prospects. Without a powerful articulation of public opinion, governments would not change.

Beyond the EU, we can point to China with its economic transformation towards more renewables and its highly developed industry of cheap production of solar panels, wind generators, and energy storage systems. Five Chinese cities and two provinces have set up an emissions trading scheme[48] and overall one gets the impression that China is more committed to a low-carbon economy than its counterpart the US. Overall, 118 countries in the world have adopted renewable energy targets through feed-in tariffs, renewable quotas, or investment programmes. China leads investment into renewables (US$ 52 billion), followed by the US (US$ 51 billion), and Germany (US$ 31 billion).[49]

Also important in this context is low-carbon technology. Countries like Germany, Finland, Sweden, The Netherlands, and, to some extent, China, Japan, and South Korea, are very much driven by technological innovation. They clearly see low-carbon technology as a win-win option: it reduces GHG emissions and increases international competitiveness.

A few examples for the sub-national level would establish this point. California recently put a cap-and-trade system in place and the Governor of the Washington state imposed an executive order to implement cap-and-trade programme, eliminate coal power, and fund green energy projects.[50] There are similar developments in other federal countries such as Canada and Germany.

Then there is the ever-increasing role that local communities and local governments play in combatting climate change. Typically, they do more than their national governments. There is a worldwide trend of big cities to improve urban transportation and reduce GHG emissions. To some extent, big cities lead the way towards a low-carbon future as they feel the pressures most directly in the form of traffic congestion and pollution.

Smaller communities have the advantage of more effective citizen engagement and can be highly innovative. Consider the example of German mid-size cities like Freiburg, Giessen, or Marburg. They have adopted local planning statutes that make it compulsory for every new built house to produce their own energy and become independent from the public grid. This is not communism, but communalism that is based on the principles and practice of communal ownership. Among the German communities, particularly in Waiheke Island, there is a very strong sense of caring for the commons. Such instances can be seen in many traditional societies. People want to have direct control over what happens in their communities and want to know how cultural commons such as schools, libraries, health institutions, or community centres, but also natural commons such as the public spaces, parks, air, water, and so on, are protected from the austerity and privatization ideology of central governments.

Overall, there has been a dramatic surge of local climate initiatives over the past decade that shifted the focus of climate policies from the global to the local. Christiana Figueres, Executive Secretary of the United Nations Framework Convention on Climate Change (UNFCCC), went so far as to speak of new metrics of the global conversation where no longer just states, but civil society and in particular, local communities engage in the actual decision-making process.[51]

Considering that nearly half of the world population is, in some form, represented in the global organization Local Governments for Sustainability, International Council for Local Environmental Initiatives (ICLEI),[52] the voice of local communities is indeed strong and able to elevate the global conversation to a new level.[53]

Overlooking the three mentioned areas of more effective climate governance, we can summarize them as playing fields for the development of governance of the atmosphere as a global commons. Each playing field has its own dynamics and we cannot be sure to what extent they are informed by a sense of actual governance of the global commons as distinct from traditional state-centred global governance. But they are relevant and deserve a lot more attention than what, for example, public media report about it.

It is simply not enough to assess global climate protection purely through the lenses of state negotiations. Somewhat cynically one could say that states are no longer in charge, they are bystanders in a world increasingly shaped by corporate forces, on the one hand, and active citizens and communities, on the other.

Trusteeship of the Global Commons

This brings us to the aforementioned transformational approach. As remarked at the beginning, governance of the global commons is inherently stewardship or trusteeship governance. As argued above, legal regimes around the commons have a long tradition in all cultures including the European history. What remains to be shown now is that trusteeship, an equally well-known concept and arguably the most effective form of caring for the global commons. Essentially, this idea promotes a shift away from state-centred to Earth-centred governance.[54] The difference between the two may appear as paradigmatic, but could also be seen as gradual, as Earth governance involves nation states, the United Nations, and global civil society alike.

Earth governance is a partnership model as perceived, for example, by the UN-initiated Commission on Global Governance. Its 1995 report *Our Global Neighbourhood*[55] calls for a 'global civic ethic' around social justice, intergenerational justice, and ecological integrity and recommends institutional reform based on a partnership between states and civil society. The Report specifically acknowledges the Earth Charter[56] as an overarching and all-inclusive framework and blames the rigidity of state sovereignty as the main reason for the failure of global governance.

It concludes, 'We propose, therefore, that the Trusteeship Council be given the mandate for exercising trusteeship over the global commons. Its functions would include the administration of environmental treaties ... It would refer any economic or security issues arising from these matters to the Economic Security Council or the Security Council.'[57]

Trusteeship over the global commons provides the basis to levy user fees, taxes, and royalties for permits to use the global commons. The global commons are defined to be 'the atmosphere, outer space, the oceans, and the related environment and life-supporting systems that contribute to the support of human life'.[58] This broad definition would give the UN authority to interfere with states if they do not comply with their duties to protect the environment within their borders. By implication, states are assumed to perform trusteeship functions with respect to both, their domestic territory that they control and to the global environment that they are part of.

Even though, the focus of this chapter is on atmosphere as global commons, atmosphere alone cannot be seen in isolation. As seen in the above definition, any global commons governance architecture should integrate the interlinking issues of atmosphere, outer space, oceans, Antarctica, and so on. As none of these global commons have any strong governance structure (barring a minimum level of oversight provided by the UN Convention on Law of the Sea, Antarctic Treaty System and the Outer Space Treaties), a holistic system has to be evolved for their governance.

The idea of trusteeship governance for the atmosphere is not new and has been further developed in a number of recent books.[59] The first author's book on the subject[60] particularly looks at the UN's own history and experience with trusteeship models to argue the feasibility of trusteeship. Apart from a possible revival of the dormant Trusteeship Council[61] that has been strongly advocated by several countries, in particular Malta, but at some point also by New Zealand, and there are other UN-affiliated bodies that perform trusteeship functions. Prominent examples include the World Health Organisation (WHO), which has its raison d'etre (reason for existence) not in serving states, but in serving human health. In this function, WHO has some far-reaching powers including setting of norms and standards and monitoring compliance by the states.

Even more powerful is the WTO with its raison d'etre of promoting free trade. It has the world's only judiciary body that can hold members states directly accountable and against their will to be tested in an international court of justice. WTO's foundational and organizational documents specifically refer to trusteeship as its key function. What this demonstrates is that the states have proven to be quite capable of conferring trusteeship functions to powerful global bodies. Obviously, the eradication and control of global diseases find the undisputed agreement of the states across the globe. Likewise, free trade is very high, if not at the top of states' concerns. It is therefore quite ironic, that the protection of the global commons—a prerequisite for human survival—seems to be of much less a concern globally.

The only plausible explanation for this could be the sheer weakness of the states to stand up against multinational corporate power. Simply because they cannot cope with the conflicting interests of climate protection, on the one hand, and economic dependence on corporates, on the other. Precisely, this weakness makes it unlikely for states to accept a new global body acting as a trustee for the atmosphere (or any other global commons). The US, in particular, is known for blocking every single initiative towards trusteeship governance, except of course, for those performed by the World Bank, IMF, and the WTO.

***

Global commons such as the earth, water, air, and energy are the foundation of everything on this planet as they constitute and sustain all animate and inanimate things. Therefore, it is suggested that a strong normative basis has to be evolved to safeguard the global commons on an urgent basis before it gets too late. Again, the common connection among the other global commons has to be understood well instead of looking at atmosphere in isolation. As sketched above, the trusteeship governance could turn out to be a viable option to effectively govern the global commons. We can either hope for gradual improvement of climate governance, even if it means that any real action may come too late, or we could insist on a radical turnaround now, despite the ongoing resistance of national governments. In practical terms, both strategies can possibly go hand-in-hand and it is observed that all the genuine reform proposals are rooted in the idea of stewardship or trusteeship for the global commons. Ultimately, however, it falls upon us, the members of the civil society—including the academic community—to push for a bottom-up transformation of the system of global climate governance. The global atmosphere is too important to leave its governance to politicians.

In this context, the choice and use of energy sources and their close linkages with the health of atmosphere cannot be underestimated.

The contemporary model of energy use has resulted in unmanageable climate crisis. This has ultimately brought a grave adverse impact on the atmosphere. New options of clean and green energy sources and a strong climate governance structure are imperative for the future of humanity. As discussed above, different new renewable energy options and carbon tax mechanisms provide hope for a healthy future. However, one needs to be too cautious in inventing and introducing new energy options into the global commons. As Garret Hardin put it, 'Every new enclosure of the commons involves the infringement of somebody's personal liberty.'[62] The present generation has to be vigilant in choosing energy options as they directly affect the atmosphere and global climate that will have significant effect on the needs of present and future generations. Needless to say that the first and primary source of life, air, in a broader sense, atmosphere, needs to be clean and safe if the states want to keep its citizens healthy and productive.

Notes and References

1. S.J. Buck, *The Global Commons: An Introduction*, 1st Edition. (London: Earthscan Publications Ltd., 2006), p. 111.

2. S.J. Mojzsis, 'Life and the Evolution of Earth's Atmosphere', in E.A. Mathez (ed.), *Earth Inside and Out* (New York: American Museum of Natural History, New Press, 2001), pp. 32–9, 33.

3. See, for example, 'The Global Commons' in *World Conservation Strategy*, Report by IUCN/UNEP/WWF (UNESCO, 1980), ch. 18; E. Ostrom, *Governing the Commons: The Evolution of Institutions for Collective Action* (Cambridge: Cambridge University Press, 1990); K. Bosselmann, *When Two Worlds Collide: Society and Ecology* (RSVP, 1995), pp. 275–81; P. Sand, 'Trusts for the Earth: New Financial Mechanisms for International Environmental Protection', in W. Lang (ed.), *Sustainable Development and International Law* (Amsterdam: Martinus Nijhoff Publishers, 1995), pp. 167–84; Kemal Baslar, *The Concept of the Common Heritage of Mankind in International Law* (Amsterdam: Martinus Nijhoff, 1998); K. Bosselmann, R. Engel, and P. Taylor, 'Governance for Sustainability: Issues, Challenges, Successes', *IUCN* (2008).

4. P. Linebaugh, *The Magna Carta Manifesto: Liberties and Commons for All* (California: University of California Press, 2008); J.M. Neeson, *Commoners: Common Right, Enclosure and Social Change in England* (Cambridge: Cambridge University Press, 1996), pp. 1700–820.

5. The Paris Agreement on Climate Change has been dealt in detail, later in this chapter under the heading 'Paris Agreement on Climate Change'.

6. See, UNGA RES/43/53 (6 December 1988), UNGA RES/45/212 (21 December 1990), UNGA RES/46/169 (19 December 1991), UNGA RES/47/195 (22 December 1992).

7. See, M. Barlow, 'The Growing Movement to Protect the Global Water Commons', *The Brown Journal of World Affairs*, 17(1) (2010), pp. 181–95.

8. Barlow, 'The Growing Movement to Protect the Global Water Commons', p. 184.

9. *Barcelona Traction, Light and Power Company Limited (New Application, 1962), Belgium v. Spain*, Judgment, Merits, Second Phase, ICJ GL No 50, [1970] ICJ Rep 3, (1970) 9 ILM 227, ICGJ 152 (ICJ 1970), 5th February 1970, International Court of Justice [ICJ].

10. *Barcelona Traction* case, p. 32.

11. *Gabc̆íkovo-Nagymaros Project (Hungary/Slovakia)*, Separate Opinion of Justice Weeramantry, [1997] ICJ Rep, 117–18, available http://www.icj-cij.org/docket/index.php?p1=3&p2=3&case=92&p3=4, accessed on 10 July 2016.

12. J.A. Frowein, *Obligations erga omnes*, Max Planck Encyclopedia of Public International Law, available www.mpepil.com, accessed on10 August 2016.

13. Order of the German Federal Constitutional Court of 26 October 2004, 2 BvR 955/00, Deutsches Verwaltungsblatt (2005), pp. 175–83, 178, quoted by S. Talmon, 'The Duty Not to "Recognize" as Lawful a Situation Created by the Illegal Use of Force or Other Serious Breaches of a Jus Cogens Obligation: An Obligation without Real Substance?', in Christian Tomuschat and Jean Marc Thouvenin (Amsterdam: Martinus Nijhoff Publishers, 2006), pp. 99–126, 100.

14. Barlow, 'The Growing Movement to Protect the Global Water Commons', p. 190.

15. J.L. Sax, 'The Public Trust Doctrine in Natural Resource Law: Effective Judicial Intervention', *Michigan Law Review*, 68(47) (1970), pp. 471–566, 477.

16. See, N. Oreskes, 'The Scientific Consensus on Climate Change: How Do We Know We're Not Wrong?' in J. DiMento and P. Doughman, (eds) *Climate Change: What It Means for Us, Our Children, and Our Grandchildren*, (Cambridge, MA: MIT Press, 2007), p. 105.

17. G. Bryner and R.J. Duffy, *Integrating Climate Energy and Air Pollution Policies* (Cambridge, MA: MIT Press, 2012), pp. 6–7, arguing that a scientific consensus has taken place already and the precautionary principle has become 'largely irrelevant'.

18. B. Sudhakara Reddy, Gaudenz B. Assenza, Dora Assenza, and Franziska Hasselmann, *Energy Efficiency and Climate Change: Conserving Power for a Sustainable Future* (New Delhi: SAGE Publications, 2009), p. 1.

19. Intergovernmental Panel on Climate Change (IPCC), 2013. *Climate Change 2013: The Physical Science Basis. Working Group I contribution to the IPCC*

Fifth Assessment Report (Cambridge, United Kingdom: Cambridge University Press), available www.ipcc.ch/report/ar5/wg1, accessed on 18 July 2016.

20. Intergovernmental Panel on Climate Change (IPCC), 2013, Climate Change 2013.

21. This chapter looks at the impacts of energy sources on the atmosphere and does not cover the pollution caused by ozone depleting substances.

22. Human Genome Project, Stanford Encyclopaedia of Philosophy, available http://plato.stanford.edu/entries/human-genome/, accessed on 10 August 2016.

23. S. Tripp and M. Grueber, 'Economic Impacts of the Human Genome Project', *Battelle Memorial Innstitute*, (2011), ES-3.

24. UNEP, *The Adaptation Finance Gap Report* 2016 (Nairobi, Kenya: United Nations Environment Programme, 2016), p. xii.

25. 'The Energy Report: 100% Renewable Energy by 2050' (WWF, Ecofys, OMA, 2011), p. 13.

26. 'The Energy Report: 100% Renewable Energy by 2050', p. 13.

27. Sustainable Development Goals, Target 7.1, 7.2, and 7.3, available http://www.un.org/sustainabledevelopment/energy/, accessed on 12 August 2016.

28. UN-Energy Knowledge Network, *About Energy*, available http://www.un-energy.org/about, accessed on 12 August 2016.

29. S. Droge, et al., 'National Climate Change Policies and WTO Law: A Case Study of Germany's New Policies', *World T.R.*, 3(2) (2004), pp. 161–87, 161.

30. See, UNFCCC, Synthesis Report on the aggregate effect of the Intended Nationally Determined Contributions, FCCC /CP/2015/7, 30 October 2015, paras 23 and 154, pp. 8, 34.

31. See, Gary Bryner and Robert J. Duffy, *Integrating Climate Energy and Air Pollution Policies*, 1st Edition (Cambridge, MA: The MIT Press, 2012), pp. 17–39.

32. O. Edenhofer, Ch. Flachsland, M. Jakob, and K. Lessmann, 'The Atmosphere as a Global Commons–Challenges for International Cooperation and Governance', The Harvard Project on Climate Agreements (Harvard Kennedy School of Government, 2013).

33. O. Edenhofer, et al., *The Atmosphere as a Global Commons–Challenges for International Cooperation and Governance*, p. 10.

34. H. Varden, 'The Lockean "Enough-and-as-Good" Proviso. An Internal Critique', *Journal of Moral Philosophy* 9 (2012), pp. 410–42.

35. Article 19 and 300-A, The Constitution of India, available http://lawmin.nic.in/coi/coiason29july08.pdf, accessed on 12 July 2016. Article 300-A reads as follows, 'No person shall be deprived of his property save by authority of law.'

36. K. Bosselmann, 'Property Rights and Sustainability: Can they be Reconciled?', in D. Grinlinton and P. Taylor (eds), *Property Rights and Sustainability:*

The Evolution of Property Rights to Meet Ecological Challenges,(Amsterdam: Martinus Nijhoff Publishers, 2011), pp. 23–42.

37. K. Bosselmann, 'Losing the Forest for the Trees: Environmental Reductionism in Law', *Environmental Laws and Sustainability*, Special Issue of *Sustainability*, 2(8) (2010) p. 2428, available http://www.mdpi.com/2071-1050/2/8/2424/2424-48, accessed on 12 July 2016; R. Kim and K. Bosselmann, 'International Environmental Law in the Anthropocene: Towards a Purposive System of Multilateral Environmental Agreements,' *Transnational Environmental Law*, 2 (2013) pp. 285–309, available www.journals.cambridge.org/action/disp layAbstract?fromPage=online&aid=8943677, accessed on 12 July 2016; R. Kim and K. Bosselmann, 'Operationalizing Sustainable Development: Ecological Integrity as a *Grundnorm* in International Law', *Review of European Community and International Environmental Law*, 24(2) (2015)1–18.

38. I. Kaul, at the launch of *The Governance Report*, Hertie School of Governance (Oxford University Press, 2014), on 23 February 2014, available http://www.governancereport.org/media/news/the-sovereignty-paradox/, accessed on 10 July 2016.

39. Such as the recently signed Trans-Pacific Partnership (TPP) free trade agreement between 12 Pacific-Rim countries.

40. Kaul, at the launch of *The Governance Report*.

41. M. Weitzman, 'On Modeling and Interpreting the Economics of Catastrophic Climate Change', *The Review of Economics and Statistics XCI* (2009) pp. 1–19.

42. A. Lange and C. Voigt, 'Cooperation in International Environmental Negotiations Due to a Preference for Equity', *Journal of Public Economics*, 87 (2003) pp. 2049–67.

43. C.D. Siniscalco, 'Strategies for the International Protection of the Environment', *Journal of Public Economics*, 52 (1993), pp. 309–28.

44. K. Pittel and D. Rübbelke, 'Transitions in the Negotiations on Climate Change: From Prisoner's Dilemma to Chicken and Beyond', *International Environmental Agreements, Politics, Law and Economics*, 12 (2012), pp. 23–39.

45. M. Hoel, 'International Environmental Conventions: The Case of Uniform Reductions of Emissions', *Environmental Resource Economics*, 2 (1992), pp. 141–59.

46. S. Barrett, 'Rethinking Climate Change Governance and its Relationship to the World Trading System', *The World Economy*, 34 (2011), pp.1863–82.

47. See generally, S. Oberthür and M. Pallemaerts (eds), *The New Climate Policies of the European Union: Internal Legislation and Climate Diplomacy* (Brussels: VUB Press, 2011).

48. O. Edenhofer et al., 'The Atmosphere as a Global Commons–Challenges for International Cooperation and Governance' (Cambridge, Massachusetts: Harvard Project on Climate Agreements, 2013), p. 15.

49. Edenhofer et al., 'The Atmosphere as a Global Commons', The Harvard Project on Climate Agreements, p. 15.

50. M. Bastasch, 'Washington Governor Imposes Cap-And-Trade Through Executive Order', available http://dailycaller.com/2014/04/29/washington-governor-imposes-cap-and-trade-through-executive-order/, accessed on 10 August 2016.

51. Address to the C40 City Mayors Summit, Johannesburg, 5 February 2014, available http://unfccc.int/files/press/statements/application/pdf/20140502_c40_check.pdf, accessed on 10 August 2016.

52. ICLEI, Local Governments for Sustainability, available http://www.iclei.org/index.php?id=9, accessed on 20 July 2016.

53. E. Ostrom, 'A Polycentric Approach to Cope with Climate Change', *Policy Research Working Paper 5095*, World Bank (2009); E. Ostrom, 'Polycentric Systems for Coping with Collective Action and Global Environmental Change', *Global Environmental Change*, 20 (2010), pp. 550–7.

54. K. Bosselmann, *Earth Governance: Trusteeship of the Global Commons* (US: Edward Elgar, 2015).

55. Commission on Global Governance, *Our Global Neighbourhood* (Oxford: Oxford University Press, 1995).

56. The Earth Charter was adopted in 2000 in the Peace Palace in The Hague, is the most inclusively negotiated global document to-date and represents a consensus across cultures, religions and peoples. It can be perceived as having constitutional founding document for global civil society. See, K. Bosselmann and R. Engel (eds), *The Earth Charter: A Framework for Global Governance* (Amsterdam: KIT, 2010).

57. *Our Global Neighbourhood*, p. 252.

58. *Our Global Neighbourhood*, p. 252.

59. For example, P. Barnes, *Capitalism 3.0: A Guide to Reclaiming the Commons* (Berrett-Koehler, 2006); K. Coghill, Ch. Sampford, T. Smith (eds), *Fiduciary Duty and the Atmospheric Trust* (Ashgate, 2012); M. Blumm and M. Wood, *The Public Trust Doctrine in Environmental and Natural Resource Law* (Durham: Carolina Academic Press, 2013); M. Wood, *Nature's Trust: Environmental Law for a New Ecological Age* (Cambridge: Cambridge University Press, 2014).

60. Bosselmann, *Earth Governance*, 2015.

61. See, for example, C. Redgwell, 'Reforming the UN Trusteeship Council', in W. Chambers and J. Green (eds), *Reforming International Environmental Governance: From Institutional Limits to Innovative Reforms* (UN University Press, 2005), pp. 66–92.

62. G. Hardin, 'The Tragedy of Commons', *Science*, 162 (1968), pp. 1243–8, esp. p. 1248.

2

UPENDRA BAXI

Making Sense of Energy and Natural Resource Law in the Anthropocene Epoch

The conversation on 'Anthropocene' crosses many a disciplinary border and boundary with multitude of difficulties. This term has to be officially certified by the verdict of an Anthropocene Working Group of the Sub-commission on Quaternary Stratigraphy, but very few doubt that a profound change popularly known as 'global warming' and 'climate change' is 'occurring and that is anthropogenic'.[1] It is also certain that the decades of climate scepticism are over, giving way to political and popular conviction that the 'weight of evidence', on which science forever relies for its progress, overwhelmingly demonstrates today, that climate change has already happened and will occur at a rapid pace in near future.[2] Even as I write this, and you read it, a number of Solomon Islands are actually disappearing as sea levels rise, glaciers have started melting, the phenomenon of climate refugees is fast becoming a reality, global temperatures are soaring, forest cover is disappearing, desertification advances, and some unpredictable changes in Earth's behaviour are happening.

But the Anthropocene is more than climate change, which is 'only the tip of the iceberg (how ironic this expression becomes now!)' it also relates to the fact that,

[H]umans are (i) significantly altering several other biogeochemical, or element cycles, such as nitrogen, phosphorus and sulphur, that are fundamental to life on the Earth; (ii) strongly modifying the terrestrial water cycle by intercepting river flow from uplands to the sea and, through land-cover change, altering the water vapour flow from the land to the atmosphere; and (iii) likely driving the sixth major extinction event in Earth history [...] Taken together, these trends are strong evidence that humankind, our own species, has become so large and active that it now rivals some of the great forces of Nature in its impact on the functioning of the Earth system.[3]

In this sense, we are entering in the Anthropocene a 'geological age of our own making'.[4] It may also be the time of our own unmaking if the urgency of human action is not sufficiently realized.[5] And realizing it also includes not merely the law as we transmit it through our teaching, learning, and research, but also developing transformations in thinking and imagining the law in terms of a new climate change justice theorizing.[6] The task is not so much that of chasing continuities and discontinuities between (and of) law and justice or thinking out of the box as in case of 'sustainable development'.[7] Rather, it is one of smashing the box itself!

It is not difficult to identify the old box. Most generally, it allowed for changes of players but not the rules of the game whereas the question now in the Anthropocene is how to change the 'game' itself. The common law—statutory as well adjudicatory—enshrines right to do a lawful harm unto others[8] and environment, whereas arresting the Anthropocene entails at least following the principle of *Primum non nocere*, meaning 'first, do no harm'.[9] The existing international law allows immunity of multinational corporations and other entities from human rights obligations and responsibilities,[10] and allows even greenwashing[11] in geoengineering, or climate engineering, called by common consent as the field of deliberately (with intent) manipulating the Earth's climate to alleviate climate change.[12] The ideas of civilization as a mastery over natural resources[13] and law as a means of promoting world capitalism through chattel slavery, colonization, apartheid have come under severe criticism and are no longer normatively valid. These ideas, in various historical incarnations, served to organize a mode of production and an accompanying social formation; and drove humankind to accelerated destruction of (what is named now) the 'Mother

Earth'. The Cochabamba Declaration, 2010, heralds an altogether different way of thinking about human rights and responsibilities.[14] Moreover, the transition from universal human rights of all individual human beings to TRMF (trade-related, market-friendly) human rights of multinationals and their affiliates and cohorts[15] is now being severely challenged, as are the notions and structures of law and jurisprudence of environmental rights.[16] We clearly need to rethink law and justice from the Anthropocene standpoint.

To Whom Does the Sovereignty over Natural Resources Belong?

This distinctively postcolonial question is answered in a revolutionary mode by the landmark UN Declaration of Permanent Sovereignty over Natural Resources (PSNR) and its eight core principles.[17] Principle 1 of which declares that the right of peoples and nations to permanent sovereignty over their natural wealth and resources must be exercised in the interest of their national development and of the well-being of the people of the State concerned. The exploration, development, and disposition of such resources, as well as the import of the foreign capital required for these purposes, is mandated by Principle 2 to be in conformity with the rules and conditions which the peoples and nations freely consider to be necessary or desirable with regard to the authorization, restriction, or prohibition of such activities. Principle 3 proclaims that in cases where authorization is granted, the capital imported and the earnings on that capital shall be governed by the terms thereof, by the national legislation in force, and by international law. The profits derived must be shared in the proportions freely agreed upon, in each case, between the investors and the recipient state, due care being taken to ensure that there is no impairment, for any reason, of that State's sovereignty over its natural wealth and resources. Principle 4 states that nationalization, expropriation, or requisitioning shall be based on grounds or reasons of public utility, security, or the national interest which are recognized as overriding purely individual or private interests, both domestic and foreign. In such cases the owner shall be paid appropriate compensation, in accordance with the rules in force in the State taking such measures in the exercise of its sovereignty and in accordance with international law. In any case where the question of compensation gives rise to a controversy, the national jurisdiction of the State taking such measures shall be exhausted. However, upon agreement by sovereign States and other parties concerned, settlement of the dispute should be made through arbitration or international adjudication.

The free and beneficial exercise of the sovereignty of peoples and nations over their natural resources must be furthered by the mutual respect of States based on their sovereign equality as per Principle 5 and Principle 6 declares that international co-operation for the economic development of developing countries, whether in the form of public or private capital investments, exchange of goods and services, technical assistance, or exchange of scientific information, shall be such as to further their independent national development and shall be based upon respect for their sovereignty over their natural wealth and resources. Principle 7 emphasizes that the violation of the rights of peoples and nations to sovereignty over their natural wealth and resources is contrary to the spirit and principles of the Charter of the United Nations and hinders the development of international co-operation and the maintenance of peace and the last, Principle 8 proclaims that 'the foreign investment agreements freely entered into by or between sovereign States shall be observed in good faith; States and international organizations shall strictly and conscientiously respect the sovereignty of peoples and nations over their natural wealth and resources in accordance with the Charter and the principles set forth in the present resolution'.

The contradictory unity of these Principles has been well analysed in the context of their emergence.[18] We also note the unsustainability of 'sustainable development'.[19] And these have matured, as it were, in the transition of UN Millennial Development Goals to Sustainable Development Goals, with particular reference to 'climate change' situation.[20] There is every reason to think why the PSNR principles should place 'beyond obvious that globalization, modernization, and development not observing responsible use and environmental protection will further undermine the capacity of life-supporting ecological systems to sustain themselves and hence provide ecosystem services for humans'; that is to be questioned in the interpretations placed on the concept rather than PSNR principles as such.[21] Gümplová, interestingly, maintains that,

[T]here are good reasons not to refuse state sovereignty as a framework for global environmental justice, or for other dimensions of global justice, for that matter. However, the viability of resource sovereignty, both in theory and in practice, depends on the very interpretation of the concept of sovereignty and the way it incorporates self-limiting standards in its exercise. A parallel already exists. Resource sovereignty can be limited by environmental sustainability and ecological stewardship standards in the same way that human rights constrain the exercise of state power over its citizens.[22]

One now hears even of 'sustainable globalization'![23] However, neither international environmental law, nor regional, supranational, and domestic constitutions (and interpretation), though necessary, are sufficient to cope with the new concerns of climate change and climate justice theorizing. Sir Robert Jennings pointed acutely, though not in terms of the Anthropocene concerns, when he said that humanity 'is faced with a multifaceted dilemma. There seems to be an urgent need for more and more complex regulation and official intervention; yet this is, in our present system of international law and relations, extremely difficult to bring about in a timely and efficient manner'. He said further:

> The fact of the matter surely is that these difficulties reflect the increasingly evident inadequacy of the traditional view of international relations as composed of pluralistic separate sovereignties, existing in a world where pressures of many kinds, not least of scientific and technological skills, almost daily make those separate so-called sovereignties, in practical terms, less independent and more and more interdependent. *What is urgently needed is a more general realisation that, in the conditions of the contemporary global situation, the need to create a true international society must be faced. It needs in fact a new vision of international relations and law* (Jennings 2003: xxiii; emphasis added).[24]

That 'vision' for alternate law (and we may add jurisprudence) is a pressing question for all the makers and students of a new world order geared to provide expeditious and equitable world order in the Anthropocene. In particular, alternate futures entail post-Carbon economies, where natural resources are renewable-mostly solar, wind, and tidal energy and power—and geo-engineering, no matter how contestable these may be. We need also to recognize that while production based on new technologies is necessary, it is always subject to corporate takeover[25] which state regulation or international 'soft' law may effectively govern. In other words, 'Making peace with the Earth was always an ethical and ecological imperative. It has now become a survival imperative for our species.'[26]

The 2G Decision and Opinion

Instead of focusing further on this discourse, we look basically at the 2G decision and the subsequent Reference for Advisory Opinion of the Supreme Court of India.[27] Here, we particularly engage the question raised by way of constitutional doubt whether 'the Government has the right to alienate, transfer, or distribute natural resources/national assets otherwise than by following a fair and transparent method consistent with the fundamentals of the equality clause enshrined in the Constitution...'.[28]

The Court considered what the natural resources are. Both the 2G Case and 2G Reference acknowledge that though there is 'no universally accepted definition of natural resources' these designate 'elements having intrinsic utility to mankind'. As renewable or non-renewable, these 'are thought of as the individual elements of the natural environment that provide economic and social services to human society and are considered valuable in their relatively unmodified, natural form'. Their 'value rests in the amount of the material available and the demand for it' and the demand 'is determined by its usefulness to production'.[29] It is true that no 'comprehensive definition' of natural resources exits in the Indian and many other legal systems, and 'nature' cannot simply be treated as a warehouse of resources for the humans. Furthermore, the Court's perspective is deeply disappointing as it ignores use value altogether and engages only the exchange value of natural resources; it is excessively liberal economistic geared only to demand, supply, and production.

The productivity predisposition creeps through the judicial fashioning of a response to the question whether natural resources belong to people or the state as a complex order of their personification. Following some precedents,[30] the 2G Reference says boldly that 'as far as "trusteeship" is concerned, there is no cavil that the State holds all natural resources as a trustee of the public and must deal with them in a manner that is consistent with the nature of such a trust',[31] it simultaneously holds that 'the public trust doctrine is a specific doctrine with a particular domain and has to be applied carefully'.[32]

What does 'carefully' signify in the context of natural resources? The Court is quite clear that the notion of popular sovereignty is creatively ambiguous. One many take the view that the invocation of 'people' here designates parliamentary sovereignty, which cannot and must not be limited by courts through interpretation. The opposed view is that 'sovereignty' of Parliament and state legislatures is curtailed by responsible crafting of the notion or doctrine of public trust. Certainly, the Supreme Court has imposed this limitation through acts of disciplined constitutional interpretation. This means that the 'supremacy' of Parliament and state legislatures is not curtailed by any notion or doctrine of public trust![33] It may at times be curtailed by judicial process, power, and review.

What 'carefully' seems to signify is the strict constitutional scrutiny at the bar of equality; citing mainly the Praful Bhagwati jurisprudence (in Paras 93–105 of the Reference), the Court insists that a showing of 'constitutional infirmity' remains necessary to invoke its supreme jurisdiction, 'law may not be struck down for being arbitrary without pointing out a constitutional infirmity...' (Paragraph 105). If no such 'infirmity' is shown at the threshold, the legislation, or even the exercise of administrative power, prevails. The question then is, 'How do we distinguish between the rhetoric and reality of dispossession? Rhetorically it refers to natural resources belong to the people' but the Court immediately adds that 'the State legally owns them on behalf of its people'.[34]

***

Leaving perforce the vexed theoretical concern about the distinction between 'legal' and 'political' sovereign,[35] what may this form of ownership diarchy signify? It is true that the 'the State is deemed to have a proprietary interest in natural resources and must act as guardian and trustee in relation to the same' and the right of peoples and nations to permanent sovereignty over their natural wealth and resources must only extend to the interest of their 'development' and 'well-being'.[36] But how does one ascertain that the people's sufferings are not aggravated by judicial insistence on executive/legislative supremacy in macroeconomic development policies which are thus held beyond constitutional judicial review? Ultimately, the operations of a free market state prevail over human rights enshrined in the Constitution. The 2G reference says, 'There is no constitutional imperative in the matter of economic policies—Article 14 does not pre-define any economic policy as a constitutional mandate. Even the mandate of 39(b) imposes no restrictions on the means adopted to subserve the public good and uses the broad term 'distribution', suggesting that the methodology of distribution is not fixed'.[37]

This escape in the *demosprudential* constitutional leadership of India[38] illustrates a wider point. How do we go beyond the commons and its tragedies? First, we need to have theory of regulation which takes communities seriously as a way of taking governance by human rights seriously. How to escape the multiple tyrannies of state and market failures (too often acting in disguised unison) is a major question,[39] which cannot be handled so long as we persist with the negative law and jurisprudence. Second, in the Anthropocene era, we need to consider how to make impossible and illegal 'contemporary enclosures' (and accompanying 'monocultures of the mind', as Vandana Shiva calls these)[40] that privatize the erstwhile commons in the name of development, alias 'creative destruction'.[41]

Third, ways need to be found to show how the public trust doctrine actually serves as eminent domain doctrine; or in other words, how the law of 'takings' evolves to reinforce the state or governmental control over private property (or more crucially as living within the contradictions of the global, national, and the local commons).[42]

Fourth, we need to think about the foundations of a new morality of justice; in particular, we need to move beyond environmental justice, sustainable justice, and global justice to some new ways of articulating justice for the Anthropocene.[43] We should not merely adjudge how the extant human rights regimes are actually crystallized in environmental law[44] but also fundamentally rethink justice notions in the contexts of extinction provided by climate change.[45]

Fifth, we somehow ought to erase the institutional apartheid between legal scholars and policy-makers.[46] Climate change justice may not emerge when political and policy actors continue to remain functionally illiterate of the academic contribution and when the academics remain unaware of the dilemmas of reasonable action confronted by political actors and both these altogether remain disinclined to take people's sufferings seriously.

Sixth, (and without being exhaustive) is the question of what may we as policy, political, academic, or scientific actors learn from social activism and movements? How may our understandings of energy and natural resources law be transformed by the notion of 'climate debt' and even 'reparations'? How may some fresh understandings of historical justice be fashioned?[47] Furthermore, how may we understand respect for human dignity and climate justice to be the same? May we all relearn from an eco-activist who said that *'the only true green and sustainable things in life is how we treat each other'*.[48]

Notes and References

1. M. Maslin, *Climate Change: A Very Short Introduction* (Oxford: Oxford University Press, 2009), p. 28.

2. Maslin, *Climate Change*, pp. 29–45.

3. W. Steffen, J. Grinevald, P. Crutzen, and J. Mcneill, 'The Anthropocene: Conceptual and Historical Perspectives', *Philosophical Transactions of the Royal Society A*, 369 (2011), pp. 842–67, 843.

4. A.C. Revkin, *Global Warming: Understanding the Forecast* (New York: Abbeville Press, 1992), p. 55.

5. N. Kline, *This Changes Everything: Capitalism versus Climate* (London: Allen Lane/Penguin, 2014). This admirable work, designed to foster creative activist knowledge, legality, justice, and solidarity among suffering and struggling peoples of the earth, is especially important in conveying a vivid description of the tactics pursued by neoliberal markets and governments, especially job blackmail, 'desperation' as a means to predation and 'total control'. See Chapters 12 and 13. See her highly popular, and assiduously accurate, work *The Shock Doctrine: The Rise of Disaster Capitalism* (New York: Metropolitan Books, 2007). In her 2014 work, Naomi Klein comes close to describing the uses of law as a 'tactic'. Michel Foucault first emphasized this dimension: see A. Hunt and G. Wickham, *Foucault and Law: Towards a Sociology of Law as Governance* (London: Pluto Press, 1994); A. Beck, 'Foucault and Law: The Collapse of Law's Empire', *Oxford Journal of Legal Studies*, 16(3), (1996), pp. 489–502. See also V. Tadros, 'Between Governance and Discipline: The Law and Michel Foucault', *Oxford Journal of Legal Studies*, 18(1), (1998), pp. 75–103. The distinctions among law, tactics, strategy, and theory are quite crucial.

6. U. Baxi, 'Towards Climate Justice Theory?' *Journal of Human Rights and the Environment*, 7(1), (2016), pp. 7–31.

7. It is without doubt that the Brundtland Report pioneered this conception well. Combining sustainably with development is hard task, even when international environmental justice is confined to a single generation. The Brundtland Report extended the notion to entire economy and society; its

notion of development is that it 'meets the needs of the present without compromising the ability of future generations to meet their own needs'. See the United Nations, *Our Common Future: World Commission on Environment and Development* (1991). See, for some critical perspectives, D. Shelton, 'Legitimate and Necessary: Adjudicating Human Rights Violations Related to Activities Causing Environmental Harm or Risk', *Journal of Human Rights and the Environment*, 6(2) (2015), pp. 139–55; A.P. Mihalopoulos, *Absent Environments: Theorising Environmental Law and the City* (Routlege-Canvendish, 2007). But see, for celebrationist perspective, V. Meg, *Sustainable Development, Energy and The City: A Civilisation of Visions and Actions* (Springer Science+ Business Media, Inc., 2005). See also, D. Schlosberg, 'Reconceiving Environmental Justice: Global Movements and Political Theories' *Environmental Politics*, 13(3) (2004), pp. 517–40; N.C. Carre, 'Environmental Justice and Hydraulic Fracturing: The Ascendancy of Grassroots Populism in Policy Determination', *Journal of Social Change*, 4(1) (2012), pp. 1–13; D. Miansom, 'Duties to The Distant: Aid, Assistance, and Intervention in the Developing World', *The Journal of Ethics*, 9 (2005), pp. 151–70. See as to the realities of war/conflict displaced people and what happens in the process to sustainable development, H. Young and L. Goldman (eds), *Livelihoods, Natural Resources, and Post-Conflict Peace building* (New York: Routledge, 2015).

8. It is amazing but true that so little in ethical or jurisprudential theory (in English) is written about this, whereas our entire liberal capitalist legality is based on the right to a lawful harm; See also, O.J. Herstein, 'Defending the Right to Do Wrong', *Law and Philosophy*, 31(3) (2012), pp.343–65; J. Waldron, 'A Right to Do Wrong', *Ethics*, 92(1) (1981), pp. 21–39. See also, U. Baxi, 'From Human Rights to a Right to be Human', in U. Baxi and G. Sen (eds), *The Right to be Human* (Delhi: Lancer International Publications, 1987), pp. 181–200.

9. See, Baxi, 'Towards Climate Justice Theory?' pp. 19–31.

10. See U. Baxi, 'Human Rights Responsibility of Multinational Corporations, Political Ecology of Injustice: Learning from Bhopal Thirty Plus?', *Business and Human Rights Journal*, 1(1) (2016), pp. 21–40.

11. See, for example, E.L. Lane, 'Greenwashing 2.0', *Columbia Journal of Environmental Law*, 38(2) (2013), pp. 279–331; I.M. Alves, 'Green Spin Everywhere: How Greenwashing Reveals the Limit of the CSR Paradigm', *Journal of Global Change and Governance*, 2(1) (2009), pp. 1–26; E.L. Lane, 'Green Marketing Goes Negative: The Advent of Reverse Greenwashing', *European Journal of Risk and Regulation*, 4 (2012), pp. 582–8; T. Choice, 'The Sins of Greenwashing: Home and Family Edition' (2010), available www.sinsofgreenwashing.org, accessed on 15 August 2016; R. Bewick, J.P. Sanchez, and C.R. McInnes, 'Usage of Asteroid Resources for Space-Based Geoengineering', in V. Badescu (ed.), *Asteroids: Prospective Energy and Material Resources* (Springer, 2013), pp. 581–603.

12. A recent report on New Biology starts with the following indictment, 'The extreme genetic engineering industry of Synthetic Biology (Syn Bio) is rapidly shrugging off earlier pretensions that it might usher in a clean, green, post-petroleum future. Instead, many Syn Bio executives and start-ups are now trying to create alliances with fracking, oil, and shale interests, which will actually increase the fossil-based extractive economy that has already brought planetary climate change and other ecological and social problems'. H.B. Stiftung, *Extreme Biotech Meets Extreme Energy*, available www.etcgroup.org/sites/www.etcgroup.org/files/files/extbio_a4, accessed on 15 August 2016.

13. The idea of mastery over, or conquest of, nature played a large role in defining civilization during the first Industrial Revolution and the Age of Enlightenment; this tendency was widely prevalent in the writings of jurisprudence. See, especially, J. Kohler, *Philosophy of Law* (Adalbert Albrecht trans., Kessinger Publishing, LLC 2007 [1914]). For discussion of Kohler's jurisprudence, see R. Pound, *Interpretations of Legal History* (New York: The Macmillan Company, 1923), pp. 141–51; see also E. Bodenheimer, *Jurisprudence: The Philosophy and Method of the Law* (Harvard University Press, 1974), pp. 113–14; J. Stone, *Human Law and Human Justice* (1965) and *Social Dimensions of Law and Justice* (Sydney: Maitland Press, 1966). But see the withering criticism of K. Marx in Marx, 'The Philosophical Manifesto of the Historical School of Law' (1842), *Karl Marx Frederick Engels Collected Works*, 16 (1975), p. 203. Marx called it the 'sole frivolous product' of the European (especially German) legal thought.

14. See, the Declaration text in readingfromtheleft.com/PDF/CochabambaDocuments.pdf., accessed on 15 August 2016. See also, Law of the Rights of Mother Earth, enacted in December 2010, available htttp://bolivia.infoleyes.com/shownorm, accessed on 15 August 2016, and the discussion in W.D. Mignolo, 'From "Human Rights" to "Life Rights"', in C. Douzinas and C. Gearty, *The Meaning of Rights: The Philosophy and Social Theory of Rights*, pp. 161–80 and I.R. Wall, 'On a Radical Politics for Human Rights' in the same volume (Cambridge: Cambridge University Press, 2014), pp. 106–20.

15. U. Baxi, *The Future of Human Rights*, Perennial Edition (Delhi: Oxford University Press, 2013). Hereafter cited as 'Baxi, *Future*'.

16. Baxi, 'Towards Climate Justice Theory'.

17. See, www.ohchr.org/Documents/ProfessionalInterest/resources.pdf, accessed on 15 August 2016.

18. See for an analysis of contexts and overall significance, S. Pahuja, *Decolonizing International Law: Development, Economic Growth, and the Politics of Universality* (New York: Cambridge University Press, 2011). See also C. Armstrong, 'Against "Permanent Sovereignty" over Natural Resources', *Politics, Philosophy & Economics*, 14(2) (2015), pp. 129–51. These include: (1) Access: 'right to interact with a resource and to enjoy "non-subtractive" benefits from it';

(2.) Withdrawal: 'right to obtain and indeed to remove resource units for one's own use'; (3) Management: 'right to regulate use patterns and to transform a resource by making improvements to it'; (4) Exclusion: 'right to determine who can access and withdraw a resource'; (5) Alienation: 'right to sell a resource, or moving up a level, the right to sell management and exclusion rights'; and (6) Derive Income: 'right to obtain proceeds from the sale of a resource, or to extract some other form of income from it'. See also U. Baxi, 'Enslavement and Environment' being a talk delivered on 5 December 1991 at the India International Centre, in *Towards Hope: An Ecological Approach to the Future* (an Indian National Trust for Art and Cultural Heritage, Delhi, 1992, edited by N. Jayal. G. Hardin has famously drawn attention to the tragedy of commons: see his 'The Tragedy of the Commons', *Science*, 162 (1968), pp. 1243–8; 'Lifeboat Ethics: The Case against Helping the Poor', *Psychology Today*, (1974), pp. 38–43, 124–6; 'The Ethical Implications of Carrying Capacity', in G. Hardin and J. Baden (eds), *Managing the Commons* (San Francisco CA: W.H. Freeman, 1977), pp. 112–25. See also for the notion of 'phantom' carrying capacity, W. Catton, *Overshoot: The Ecological Basis of Revolutionary Change* (Urbana IL: University of Illinois Press, 1980).

19. See the important work of P. Birnie and A. Boyle, *International Law & The Environment*, Second Edition (Oxford: Oxford University Press, 2002); S. Cohen, *The Resilience of the State. Democracy and the Challenge of Globalization* (London: Hurst & Company, 2003); J. Derrick, M.C. Segger, and A. Khalfan, *Sustainable Development Law. Principles, Practices and Prospects* (Oxford: Oxford University Press, 2004); P. Wetterstein, *Harm to the Environment: The Right to Compensation and the Assessment of Damages* (Oxford: Oxford University Press, 1997). See also, J.M. Harris, T.A. Wise, K.P. Gallagher, and N.R. Goodwin (eds), *A Survey of Sustainable Development: Social and Economic Dimensions* (Washington: The Global Development and Environment Institute, Tufts University, Island Press, 2011).

20. See, for an early exposition, B. Jenks. 'From an MDG World to an SDG/ GPG World: Why the United Nations Should Embrace the Concept of Global Public Goods', *Development Dialogue Paper* No.15 (2015); D. Osborn, A. Cutter, and F. Ullah, 'Universal Sustainable Development Goals: Understanding the Transformational Challenge for Developed Countries—Report of a Study By Stakeholder Forum' (2015); German Development Institute, Briefing Paper 18, 'Post 2015: How to Reconcile the Millennium Development Goals (MDGs) and the Sustainable Development Goals (SDGs)?' (2012).

See, further, Chapin. III, F. Stuart, G.P. Kofinas, and C. Folke (eds), *Principles of Ecosystem Stewardship. Resilience-Based Natural Resource Management in a Changing World* (New York: Springer, 2009); J.L. Cohen, *Globalization and Sovereignty: Rethinking Legality, Legitimacy, and Constitutionalism* (New York: Cambridge University Press, 2012); *Whose Common Future:* Reclaiming the

Commons (Pennsylvania; Phil., The Ecologist 1993); C. Singh, *Common Property and Common Poverty: India's Forests, Forest Dwellers and the Law* (Delhi: Oxford University Press, 1987).

21. See P. Gümplová, 'Restraining Permanent Sovereignty over Natural Resources', *Enrahonar: Quaderns de Filosofia*, 53 (2014), pp. 93–114).

22. Gümplová, 'Restraining Permanent Sovereignty over Natural Resources', p. 96.

23. See the Preamble to the UN *Principles for Responsible Contracts Integrating the Management of Human Rights Risks into State–Investor Contract Negotiations: Guidance for Negotiators* (Geneva: OHCR, 2015).

24. R. Jennings, 'Foreword', to P. Sand, *Principles of International Environmental Law*, Third Edition (Cambridge: Cambridge University Press, 2003).

25. V. Shiva describes all this in terms of an old and new civil war as the Ecowars of the 'environmental law' and climate change now:

> Protecting the commons is vital to making peace with the earth, and maintaining peace within and between communities. Privatization of the Earth's resources is leading to wars—privatization of water is leading to water wars, patents are leading to biodiversity wars, corporate takeover of land is leading to land wars. Corporate takeover of the atmospheric commons first through pollution which gave us climate change and then through the pseudo solution of emissions trading is creating wars over the atmosphere … this … contest between eco-imperialism and earth democracy. Eco-imperialism expands control by the powerful over the earth's resources violating the rights of the earth and people. Earth democracy is the democracy of all life, and it is based on the rights of the earth and the rights of all people.

See *Making Peace with the Earth: City of Sydney Peace Prize Lecture* (3 November 2010), in part later published as a book by Pluto Press (2013). See also V. Shiva, *Earth Democracy: Justice, Sustainability, and Peace* (Dehradun: Natraj Publications, 2010). I would add here the category of 'negative law and jurisprudence' to those of 'negative economics and negative politics' at p. 106. See also ETC Group, *Geopiracy: The Case against Geoengineering* (2010) [www.etcgroup.org]. See at a more sustained theoretical level (though he does not mention messy matters such as Biopiracy and Geopiracy), G. Agamben, *STASIS: Civil War as a Political Paradigm (Homo Sacer II, 2)* (California: Stanford University Press, 2015: Nicholas Heron trans).

26. Law as an instrument of peace is a varied and old idea (notably developed by great jurists and philosophers as wells by the old and the new indigenous people's movement and ancient inhabitants of the earth) and needs urgent revival in the Anthropocene.

27. Special Reference No. 1 of 2012 [Under Article 143(1) of the Constitution of India], per D.K. Jain, J. [For S.H. Kapadia, CJI, D. Misra, and R. Gogoi, JJ.,

with an additional concurring opinion by J.S. Khehar, J.] The case hereafter will be cited as 2G Reference by paragraph numbers.

28. The Reference arose out of the decision in *Centre for Public Interest Litigation & Ors. v. Union of India & Ors.* (2012) 3SCC 1. In that case (hereafter referred to as 2G Case) the Court formulated as a general first issue the constitutional concern set out in the text and responded to it as follows:

> [W]hile distributing natural resources the State is bound to act in consonance with the principles of equality and public trust and ensure that no action is taken which may be detrimental to public interest. Like any other State action, constitutionalism must be reflected at every stage of the distribution of natural resources. In Article 39(b) of the Constitution it has been provided that the ownership and control of the material resources of the community should be so distributed so as to best subserve the common good, but no comprehensive legislation has been enacted to generally define natural resources and a framework for their protection...'. See Para 75 of 2G Case and para 73 of 2G Reference.

29. See the 2G Reference Paras 83–9.

30. Especially, *Natural Resources Limited v. Reliance Industries Limited* (2010) 7 SCC 1.

31. 2G Reference, Para 85.

32. 2G Reference, Para 90.

33. In Para 91 of the Reference, the Court says that as early as 1951 'the notion that the Parliament is an agent of the people was squarely rebutted in *Re: Delhi Laws Act, 1912,* 1951 AIR 332, 1951 SCR 747 where it was observed that "the legislature as a body cannot be seen to be an agency of the electorate as a whole" and "acts on its own authority or power which it derives from the Constitution".

34. In 2G Case, the Court concluded that 'In conclusion, we hold that the State is the legal owner of the natural resources as a trustee of the people and although it is empowered to distribute the same, the process of distribution must be guided by the constitutional principles including the doctrine of equality and larger public good': see also Paras 75–6 of the 2G Reference.

35. See, for example, D. Dyzenhaus, 'Hobbes and the Legitimacy of Law', *Law and Philosophy* 20 (2001), p. 461; H.L.A. Hart, *The Concept of Law* (Oxford: Oxford University Press, 1994), pp. 1–4; J. Raz, *The Concept of a Legal System,* Second Edition (Oxford: Clarendon Press, 1980); N. MacCormick, *Questioning Sovereignty* (Oxford: Oxford University Press, 1999); MacCormick concludes that 'sovereignty is neither necessary to the existence of law and state nor even desirable', p. 129; J. Waldron, *Law and Disagreement* (Oxford: Oxford University Press, 1999); J. Habermas, *Between Facts and Norms* (Cambridge: Polity Press, 1996; trans. by William Rehg); A.R. Amar, 'The Central Meaning of Republican

Government: Popular Sovereignty, Majority Rule, and the Denomination Problem', *Colorado Law Review*, 9 (1994), pp. 65–74; 'The Consent of the Governed: Constitutional Amendment Outside Article V', *Columbia Law Review*, 94 (1994), p. 457; R. Dworkin, *Law's Empire* (London: Fontana, 1986); N.E. Simmonds, *Law as a Moral Idea* (Oxford: Oxford University Press); J. Rawls, *Lectures in the History of Political Philosophy* (Cambridge, MA: Harvard University Press, 2007); C. Schmitt, *Constitutional Theory* (Durham: Duke University Press, 2008); trans. and edited by J. Seitzer.

36. The 2G case at Para 64; see Para 65–73 for further elucidation.

37. 2G Reference at para 120.

38. See U. Baxi, 'Introduction' to M. Suresh and S. Narrain (eds), *The Shifting Scales of Justice: The Supreme Court in Neoliberal India* (Delhi: Orient Blackswan, 2014); U. Baxi, 'Demosprudence v. Jurisprudence? The Indian Judicial Experience in the Context of Comparative Constitutional Studies', *Macquarie L.J.* 14 (2015), pp. 1–13; U. Baxi, 'Demosprudence and Socially Responsible / Response-able Criticism: The NJAC Decision and Beyond', the Ninth D.D. Basu Memorial Lecture WBNAJS, Kolkata (forthcoming *NUJS Law Rev*, 2016); P.B. Mehta, 'The Rise of Judicial Sovereignty', *Journal of Democracy*, 18(2) (2007), pp. 70–83.

39. E. Ostrom, With Contributions from C. Chang, M. Pennington, and V. Tarko, *The Future of the Commons: Beyond Market Failure and Government Regulation* (London: The Institute of Economic Affairs, Westminster, in association with Profile Books Ltd, 2012) pp. 68–83. Ostrom proposes several 'design 'principles, among which are: (a) a clear definition of natural resources; (b) clear rules 'governing the use of resources should be adapted to local condition'; (c) 'enforcement of the governance system: monitoring, conflict resolution, and sanctions for those violating rules'; and (d) ways in which the 'community's own organisation fits within a broader reality', see especially, the contribution of Christian Chang in this volume. She says, and I cannot agree more, 'Ostrom's design principles also involve 'the distillation of the practical tools that are actually used to promote productivity and sustainability in a very unpromising environment'.

40. Vandana Shiva, *Monocultures of the Mind: Perspectives on Biodiversity and Biotechnology* (London: Zed Books, 1993).

41. Commonly called J. Schumpeter's gale, the 'gale of creative destruction' describes the processes of industrial change or mutation which 'incessantly' revolutionizes the economic formation from within, incessantly destroying the old one, incessantly creating a new one', see J.A. Schumpeter, *Capitalism, Socialism and Democracy* (London: Routledge, 1979). The crucial word here is 'incessant' but this 'prophet of innovation' often forgets what Lord Buddha names as the law of impermanence!

42. The widely read book, even now, is R.L. Epstein, *Takings: Private Property and the Power of Eminent Domain* (Cambridge MA: Harvard University

Press, 1985). But see M. Kelman, 'Taking Takings Seriously: An Essay for Centrists', *California Law Review*, 74 (1986), pp. 1829–63; see also A. Bell and G. Parchomovsky, 'The Uselessness of Public Use', *Columbia Law Review*, 106(6) (2006), pp. 1412–49. V. Shiva makes a deeper point about the erasure by abstract legal doctrines of the lived histories of colonialism, see note 23, *Earth Democracy* at 44–8.

On another note, see the recent study of the way in which the Indian Supreme Court and the High Courts are struggling to install a constitutional conception of the commons, see S. Bhutani and K. Kohli, *The Case for the Commons: Lessons from Implementation of the Landmark Judgment of the Supreme Court of India* (Delhi: Supported by Foundation for Ecological Security, 2015).

43. Baxi, 'Towards Climate Justice Theory', pp. 19–31.

44. A. Dias, 'Human Rights, Environment And Development: With Special Emphasis on Corporate Accountability', UNDP Human Development Report 2000 Background Paper, available hdr.undp.org/sites/default/files/ayesha-dias.pdf, accessed on 1 August 2016. See also, the additional materials referred to in Baxi, 'Towards Climate Justice Theory'.

45. C. Colebrook, *Death of the PostHuman: Essays on Extinction*, Vol. 1 (Open Humanities Press, 2014) available http://dx.doi.org/10.3998/ohp.12329362.0001.001; S. Shaviro, *No Speed Limit: Three Essays on Accelerationism* (Minn., University of Minnesota Press, 2016); R. Lee, 'Everybody's Novel Protist: Chemeracological Entanglements in A. Ghosh's Fiction', in *The Exquisite Corpse of Asian America: Biopolitics, Biosociality, and Posthuman Ecologies* (New York: New York University Press, 2014); M. Cooper, *Life as Surplus: Biotechnology & Capitalism in the Neoliberal Era* (University of Washington Press, 2008).

46. This is a general point as law and Anthropocene scholarship is in relative infancy in India and even today most law schools do not offer a full conventional course on natural resources and the law. But at last, one Minister in India candidly acknowledges the non-conversation between legal academics and political/policy actors in his autobiography: see J. Ramesh, *Green Signals: Ecology, Growth, and Democracy in India* (Delhi: Oxford University Press, 2015), pp. 186–8. Ramesh there bemoans, refreshingly, the 'institutional monocultures' of policy and political actors.

47. See Klein, *This Changes Everything*, the entire Chapters 12 and 13.

48. Klein quotes the above (at p. 407) a Greensburg Mayor, Bob Dixon a 'former postmaster who comes from a long line of farmers'.

3

USHA TANDON[*]

Regulation of Nuclear Energy for Sustainable Development

A Critical Overview of International Regime with Special Reference to IAEA

Various principles[1] and models of Sustainable Development[2] have been projected to explain the concept of Sustainable Development. However, the underlying idea of this concept points to the confluence of three dimensions of development—economic, environmental, and social.[3] All the three aspects of Sustainable Development are inter-related and affect each other. For instance, a state cannot afford to protect environment at the cost of the basic needs of its masses as poverty itself has been considered as the biggest polluter.[4] Similarly, undue emphasis on economic development without any regard to environmental concerns and social justice cannot go a long way to achieve sustainable developmental goals.[5] For economic development, sources of energy have played a pivotal role. It is worth mentioning that major economic progress, in the last century, was due to electric energy.

[*] I acknowledge the wonderful research assistance rendered by Neeraj Gupta, Research Scholar, Faculty of Law, University of Delhi, New Delhi.

Energy including electric energy is the wheel which has allowed the vehicle of development to move forward smoothly.[6]

The nuclear energy discovered by the scientists, in mid twentieth century, had the potential to provide solutions to energy security issues. The scientists were well aware that the discovery of nuclear technology has great potential to serve the humanity, if utilized wisely, as they were conscious, that any misuse of it may prove catastrophic for the entire world.[7] Hiroshima and Nagasaki nightmare has led to the realization of the world community that nuclear energy must be used only for peaceful purposes and any military use must be eliminated. The peaceful use of nuclear energy is dependent on peaceful coexistence. Conflicts and sustainable development are antithetical.[8] The international community must ensure that conflicts, being an inevitable part of any society, must be resolved with peaceful means alone. All this requires that the arms race, which was very prevalent in the cold war period, is given a good bye.[9]

The present chapter accepts, having put forward the arguments in favour and against the use of nuclear energy, the relevance of nuclear energy for sustainable development. Having provided a brief analysis of the concept of sustainable development, it outlines global legal framework for nuclear energy's safe and peaceful development. It explains that the focus of the international community has always been to ensure that nuclear energy is used peacefully and safely. Though, the primary responsibility for the regulation of the use of nuclear energy rests with national authorities, however as other countries may be affected as well, it has been recognized that the regulation of nuclear energy necessitates the international regulation to ensure, inter alia, uniformity of standards, co-ordination, as well as compliance. In this respect, International Atomic Energy Agency (IAEA), has served as a focal point. It provides an overview of the Statute of International Atomic Energy Agency, discusses various Conventions for the safe use and handling of Nuclear Materials, and critically analyses IAEA in terms of its leadership, and absolute efficiency. It argues that nuclear safety measures must be carried out with utmost transparency and coordination among the countries and international organization and all states using nuclear energy must give due regard to the provisions contained in various conventions.

Sustainable Development: A Conceptual Analysis

The idea of sustainable development in modern international law is not very old. The term was firstly used in the year 1987 in the modern sense by the World Commission on Environment and Development.[10] The emphasis on the concept was the result of economic policies which were followed by the imperial states at national and international level.[11] The economic policies of the major industrialized countries, before the beginning of the eighth decade of the last century, were guided towards indiscriminate exploitation of natural resources for the purposes of higher economic growth.[12] The only aim of these countries was attaining the highest possible Gross National Product (GNP) for which they were indiscriminately exploiting the natural resources. The continuous exploration and mining of natural resources, on the one hand, led to depletion of natural sources and on the other hand, resulted in the creation of toxic and non-toxic wastes leading to various types of pollution.[13] This cumulative pollution resulted on the adverse impact on health of people at large, especially poor.[14] Further, the emphasis on the GNP led to the ignorance of certain other important aspects of societal balance. The rich were becoming richer and the poor were becoming poorer.[15] The indiscriminate use of natural resources, the ever expanding gulf between the rich and the poor and demands of the newly formed independent states for economic development formed the basic idea behind the concept of sustainable development.[16]

The concept of sustainable development, thus, is considered to be a compromise of the two conflicting course of action propagated by the world community.[17] While the developed world were more concerned with the deteriorating environmental conditions and climatic change, the newly formed states, who had recently gained independence from the colonial system, were yearning for the economic development so that they could provide their population the basic minimum conditions of living.[18]

At the United Nations Conference on the Human Environment, which was held at Stockholm in the year 1972, the developed and developing countries had already agreed to take measures to protect and improve the environment but without compromising the developmental needs of the developing countries.[19] However, this Conference is also known as the conference where the difference in the agenda of developed and developing countries became very prominent. Developed countries were vociferously arguing that any developmental activity now onwards must take into account the environmental aspect. The developing countries' robust stand was that the environment pollution existing at the date was the result of the indiscriminate economic activities of the industrialized countries and the new countries also have the right to develop and they cannot be made liable for wrong done to environment by the developed countries.[20]

The Stockholm Conference, which is the first universal conference, can also be considered as the beginning of International Environmental Law in real sense, as the earlier efforts relating to environment protection were restricted to industrialized countries alone.[21] This universal conference was the beginning of other steps which were to be taken in the future by the United Nations such as establishment of UNEP and adoption of World Charter for Nature by the United Nations General Assembly (UNGA) in the year 1982.[22] In the year 1983 UNGA established the commission known as the World Commission on Environment and Development (WCED).

These conflicting demands of developed and developing countries found place in the report of the World Commission on Environment and Development, also known as Brundtland Report. The report defined Sustainable Development as the kind of development that 'meets the needs of the present without compromising the ability of future generations to meet their own needs'.[23] The concept contains within it two key concepts—the concept of 'needs', in particular the essential needs of the world's poor, to which overriding priority should be given and the idea of limitations imposed by the state of technology and social organization on the environment's ability to meet present and future needs.[24] The Report clearly lays down the crucial relationship between environment and development by emphasizing that 'Environment and development are not separate challenges; they are inexorably linked'. Development cannot subsist upon a deteriorating environmental resource base; the environment cannot be protected when growth leaves out of account the costs of environmental destruction. These problems cannot be treated separately by fragmented institutions and policies. They are linked in a complex system of cause and effect'.[25]

The Rio Declaration on Environment and Development and other binding and non-binding documents which were adopted in the year 1992 were also based on the concept of Sustainable Development. Out of the 26 Principles adopted in Rio Declaration, 11 Principles expressly incorporate the phrase Sustainable Development. The Forest Principles, Agenda 21, and UNFCCC were also adopted simultaneously with Rio Declaration which is also based on the concept of Sustainable Development. The concept of sustainable development has been emphasized and reemphasized in future conferences related to environmental law post Rio. The World Summit on Sustainable Development (WSSD) held at Johannesburg highlighted the role of private corporations in attaining sustainable development, it was observed in the summit that, apart from the role of the government in attaining sustainable development the role of corporate world also plays a pivotal role. In other words, the corporate social responsibility was recognized as an important tool for attaining sustainable development.[26] Similarly, the United Nations Conference on Sustainable Development, held at Rio de Janeiro (popularly known as Rio+20) made efforts to highlight the impact of non-governmental organizations in achieving sustainable development. Thus, it can be said that the concept of Sustainable Development is not against economic development. Along with economic development, it advocates for the protection and preservation of environment as well as social development; and any effort in this area requires participation from every segment of society.

Energy and Sustainable Development

Energy is the prerequisite for humans to grow and develop. Supply of energy has always been at the core of human progress. The Brundtland Report also emphasized the need of supply of energy, when it observed that, 'Energy is necessary for daily survival. Future development crucially depends on its long-term availability in increasing quantities from sources that are dependable, safe, and environmentally sound. At present, no single source or mix of sources is at hand to meet this future need [...] energy provides "essential services" for human life—heat for warmth, cooking, and manufacturing, or power for transport and mechanical work.'[27]

The current world population is beyond seven billion and is projected to touch the mark of ten billions by the year 2050.[28] Providing them basic necessaries of life including gainful employment requires faster pace of economic progress, which is dependent on the energy security. Apart from the industrial energy requirements, the domestic consumption of energy also needs to be addressed.[29] According to a report published by the International Union for Conservation of Nature (IUCN), one tonne of oil equivalent (toe)/capita per year is the minimum energy needed to guarantee an acceptable level of living as measured by the Human Development Index, despite many variations of consumption patterns and lifestyles across countries.[30]

The level of economic progress achieved so far, is still insufficient and the world needs to continue with the economic activities at a fast pace. The last century's economic development was fuelled by the fossil fuels such as coal, petroleum, and natural gas.[31] The indiscriminate and inefficient use of these fossil fuels has created adverse impact on the environment. The emissions of carbon and other gases due to the burning of fossil fuels have affected the temperature of the atmosphere which has resulted in the change in weather pattern, described as climate change.[32] Climate change has led to melting of glaciers and early flowering. Various species of animals and living creatures have become extinct and there is a huge threat to biodiversity and ecological balance.[33]

All this, cumulatively, requires finding the alternatives to fossil fuels. Various alternatives, in the form of solar power, wind energy, and tidal energy have been discovered and used for electricity generation. Nuclear energy is also one of the potential energy sources which may help in lightening the future of the world's poor. Around 11 per cent of total electricity production of the world is generated with the help of nuclear energy. And, the developed countries are producing 40 per cent of their electricity with the help of nuclear energy.[34] At present, as of May 2016, 30 countries worldwide are operating 444 nuclear reactors for electricity generation, and 63 new nuclear plants are under construction in 15 countries.[35] The major argument in favour of the use of nuclear energy for electricity generation is that energy released by nuclear fission of each atom is about 60 million times larger than the energy released per atom by burning carbon and hydrogen. It is argued that nuclear energy is the answer to the worries of greenhouse gases and it provides the energy security as well.[36]

However, the inherently dangerous nature of nuclear material being used for electricity generation has met with stiff resistance worldwide.[37]

The opponents assail it on various grounds like the process of obtaining uranium, which is used as nuclear fuel, causes grave environmental pollution. Further, the nuclear power plant construction also requires high-end expertise and precautions, which makes establishing it very cumbersome demanding more resources in its construction. Then, the spent nuclear fuel requires high degree of precaution while disposing and decommissioning it is the process which requires high level planning and expertise. Any error may prove fatal to the workers involved and the site may get affected with radioactivity.[38]

The above discourse leads to the inference that nuclear energy, though having the potential to brighten the future of the masses as an efficient alternative source of energy, however, has its own inherent challenges. These challenges include, first, the utilization of nuclear energy only for peaceful purposes; second, the high costs involved in the construction of nuclear reactor for power plants and the high prices of its fuels; third, construction, operation, and utilization of nuclear power plant in such a manner that it has minimal effect on nature, human, and environment. These challenges require that international legal regime must be efficiently developed to work in the holistic manner, failing which the dream of achieving Sustainable Development may remain a distant reality.

International Legal Regime

Bombing of Hiroshima and Nagasaki with nuclear weapon ended the World War II. Forthwith, international efforts to establish and maintain peace came at the centre stage. The efforts in this direction led to the establishment of United Nations with the primary objective to save succeeding generations from the scourge of war and to maintain international peace and security.[39] The UN Charter further aims to reduce the use of armed forces only as last resort in case international peace is at stake.[40] By its very first Resolution, the General Assembly of United Nations established the United Nations Atomic Energy Commission (UNAEC) to deal with the problems raised by the discovery of atomic energy.[41] It can also be said that nuclear regulation and United Nations are twins being born simultaneously.[42] Various plans were proposed by different countries in UNAEC with regard to reduction in nuclear arms at various times, some of them were directly opposed to each other.

However, in 1948, this agency declared that its function had ceased to be meaningful as the aim of the US and other allies was not restricted to reduction or prevention of nuclear arms but its total elimination.[43] Later, UNGA also adopted a resolution for peaceful use of nuclear energy for economic and social development. The resolution stated that states have the sovereign right to use nuclear energy for peaceful purposes, and free access to acquire the nuclear technology, material, and so on.[44]

In the year 1953 the US president Eisenhower's declaration of 'Atoms for Peace' is considered the basis which led to the establishment of the International Atomic Energy Agency.[45] IAEA is an international organization in the UN family. In the present international regime of nuclear energy regulation, IAEA is considered the world body which is responsible to ensure safe and secure use of nuclear energy for peaceful purposes.[46] Various conventions under the leadership of IAEA have come into existence related to safe use of nuclear energy and its regulation.[47] In all the conventions related to nuclear energy, IAEA is given a prominent role and is bestowed with various duties under them.[48] In the next few paragraphs, the powers and functions of IAEA under the Statute of International Atomic Energy Agency (hereinafter referred as the Statute) and other conventions is discussed. The chapter proceeds to examine as to what provisions have been enacted at international level to ensure safe use of nuclear energy for peaceful purposes which can help attain sustainable development.

Statute of International Atomic Energy Agency: An Overview[49]

The Statute is enacted with the aim to attain the dual object of accelerating and enlarging contribution of atomic energy to peace, health, and prosperity throughout the world, and to ensure that the atomic energy is not used for military purposes.[50] The Statute prescribes that IAEA is bound to carry out its activities in accordance with the purposes and principles of UN charter for furthering the policy of world peace and disarmament.[51] It further states that IAEA will establish control over the use of fissionable material so that it is used only for peaceful purposes.[52] IAEA is bound by the principles of efficient utilization, greatest possible general benefit of all areas of the world, having special regards to the underdeveloped areas.[53] It has to annually submit its report to UNGA and if required, also to the UNSC.[54] The Statute provides that IAEA will have due regard to the sovereignty of the states and it will not impose any conditions while granting the assistance which are contrary to the Statute.[55] IAEA has to work in cooperation with other international and expert bodies and for that purposes it may enter into any agreement with them.[56]

The functions of IAEA may be divided into four major categories. First, research and development of nuclear energy for peaceful purposes; second, acting as an intermediary in the negotiations and activities related to nuclear energy for peaceful purposes; third, setting and enforcing the standards and safeguards related to use and activities of nuclear technology; and lastly, acting as a repository of nuclear substance, equipment, facilities as discussed in detail below. Various other Conventions adopted by the international community have also conferred divergent duties on IAEA such as *Comprehensive Nuclear-Test-Ban Treaty, Convention on Nuclear Safety, Convention on Early Notification of a Nuclear Accident,* and many others. The nature of duties conferred on it by various other conventions makes IAEA a central and the most important international organization involved in the regulation of nuclear energy.

The Statute authorizes IAEA to encourage and assist research on nuclear energy and research and development of practical application of atomic energy for peaceful uses throughout the world. IAEA is also bound to foster the exchange of scientific and technical information on peaceful uses of atomic energy and encourage the exchange of training of scientists and experts in the field of peaceful uses of atomic energy.[57] It provides elaborate provisions in regard to research and development and assists the member states in carrying out the projects related to nuclear energy for peaceful purposes. It lays down sufficient safeguards to ensure that the assistance provided by it or at its request is utilized for peaceful purposes only and not diverted for any military purposes in the forms of various provisions.[58]

Article VIII.A of the Statute says that each member should make available such information as would, in the judgment of the member, be helpful to IAEA. The information may relate to anything—it may be scientific information related to nuclear technology or relating to any hidden project being carried out by any state or organization related to nuclear arms, or it may also relate to any accident which has already occurred.[59] The important thing to understand here is that this Clause is recommendatory and the information may relate to anything, which in the opinion of the member state, may prove helpful.[60]

The second clause of Article VIII is more specific and it creates a duty on the member state to provide scientific information which has been developed by such state with the assistance from IAEA under Article XI. The information given to IAEA has to be made accessible by its members. It is the bounden duty of IAEA to encourage and promote the exchange of information related to nuclear energy for its use in peaceful purposes.

Article XI provides that any member or group of member interested in a research project related to nuclear energy for peaceful purposes may seek the assistance of IAEA for supplying them with the fissionable material, source material or equipment, and so on, necessary for such research. At the request of the member, it may assist in obtaining the financial assistance as well. However, it shall not give any guarantee for such assistance or assume any financial responsibility for any project.[61]

IAEA may consider the project before providing the assistance and for this purpose it may also send experts, in the territory of the requesting member, who will assess the project.[62] The Board of Governors can approve the project only after considering various aspects, such as usefulness of the project, its scientific feasibility, and adequacy of resources including funds, human resources, safeguard standards, and any other relevant factors.[63]

Once the project is approved in the above manner, IAEA enters into an agreement with the member to supply the fissionable material, equipment, and other necessary materials.[64] The Statute provides certain important terms of the agreement which must be part of the agreement.[65] Apart from the usual clauses of charges, and so on, the agreement must contain clauses declaring that the assistance provided shall not, in any way, be used for any military purposes and the project will be carried out in accordance with the safeguards provided in Article XII, which shall be specified in the agreement.

Article XII specifies the right and responsibility of IAEA to obtain the compliance of the safeguards and examines the design of the equipment and reactor from the perspective of health and safety, and also to ensure that it will not be used for any military purposes. It is empowered to ask member to observe the health and safety measures prescribed by IAEA. It can also ask to produce the accounts related to the use of the fissionable and any other material to keep a check that the material is not being diverted to any military purposes.

The most important power of IAEA with regard to safety standards relates to sending the inspectors in the recipient states after consulting them.[66] The Statute confers the right on these inspectors to access any place, document, instrument, and so on. These inspectors are empowered to examine the equipment, plant, or any other thing relevant to ensure that the assistance given by IAEA is not being utilized for any military purposes. Inspectors are also empowered to examine the project from the perspective of safety standards. If the inspectors find any non-compliance with the safety requirements prescribed or agreed, they shall report to the Director General of IAEA, who shall convey the report to the Board of Governors. The Board of Governors shall ask the recipient state to take the corrective measures; the Board shall also report the matter to all the members of UNSC and to UNGA. In case the recipient state fails to take the remedial measures required, penal actions may be taken against it, which include curtailment or suspension of the assistance, directing the recipient state to return the materials and equipment made available to it. Further, the privileges available to a member under the Statute may also be suspended.[67]

The Statute also makes provisions with regard to supply of nuclear materials, equipment facilities, and so on.[68] It provides that members may make available to IAEA services, equipment, and facilities which may be of assistance in fulfilling IAEA's objectives and functions.[69] The members may make available fissionable and source material as per the agreement with IAEA and IAEA may use it and request the members to supply such material to other members.[70] Similarly, when IAEA assist in any project as per Article XI, they have to enter an agreement with the member whom it is assisting, regarding the supply of fissionable nuclear material and other materials.[71] Whenever IAEA takes the control of the nuclear material, equipment facilities, and so on, IAEA has to ensure that the materials are kept in proper custody and place so that there is no danger to human and environment. It has to also take prevention in regard to sabotage any terrorist activities or forcible dispossession. Further, IAEA has to ensure that the storage policy is such that it ensures that the nuclear substance is not concentrated in large quantity at one place.[72] These provisions make it clear that IAEA also acts as a repository of nuclear materials.

The most important function of IAEA relates to formulation, prescription, and enforcements of safeguards and standards related to use and handling of nuclear material, equipment, their proper storage, transportation, and so on. Article III.A.6 of the Statute empowers IAEA to establish or adopt standards of safety for the protection of health and the minimization of danger to life and property. The provision also requires that the standards must be applied to IAEA's own operations and to operations making use of materials, services, equipment, facilities, and information made available by IAEA or at its request, or under its control, or supervision. States which receive technical assistance or assistance for project must sign an agreement with IAEA in which they undertake the responsibility to apply to the assisted operations, IAEA's safety standards, and measures that are specified in the agreement. The Statute also authorizes IAEA to apply its safety standards, at the request of states, to any of their operations or activities.[73]

According to El Baradei, who was the Director General of IAEA and was conferred with Nobel Prize for peace in the year 2005, in developing safety standards IAEA takes account of the work of relevant international scientific and technical bodies, such as the International Commission on Radiological Protection (ICRP), the United Nations Committee on the Effects of Atomic Radiation (UNSCEAR), the World Health Organization (WHO), and the International Labour Office (ILO).[74] He argues that international action in this field began with the establishment of the ICRP, which has issued recommendations on radiation protection since its inception in 1928. In 1955, the UNGA established United Nations Scientific Committee on the Effects of Atomic Radiation (UNSCEAR) to evaluate doses, effects, and risk from ionizing radiation on a worldwide scale. The work of ICRP and UNSCEAR provides the basis for the standards elaborated by other international and regional organizations, such as IAEA, ILO, WHO, Euratom, and the NEA. He mentions in his work, that these organizations have built close working relationships in developing these standards. The need to establish appropriate standards designed to ensure the safe use of nuclear energy is reflected in the constituent instruments of such organizations. It is to be noted that the IAEA's safety standards are mandatory with regard to nuclear activities undertaken with IAEA assistance, but where such assistance is not provided the standards are recommendatory.[75]

Conventions for the Safe Use and Handling of Nuclear Materials

IAEA has evolved various guidelines and safety standards for the use of nuclear energy in collaboration with specialized bodies. Apart from that, it has provided a platform where international community has come together and negotiated various conventions related to safe use and handling of nuclear materials. The basic features of major conventions are discussed below.

Convention of Nuclear Safety 1994[76]

This convention aims to legally commit participating states operating land-based nuclear power plants to maintain a high level of safety by setting international benchmarks to which states would subscribe.[77] The main objective of this convention is to ensure that the use of nuclear energy is safe, well regulated, and environmentally sound.[78] The convention provides basic consideration which must be taken into account while establishing and operating nuclear installations. It also provides what measures should be taken in case of an accident. Overall, the purpose of the convention is to achieve safe and secure human health and environment. All these things could be achieved by establishing a regulatory regime which shall be responsible for the proper functioning and overseeing the operations related to nuclear energy.

It provides that states shall establish and maintain a legislative and regulatory framework to govern the safety of nuclear installations. The legislative and regulatory framework shall provide, inter alia, the safety requirements, a system of licensing, a system of regulatory inspection and assessment of nuclear installations to ascertain compliance, the enforcement of applicable regulations and of the terms of licences, including suspension, modification, or revocation.[79]

The convention goes ahead and provides that it is the duty of the state to ensure that the regulatory body envisaged under this convention is capable of performing its function and for that purposes, the regulatory body must be provided with the adequate resources, including technical experts and financial resources.[80] Further, the regulatory body must be such which is independent from the authority which is responsible for the promotion and propagation of nuclear energy in the state.[81]

The convention also provides a general safety consideration which must be implemented during the construction and operation of the nuclear installation. Article 8 provides that states shall take the appropriate steps to ensure that all organizations engaged in activities directly related to nuclear installations shall establish policies that give due priority to nuclear safety. Similarly, it is the duty of the state to ensure that there are adequate financial and human resources available to support the safety of each nuclear installation throughout its life.[82] States should ensure that they have adequate number of trained and experienced human resource for ensuring that the safety is not compromised at all.[83] Further, the policies and guidelines with regard to nuclear installation must take into account human factors.[84] States are bound to run quality assurance programme effectively, which can assure that safety concerns are addressed properly.[85] They need to take steps to ensure that comprehensive safety assessment is carried prior to and during the operation of the nuclear installation throughout its life. To ensure that above-mentioned duty is performed properly, the convention provides that the activities undertaken must be well documented and annual reporting under the convention is mandated under it. It is the duty of the state under this convention to ensure safety and health of the persons generally and staff involved in the operation.[86] The convention prescribes that the state must prepare the emergency plan and it must be tested frequently, so that in case of emergency, remedial steps can be taken in the right direction.[87]

The convention provides that before finalizing the site where the proposed nuclear installation has to be erected, various steps related to safety assessment must be carried out which must take into account all the relevant factors including the consultation of the neighbouring state which might be affected by such nuclear installation.[88] The design and technology used in the nuclear installation must be safe and reliable. The design must be such which provides various levels of safety measures.[89] The operation of the nuclear installation must be in conformity of the legal regime set up under this convention. The initial operation must be allowed only after ensuring that safety measures and prescribed standards have been complied. It must be ensured that the operation of the installation must be restricted to the prescribed limit as deciphered from the testing period. Continuous inspection and regular checks of the safety measures must be ensured during the whole life cycle of the nuclear installation.[90] It emphasizes that the waste generated from the installation must be kept at the minimum and it must be ensured that it is disposed in proper manner.

Convention on the Physical Protection of Nuclear Material and Nuclear Facilities 1979[91]

The convention is intended to achieve and maintain worldwide effective physical protection of nuclear material and nuclear facilities used for peaceful purposes. It also aims to prevent and combat offences relating to such material and facilities worldwide and to facilitate co-operation among states parties to those ends.[92] The convention is restricted to use, storage, and transport of nuclear material and nuclear facilities used for peaceful purposes only.[93] The convention expressly provides that the convention does not affect the activities undertaken by the armed forces during the armed conflict.[94] The convention establishes the norm that the responsibility of physical protection of the nuclear material and nuclear facility vests in the state under whose jurisdiction such material or facility is located.[95]

Article 2A of the convention provides that the state has to ensure the physical protection of the nuclear material and nuclear facility. For this purpose, state parties have to establish, implement, and maintain a protection regime which ensures protection against theft or unlawful taking, its immediate recovery in case of theft, preventing any attempt of sabotage, and mitigating and minimizing the impact of sabotage if it occurs. For the above-mentioned purposes, the states need to establish and maintain an effective legislative and regulatory mechanism. The convention requires the states to establish a competent authority or designate one for the proper implementation of such legislative and regulatory mechanism. The Article also incorporates various principles which have to be incorporated while framing and implementing the measures related to nuclear materials.

Article 4 of the convention prescribes that the state parties shall not indulge in the export and import of nuclear material unless they have received the assurance from the other party that the nuclear material will be provided protection as per the Annex 1 of the this convention.[96] However, a state exporting the nuclear material may impose a term on the importing party, with the mutual agreement, for the protection of nuclear material.[97] In case, the nuclear material is in transit, from a member state of this convention, and the state from which, and to which the material is transported is not the party, in such situation, the transit state shall not allow such transit without the assurance being obtained that transport is protected as per the mandate given under Annex 1. It is also the duty of the state to identify and inform the states to which the nuclear transport shall transit during the import, export, or transit.[98] The standards laid down in the Annex 1 have also to be followed by the states in intra-country transportation of the nuclear materials.[99]

The convention also provides for cooperation in cases of theft or any unlawful taking of the nuclear material or in case of any credible threat in this respect. It provides that the states shall cooperate and give assistance for discovering and recovering the stolen material as much as possible for them.[100] States have a duty to take steps to inform other states as soon as possible, of any theft, robbery, or other unlawful taking of nuclear material or credible threat thereof, and to inform, where appropriate, IAEA and other relevant international organizations. Concerned states parties are required to exchange information with each other, with a view to protecting threatened nuclear material, verifying the integrity of the shipping container, or recovering unlawfully taken nuclear material. Furthermore, they are obliged to co-ordinate their efforts through diplomatic and other agreed channels, render assistance, if requested and ensure the return of recovered nuclear material stolen or missing as a consequence of the above-mentioned events. The means of implementation of this co-operation are left to be determined by the state parties concerned.[101]

Convention on Early Notification of Accident[102]

The radiation emitted by the nuclear energy is such that it can come into contact with water and air and spread from one place to other at a very fast pace. Thus, it may have harmful effects at places far beyond the place at which the accident occurs. The primary objective of this convention is to minimize the resultant harm which may occur in case of nuclear accident. The main provisions of the convention revolve around notifying IAEA and the neighbouring or adjacent countries about the occurrence of an accident.

Article 1 of the convention, while defining its scope, provides that the convention is applicable to any accident, occurring in the jurisdiction of any state party, in any facility or activity from which release of radioactive material occurs or is likely to occur and which has resulted or which may result in an international trans-boundary release that could be radiological. The words 'facility' and 'activity' have been defined in the broadest possible manner, so as to include every possible handling in nuclear energy.

The convention prescribes that the state parties shall provide IAEA and state parties, the details related to the competent authority and point of contact for the exchange of information, in case any nuclear accident occurs. The parties are required to ensure that they make available the updated information in case of any change in the above-mentioned details. Further, IAEA is also obliged to ensure that it keeps ready the above information so that it can be communicated at the time of emergency in the quickest possible manner. IAEA will also keep ready the information of above mentioned nature with regard to relevant international organizations.[103]

Article 2, which is the fulcrum of this convention, provides that the state where the nuclear accident has occurred shall, immediately inform the adjacent state which is or which may be affected by such an accident. The information may be given to the state directly or through IAEA. In any case, the state must also inform IAEA about the occurrence of the accident. The information given by the state to IAEA must contain the details, as far as possible, of accident such as the exact location, magnitude, and the exact time of the accident. The state, where the accident has occurred, is also bound to provide the necessary information which is relevant in minimizing the consequences of accident to IAEA as well as the state.

The information for minimization of the consequences of the accident include the information such as, exact time, location, and nature of the accident, the facility or the activity in which the accident has occurred, the assumed or established cause of the accident and the foreseeable development of it. The state will also provide the technical information which may aggravate or reduce the impact of the accident, such as meteorological prediction and forecast, the action taken by the state itself regarding the reduction of the impact. The state is also bound to keep providing the updated data related to all the above factors which have come to their knowledge.[104]

Once the information has been given to IAEA about the accident, IAEA has to act promptly, to inform the international organizations and states which may be affected by such an accident.[105] A state party which has no nuclear activities, but which is adjacent to a state which has and which is not party to this convention, may request IAEA to assist in establishing a radiation monitoring systems so that the objectives of the convention may be achieved.[106]

Convention on Assistance[107]

This convention must be read in continuation to the convention on Early Notification of Accident. The convention sets out an international framework for co-operation among state parties and with IAEA to facilitate prompt assistance and support in the event of nuclear accidents or radiological emergencies. It requires states to notify IAEA, of their available experts, equipment, and other materials for providing assistance. In case of a request, each state party decides whether it can render the requested assistance as well as its scope and terms. Assistance may be offered without costs taking into account, inter alia, the needs of developing countries, and the particular needs of countries without nuclear facilities. IAEA serves as the focal point for such cooperation by channelling information, supporting efforts, and providing its available services.[108]

The convention provides a mechanism where the state parties can identify and notify about the agencies, experts, materials, and so on, which they are willing to provide in the event of nuclear accident or radiological emergency. They have also to specify the terms, especially financial, under which the assistance will be made available.[109] This mechanism is intended to create a pool, which can be utilized in the event of emergency without any delay. Further, the states have to provide to IAEA the details about the authority and the point of contact as discussed in the convention of Early Notification of Accident.[110]

Article 2 provides that in case of a nuclear accident or radiological emergency, the state party may call other state parties, IAEA, and any other international organization for assistance. The state seeking assistance may directly call upon other state or it may request IAEA to convey the message. The state seeking the assistance shall specify the scope and the extent of the assistance and it will also provide the necessary information so that the assisting party may decide whether it is able to assist as requested. In case the requesting state is not able to provide the necessary information about the extent of assistance, in such situation, both the parties will mutually decide the extent of assistance. The nature of the request is inclusive of medical treatment of the victims and their temporary relocation in the assisting state. It is the duty of the state party to respond to the request promptly. They are bound to notify the state requesting assistance, about its capability or incapability to meet the request, as soon as possible along with the terms of assistance, in case it is capable to assist.

Article 3 of the convention creates a balance between the sovereign right of the requesting state and freedom and expertise required for the effective assistance. It provides that, unless otherwise agreed, requesting state shall be responsible for the direction, control, coordination, and supervision of assistance in its territory or the place under its jurisdiction. The assisting state is bound to designate competent personnel who shall be in charge of the overall functioning of the assistance; the personnel will work in coordination with the requesting state's competent authorities. The requesting state, on the other hand, has to provide local facilities and services for effective implementation and execution of the assistance needed. They are also bound to ensure safety and security of the life and property of the assisting state. The convention also prescribes various duties which IAEA has to perform under it. The first and foremost duty of IAEA relates to ensure effective coordination and cooperation in case of nuclear emergency. The role of IAEA is of an expert intermediary and repository of resources, which is expected to perform expert tasks in case the states are facing emergent situation.[111]

Apart from the above-mentioned functions, IAEA has certain duties relating to the training of personnel and research in the field of nuclear emergency. It has to lay down standards related to radiation monitoring, and so on. The convention is obliged to protect confidential information, which has come in the hands of the assisting state and requesting state, while performing its function under the convention.[112] The convention also deals with the reimbursement of the costs incurred by the state as well as IAEA while assisting.[113]

Joint Convention on Spent Fuel and Radioactive Waste Management, 1997[114]

This convention is applicable to the waste generated by the use of nuclear technology for civilian purposes and radioactive waste generated from civilian use. The radioactive waste occurring naturally is not covered by the convention. The convention specifically provides that the provisions under it are not applicable to the waste generated from military use. The aim of the convention appears to be to ensure that peaceful use of nuclear energy does not adversely affect the environment, thus, the risk of any adverse impact of it on human health, and environment is sought to be eliminated to the maximum possible extent.[115]

The convention requires the states to ensure that at all stages of spent fuel management, individuals, society, and the environment are adequately protected against radiological hazards.[116] The state has to take appropriate steps to ensure that criticality and removal of residual heat generated during spent fuel management are adequately addressed; generation of radioactive waste is kept to the minimum practicable and take into account inter-dependencies among the different steps in spent fuel management. The states have to provide for effective protection of individuals, society, and the environment, by applying at the national level, suitable protective methods, as approved by the regulatory body, in the framework of its national legislation, which has due regard to internationally endorsed criteria and standards. The states must take into account the biological, chemical, and other hazards that may be associated with spent fuel management and strive to avoid actions that impose reasonably predictable impacts on future generations greater than those permitted for the current generation, and aim to avoid imposing undue burdens on future generations.[117]

After providing general safety requirements, the convention prescribes the various measures which have to be taken with regard to proposed site of the facility, design of the facility, safety of the facility, operation of it, and disposal of the spent fuel. The state parties are mandated to establish a procedure related to finalization of the site, where spent fuel management facility is to be established. The procedure must contain the aspects related to evaluation of all relevant site-related factors likely to affect the safety of such a facility during its operating lifetime, the likely safety impact of such a facility on individuals, society, and the environment. In addition, the procedure must be such which ensures that the information on the safety of such a facility is available to members of the public. Further, parties are also required to consult other parties in the vicinity of such a facility, insofar as they are likely to be affected by that facility, and provide them, upon their request, with general data relating to the facility to enable them to evaluate the likely safety impact of the facility upon their territory. In any case the parties need to take all the efforts to ensure that its activity does not result in such consequences which are unacceptable to other party and it will take every effort in this direction.[118]

The next Article deals with the design of the nuclear facility. It provides that the parties shall take appropriate steps to ensure that the design and construction of the facility must be such which provide suitable measures to limit possible radiological impacts on individuals, society, and the environment including those from discharge or uncontrolled release. The design must be such which is supported by experience, testing, or analysis.[119] The convention requires that the facility must be assessed for safety and environment before the construction and before the operation starts in the facility.[120]

After prescribing about the design of the facility, the convention mandates the procedure which must be followed with regard to the operation of the facility. It provides that the licence to operate a spent fuel management facility is based upon appropriate assessments as specified in Article 8 and is conditional on the completion of a commissioning programme demonstrating that the facility, as constructed, is consistent with design and safety requirements. Further, the operation of the facility must be limited to conditions derived from tests, operational experience, and the assessments, as specified in Article 8. It must be ensured that engineering and technical support in all safety-related fields are available throughout the operating lifetime of a spent fuel management facility. It ensures that the incidents significant to safety are reported in a timely manner by the holder of the licence to the regulatory body. It also puts a duty on the state party to organize the programmes to collect and analyse relevant operating experiences.[121] The safety provisions related to radioactive waste management are also similar to the above provisions.

Critical Assessment of IAEA

In the sphere of international regulation of nuclear energy, the massive and unprecedented role of IAEA is evident from the fact that one of the Director Generals of IAEA and IAEA were jointly conferred the Nobel Prize for peace in the year 2005. IAEA has worked as norm creator for the world with regard to safety and security of nuclear materials.[122] The relevance of the work done by IAEA can be assessed from the fact that the standards and guidelines, though recommendatory in nature, have been considered as crucial by most of the countries.[123] It must also be noted that the non-use of nuclear weapon since the Second World War in itself, to a certain extent, is the evidence of the effective role played by IAEA along with other international organizations. Voices have been raised to provide more power to IAEA in the sphere of nuclear regulation so that it can play a more effective role.[124]

However, there are various fronts on which IAEA and its mandate are seen with suspicion. It is argued that the Statute was drafted and negotiated by few countries in the leadership of the US. The states which were involved in the drafting of the Statute were hand-picked by the US.[125] In such a situation, question may arise whether the Statute is meant to serve the interest of those few countries only. Further, it is also said that the objective of IAEA, as mandated in the Statute is contradictory. It is difficult to differentiate peaceful and military purposes of nuclear energy and in the light of this, the restricted role of IAEA under the Statute does not serve the purpose.[126] Also, the inspection provision under the Statute, if enforced in the proper spirit, will go against the sovereignty of a country party to the Statute.[127]

IAEA is criticized for failure to detect the use of nuclear material by Iraq, Iran, Syria, and Libya for military purposes. Further, in the case of Fukushima disaster, the functioning of IAEA came under severe criticism as it could not perform the basic function of providing information with regard to accident with promptness to the world at large.[128] Post-Fukushima, another criticism which has erupted is the agreement which has been signed by IAEA with WHO.[129] According to this agreement, both the parties to the agreement shall share the information with each other in their respective fields. On the basis of this agreement, the WHO could not conduct the assessment of the impact of Fukushima disaster on the health of the person and IAEA supplied the required information to WHO.[130] It is argued that as IAEA is the body which is indulged in the programme of nuclear promotion, it might have under-supplied the information.[131] Terrorism has also added to the concerns of IAEA, security of nuclear materials from the terrorist organizations has come up as one of the major concerns for IAEA. International Convention for the Suppression of Acts of Nuclear Terrorism[132] aims to deal with the problem of security of nuclear material from terrorist activities.

As far as the functioning of IAEA is concerned, with the emergence of new independent states, which also happen to be the developing countries, the politicization of the function of IAEA was bound to happen, as is the case with other international bodies. The developing countries have their own aims and aspirations and they voice their concerns on every platform available to them, this is true with regard to nuclear energy as well. They demand that developed countries must assist them technologically and financially for the research and development of nuclear power plants and other use of nuclear energy, and in this direction IAEA must act as the mediator. However, the reality is otherwise, the developed countries have various apprehensions with regard to transfer of technology related to nuclear energy.

***

The above critical analysis shows that economic development is not antithesis to sustainable development and one cannot simply deny the rights of the poor to develop. The pace of economic development, especially in the least developed countries, needs to be accelerated. The economic development, however, is largely dependent on the availability of energy resources and electricity generation. Even the social and political upliftment of the masses is dependent on electricity in the modern world of digitization. All this requires that states take efforts in cooperation with each other to make available the latest technology in the field of power generation and electricity production. Nuclear energy has the potential to address the energy security concerns of the world provided it is utilized in the proper direction. One cannot simply avoid use of nuclear energy without compromising on the economic front. However, nuclear energy can be only a part of the energy mix and one cannot be totally dependent on it.

The inherent dangerous nature of the nuclear technology requires it to be handled with due care. Further, the high cost of the installation of nuclear power plant, the process of extraction of uranium being considered as highly polluting activity, and the issue of waste generated out of nuclear reactors are the core areas which compels one to ponder over the issue whether the use of nuclear energy really helps in attaining the sustainable development goals or not.

The international efforts in the field of proper handling of nuclear energy, through various conventions, as discussed above, have made a mark and it is the need of the hour that all the states which are using nuclear energy must give due regard to the provisions contained in these conventions. The Statute and other conventions clearly prescribe that the least developed countries must be kept in mind while deciding on the use of nuclear energy. This imposes an obligation on the developed states using nuclear technology that they provide technological and financial assistance to the poorer countries to establish and set up nuclear power plants. On the other hand, it is the duty of the countries seeking assistance to comply with the international norms related to safeguards attached with the use and handling of nuclear technology. The Fukushima disaster of 2011 highlights the fact that even the most advanced country like Japan may lack in handling the accident properly, demonstrating various lapses in coordination between the state and international organization. Thus, it becomes important that nuclear safety measures must be carried out with utmost transparency and coordination among the countries and international organization.

Notes and References

1. See United Nations Conference on Environment and Development (UNCED), Rio de Janeiro, 3–14 June 1992, UN Doc. A/CONF.151/26 (Vol I) / 31 ILM 874 (1992), Rio Declaration Arts 15, 16. These principles lay emphasis on precautionary measures and polluter pays principle, available http://www. unep.org/Documents.Multilingual/Default.Print.asp?documentid=78&articl eid=1163, accessed on 25 August 2016.

2. For details see M. Keiner, 'History, Definition(s) and Models of Sustainable Development,' available http://e-collection.library.ethz.ch/eserv/ eth:27943/eth-27943-01.pdf, accessed on 21 August 2016.

3. J.M. Harris, 'Basic Principles of Sustainable Development', Global Development and Environment Institute Working Paper 00-04, 2000, available http://www.ase.tufts.edu/gdae/publications/working_papers/Sustainable%20 Development.PDF, accessed on 21 August 2016; D. Tladi, 'Sustainable

Development in International Law: An Analysis of Key Enviro-economic Instruments', Pretoria University Law Press, 2007, available http://www.pulp. up.ac.za/pdf/2007_03/2007_03.pdf, accessed on 26 August 2016.

4. See speech of Indira Gandhi, the then Prime Minister of India delivered at the Plenary Session of United Nations Conference on Human Environment, Stockholm, 14 June 1972, available http://lasulawsenvironmental.blogspot. in/2012/07/indira-gandhis-speech-at-stockholm.html, accessed on 25 August 2016.

5. UNGA /RES/70/1, 21 October 2015, Sustainable Development Goals, available http://www.un.org/ga/search/view_doc.asp?symbol=A/ RES/70/1&Lang=E, accessed on 21 August 2016.

6. See generally, M. Toman and B. Jemelkova, 'Energy and Economic Development: An Assessment of the State of Knowledge', Discussion Paper 03–13, Resources for the Future, Washington, D.C., 2003, available http:// www.rff.org/files/sharepoint/WorkImages/Download/RFF-DP-03-13.pdf, accessed on 05 September 2016.

7. The first use of nuclear technology created a deep scar on the humanity's face. The nuclear bombing at Hiroshima and Nagasaki and its consequences has given the real time picture as to what nuclear energy is capable of doing to the humanity if not handled properly.

8. See generally, *Our Common Future*, Report of the World Commission on Environment and Development, p. 197, available http://www.un-documents. net/our-common-future.pdf, accessed on 25 August 2016.

9. *Our Common Future*, p. 201.

10. There are scholars who argue that sustainable development is a deep rooted concept and it existed in various forms even prior to the modern emphasis given by *Our Common Future*. See generally, U. Grober, 'Deep Roots–A Conceptual History of 'Sustainable Development (Nachhaltigkeit),' available https://bibliothek.wzb.eu/pdf/2007/p07-002.pdf, accessed on 26 August 2016.

11. The word 'international' is used for the simple reason that the entire globe was ruled by few major powers by the method of colonialism and imperialism.

12. Harris, 'Basic Principles of Sustainable Development'.

13. *Our Common Future*, p. 17.

14. *Our Common Future*, p. 17.

15. Harris, 'Basic Principles of Sustainable Development'.

16. See the Chairman's Foreword to *Our Common Future*, pp. 6–9.

17. J.A.D. Pisani, 'Sustainable Development—Historical Roots of the Concept', *Environmental Sciences*, 3(2) (2006), pp. 83–96, DOI: 10.1080/ 15693430600688831

18. See speech of Indira Gandhi, the then Prime Minister of India delivered at the Plenary Session of United Nations Conference on Human Environment, Stockholm, 14 June 1972.

19. The Conference resulted in two non-binding documents. Stockholm Declaration and Action Plan. See U. Beyerlin and T. Marauhn, *International Environmental Law* (Oxford: Hart Publishing, 2011), p. 7.

20. For details, see generally, Beyerlin and ThiloMarauhn, *International Environmental Law.*

21. Beyerlin and Marauhn, *International Environmental Law*, p. 6.

22. Beyerlin and Marauhn, *International Environmental Law*, p. 7.

23. *Our Common Future*, p. 37.

24. *Our Common Future*, p. 37.

25. *Our Common Future*, p. 32.

26. Report of the World Summit on Sustainable Development, held on 26 August to 4 September 2002, A/CONF.199/20, available http://www.unmil-lenniumproject.org/documents/131302_wssd_report_reissued.pdf, accessed on 25 August 2016.

27. *Our Common Future*, p. 119.

28. See World Population Prospects: Key Findings and Advanced Tables, 2015 Revision, available https://esa.un.org/unpd/wpp/publications/files/key_findings_wpp_2015.pdf, accessed on 5 September 2016.

29. Nuclear Power and Sustainable Development, IAEA, Austria, April 2006, available https://www.iaea.org/OurWork/ST/NE/Pess/assets/06-13891_NP&SDbrochure.pdf, accessed on 3 September 2016.

30. A. J. Bradbrook and R.L. Ottinger, *Energy Law and Sustainable Development* (Gland, Switzerland and Cambridge, UK: IUCN, 2003), p. 7, available http://cmsdata.iucn.org/downloads/eplp47en.pdf, accessed on 21 August 2016.

31. *Our Common Future*, p. 119.

32. M. Venkataramanan and Smitha, 'Causes and Effects of Global Warming', *Indian Journal of Science and Technology*, 4(3), pp. 226–9, available http://www.indjst.org/index.php/indjst/article/viewFile/29971/25926, accessed on 5 September 2016.

33. Venkataramanan and Smitha, 'Causes and Effects of Global Warming'.

34. 'Nuclear Power and Sustainable Development'; see also, write up on 'Atomic Energy', available http://www.un.org/en/globalissues/atomicenergy/index.shtml, accessed on 3 September 2016.

35. As per the information available at http://www.nei.org/Knowledge-Center/Nuclear-Statistics/World-Statistics, accessed on 5 September 2016.

36. W. Barletta, et al., 'Clean, Sustainable, Responsible: Nuclear Power for the U.S.', available https://www.innovation.ch/personal/chris/Nuclear%20Power%20paper_def.pdf, accessed on 21 August 2016.

37. B.K. Sovaccol and C. Cooper, 'Nuclear Nonsense: Why Nuclear Power Is No Answer to Climate Change and the World's Post- Kyoto Energy Challenges', *Williamm & Mary Environmental Law and Policy Review,* 33(1) (2008) Article 2, available http://scholarship.law.wm.edu/wmelpr/vol33/iss1/2, accessed on 05 September 2016.

38. See generally, *The Regulatory Challenges of Decommissioning Nuclear Reactors,* Nuclear Energy Agency, 2003, available http://www.oecd-nea.org/nsd/reports/nea4375-decommissioning.pdf, accessed on 5 September 2016.

39. Preamble, Charter of The United Nations, United Nations, Charter of the United Nations, 24 October 1945, 1 UNTS XVI, available https://treaties.un.org/doc/publication/ctc/uncharter.pdf, accessed on 26 August 2016.

40. Preamble, Charter of the United Nations.

41. See, Resolutions Adopted on the Reports of the First Committee UNGA RES/1/ (I), available http://www.un.org/ga/search/view_doc.asp?symbol=A/RES/1(I), accessed on 21 August 2016.

42. See the Official Website at http://www.un.org/en/globalissues/atom-icenergy/, accessed on 21 August 2016.

43. D. Fischer, *History of International Atomic Energy Agency: the First Forty Year,* available http://www-pub.iaea.org/mtcd/publications/pdf/pub1032_web.pdf, accessed on 21 August 2016.

44. UNGA/RES/32/50, 08 December 1977

45. UNGA/RES/32/50.

46. E. Benz, 'Lessons from Fukushima: Strengthening the International Regulation of Nuclear Energy', *William. & Mary Environmental Law and Policy Review,* 37 (2013), available http://scholarship.law.wm.edu/wmelpr/vol37/iss3/8, accessed on 5 September 2016.

47. See generally, *Convention on Nuclear Safety, Convention on Early Notification of a Nuclear Accident,* and so on.

48. See, Website of United Nations.

49. 1956, entered into force on 29 July 1957.

50. Art II of Statute of IAEA

51. Art III.B.1.

52. Art III.B.2.

53. Art III.B.3.

54. Art III.B.4.

55. Art III.B.5.

56. Art XVI.

57. Art III.A.

58. See generally, Art XII and XIX.

59. Art VIII, clause A.

60. Art VIII, clause A.

61. Art XI, clause B.
62. Art XI, clause D.
63. Art XI. clause E.
64. Art XI, clause F.
65. Art XI, clause F.
66. Art XII, clause 6.
67. Art XII read with Art XIX.
68. Art IX.
69. Art X.
70. Art IX.
71. See, Art XI.
72. Art IX.
73. Art III.A.5.

74. M.E. Baradei et al., 'International Law and Nuclear Energy: Overview of the Legal Framework', available https://www.iaea.org/sites/default/files/37302081625.pdf, accessed on 5 September 2016.

75. Baradei et al., 'International Law and Nuclear Energy: Overview of the Legal Framework'.

76. 1963 UNTS 293.

77. For more information, see http://www-ns.iaea.org/conventions/nuclear-safety.asp, accessed on 5 September 2016.

78. Preamble, Charter of the United Nations.
79. Art 7.
80. Art11.
81. Art 8.
82. Art 11.
83. Art 11.
84. Art12.
85. Art 13.
86. Art15.
87. Art 16.
88. Art17.
89. Art 18.
90. Art 19.

91. INFCIRC/274. The convention prior to the amendment was called Convention on the Physical Protection of Nuclear Material. The amendment was made in the year 2005 which came into force on 8 May 2016.

92. Art 1A.
93. Art 2 (1).
94. Art 2(4)(b).
95. Art 2(2).

96. The Annex provides different levels of protection of nuclear material as per the quantity of it. The greater the quantity the greater the protection is the idea. For details, see the original document.

97. Art 4(6).

98. Art 4(5).

99. Art 4(4).

100. Art 5, clause 2.

101. Art 5, clause 2.

102. INFCIRC/ 335. Convention on Early Notification of a Nuclear Accident, 1986.

103. Art 7.

104. Art 5.

105. Art 7.

106. Art 8.

107. INFCIRC/336. Convention on Assistance in the Case of a Nuclear Accident or Radiological Emergency, 1986.

108. Available at https://www.iaea.org/publications/documents/treaties/convention-assistance-case-nuclear-accident-or-radiological-emergency, accessed on 05 September 2016.

109. Art 2, para 4.

110. Art 4.

111. Art 2, para 6.

112. Art 6.

113. Art 7.

114. INFCIRC/546.

115. Art 1.

116. Art 1.

117. Art 4.

118. Art 6.

119. Art 7.

120. Art 8.

121. Art 9.

122. T. Findlay, *Unleashing the Nuclear Watchdog: Strengthening and Reform of the IAEA*, The Centre for International Governance Innovation (2012), p. 23, available https://www.files.ethz.ch/isn/144984/IAEA_findlay.pdf, accessed on 3 September 2016.

123. Baradei et al., 'International Law and Nuclear Energy: Overview of the Legal Framework'.

124. Findlay, *Unleashing the Nuclear Watchdog*, p. 43.

125. Findlay, *Unleashing the Nuclear Watchdog*, p. 10.

126. Findlay, *Unleashing the Nuclear Watchdog*, p. 10.

127. Findlay, *Unleashing the Nuclear Watchdog*, p. 10.

128. Findlay, *Unleashing the Nuclear Watchdog*, p. 29.

129. 'The Role of the IAEA Today', International Association of Lawyers Against Nuclear Arms (IALANA), available http://www.ialana.de/files/pdf/ver%C3%B6ffentlichungen/TheRoleoftheIAEAtoday-02-1.pdf, accessed on 05 September 2016; see also, the website, available http://independentwho.org/en/who-and-aiea-aggreement/, accessed on 5 September 2016.

130. See the Agreement between the International Atomic Energy Agency and the World Health Organization, text available http://apps.who.int/gb/bd/PDF/bd47/EN/agreements-with-other-inter-en.pdf, accessed on 5 September 2016.

131. 'The Role of the IAEA Today', International Association of Lawyers Against Nuclear Arms.

132. Available at http://www.un.org/en/sc/ctc/docs/conventions/Conv13.pdf, accessed on 5 September 2016.

4

V. CHANDRALEKHA

International Energy Law and WTO

Issues and Challenges

The demand and supply of energy have been dynamic subjects. Scientifically, technologies change; some are entirely new and others result in improved function and efficiency. Structurally, supply organizations vary, ranging from nationalized utilities to privately owned companies. Environmentally, all energy processes have impacts; some are heavily polluting, some cause effectively no pollution and most have less polluting alternatives.[1]

A successful economy depends on both supply and use of energy which is secure, safe, and efficient. Therefore, it is the duty of the governments to have clear policies for national energy supply as administered and checked by regulatory mechanisms. Most governments link such actions with policies for industry, commerce, and environmental care. Within energy supply and use, electricity is always of major interest, especially since grid supply has monopolistic characteristics and there are always concerns for safety and security.[2]

Energy laws govern the use and taxation of energy, both renewable and non-renewable Energy laws include the legal provision for oil, gasoline, and 'extraction taxes'. The practice of energy law includes contracts for extraction, licenses for the acquisition and ownership rights in oil and gas both under the soil before discovery and after its capture, and adjudication regarding those rights.[3]

Energy is the lifeblood of modern economies. The production and transmission of energy at the international level is a complex operation which often involves both goods and services. All countries are interdependent. Yet the energy trade at the international level is mainly regulated by domestic laws. There is no international energy law as such to regulate energy trade. The World Trade Organization (WTO) rules are applicable to trade in energy goods and services. But there is no separate agreement under the institutional framework of WTO. The General Agreement on Tariffs and Trade (GATT) and General Agreement on Trade in Services (GATS) do not define energy trade and trade in energy is not expressly included in its purview. Energy trade and GATT/WTO rules seem to largely operate in isolation from one another since, for the most part, these agreements contain rules of general application. Neither refers to 'energy' or deals specifically with energy matters, although the coverage of the GATS includes market access commitments on various kinds of services equally pertinent to the energy sector. GATT 1947 did not deal with energy as a distinct sector. Many other instruments like Energy Charter Treaty, Organization of the Petroleum Exporting Countries (OPEC), International Energy Agency (IEA), Regional Agreements like North American Free Trade Agreement (NAFTA), and so on, regulate international trade in energy. There are many conventions and treaties at the international level to protect environment and prevent climate change, through energy issues. From the point of view of international law, all these agreements regulate the international energy market, but the establishment of WTO has created much confusion by ignoring the energy trade totally. Importantly, some of the energy exporting countries are not the members of WTO. Now because of globalization and privatization, many energy exporting countries have recently joined WTO. The increasing energy needs and privatization of state owned entities, the relationship of energy trade and environmental pollution, debate on sustainable development, and so on, have strongly compelled the WTO to adopt some policy frameworks towards the regulation of trade in energy.

In the light of the above situation, it is pertinent to study whether GATT 1947 covers oil and petroleum products as goods. Or, what are the restrictions imposed upon the countries carrying energy trade? Why did the WTO not adopt separate agreement on energy trade? What are the other international instruments which cover trade in energy? Apart from these issues, this article highlights the WTO approach towards environmental pollution and conservation of biodiversity and renewable energy issues and restrictions imposed by the member countries themselves upon energy trade.

Multilateral Trade System: A Brief Overview

Before addressing these issues, it is important to know about the multilateral trade system at the international level after the Second World War.

International Trade after World War II

Most of the countries suffered economically and commercially after the Second World War. It was the US which had trade relation with almost all the major countries and controlled nearly 70 per cent of the international trade at that time. By then, most of the developed and developing countries had decided to establish international economic order. As a result, International Monetary Fund (IMF) and the International Bank for Reconstruction and Development (IBRD) were established.[4] At the same time, the proposal for establishing International Trade Organization (ITO) was adopted and a preparatory committee of 19 countries constituted by the United Nations Economic and Social Council to draft a charter for the same was formed. In the first meeting of the preparatory committee, US published its Suggested Charter for ITO which was popularly called the Havana Charter 1948. But the reluctance of the less developed countries towards the suggested charter and the conflict between the private enterprises and the state control over the issue of making decision on imports and exports, led to the cancellation of the proposal for establishing ITO. The countries which showed interest in establishing ITO came together and entered GATT. In other words, the birth of GATT is essentially by way of a historical accident, for the fact that when the president of the US refused to submit the Havana Charter to Congress for ratification, the Havana Charter and the ITO collapsed and there was a virtual head on collision between those who were wedded to the idea of free trade based on multilateralism on the one hand, and those who placed the whole emphasis on full employment policies on a national basis.[5] The 23 original contracting parties had participated in the Geneva round which established GATT 1947.

Energy Law and Policies under GATT 1947

When the rules of the GATT were negotiated, world energy demand, as well as price, was a fraction of what it is today. While energy has always been a crucial factor in geopolitics, at that time liberalizing trade in energy was not a political priority. The industry was largely dominated by the state run monopolies and thus governed by strict territorial allocation. International trade in energy resources and products was heavily concentrated, centralized, and controlled by a few multinational companies. This explains why the rules of GATT[6] do not deal with energy as a distinct sector. Apart from this, the main petroleum producing and exporting countries were not the original parties to the General Agreement on Tariffs and Trade 1947 (GATT 1947)[7] and the reigning so-called 'Seven Sisters' oil company cartel dominated the petroleum industry from the 1940s until the 1970s and preferred to settle its business outside the global trading system.[8] As a strategic commodity, energy was a particularly sensitive topic in international trade, a view poignantly reflected in the 1962 UNGA Resolution 1803 on Permanent Sovereignty over Natural Resources.[9] The exploration, development, and disposition of such resources, as well as the import of the foreign capital required for these purposes, should be in conformity with the rules and conditions which the people and nations freely consider to be necessary or desirable with regard to the authorization, restriction, or prohibition of such activities.[10]

The WTO

The establishing of the WTO on 1 January 1995 marked the biggest reform of international trade. From 1948 to 1994, the GATT provided the rules for much of the world trade and presided over periods that saw some of the highest growth rates in international commerce.[11] After a long period of Uruguay trade negotiation, the WTO was established through the instrument called Marrakesh Agreement, on 1 January

1995. The WTO agreement contains a series of multilateral agreements that are binding on all WTO members. These agreements include trade, not only in goods, but also in services, IPR, government procurement, and so on. There is a separate agreement for establishing the procedures for settling disputes that arise under these multilateral agreements.

In 1995, the WTO agreement was negotiated and signed by bulk of the world's trading nations not only for liberalizing trade, but in some circumstances, for maintaining trade barriers. For example, to protect consumers and prevent the spreading of disease. However, the WTO does not have any agreement that specially deals with energy trade even though energy is one of the important parts of international trade. But all tradable energy products or raw materials like coal, oil, natural gas, and so on, are covered under GATT 1994 as goods. Even though expressly energy trade was not included under GATT, the member countries were free to follow the principles of GATT in their energy trade.

Why GATT/WTO Rules Largely Operate in Isolation from Energy?

During the trade negotiations before establishing WTO, international trade in energy was limited. Most of the imports as well as exports were carried out by the governments and privatization was not encouraged. The need for energy was less at that time.

Apart from these, the major oil exporting countries were not the original members of the WTO or GATT. The Organization of Petroleum Exporting Countries (OPEC) established in the year 1960, was regulating the trade in petroleum energy.[12] A lion's share of energy is produced from fossil fuels and petroleum products and the countries producing these were out of the agreement. That is why the WTO member countries did not accept any separate provision in relation to trade.

WTO and Other Instruments of International Energy Law

Energy Charter Treaty

A more specialized international law instrument, tailored to regulate trade and investment in the energy sector, does exist, the Energy Charter

Treaty (ECT). The Energy Charter Treaty provides a multilateral frame-work for energy cooperation that is unique under the international law. It is designed to promote energy security through the operation of more open and competitive energy markets, while respecting the principles of sustainable development and sovereignty over energy resources.[13] This treaty establishes a legal framework in order to promote long-term cooperation in the energy field, based on complementarities and mutual benefits, in accordance with the objectives and principles of the Charter.[14]

The main provisions of the treaty are the protection of foreign invest-ments, based on the extension of national treatment, or most-favoured nation treatment (whichever is more favourable), and protection against key non-commercial risks;[15] non-discriminatory conditions for trade in energy materials, products, and energy-related equipment based on WTO rules, and provisions to ensure reliable cross-border energy tran-sit flows through pipelines, grids, and other means of transportation; the resolution of disputes between participating states, and—in the case of investments—between investors and host states; the promotion of energy efficiency, and attempts to minimize the environmental impact of energy production and use.[16]

The Treaty's trade provisions were initially based on the trading regulation of the GATT and were modified by the adoption of trade Amendments to the treaty in April 1998. After this Amendment, the treaty's provisions are in conformity with the WTO rules and practices based on the fundamental principles of non-discrimination between the Most favoured Nations (MFN)[17] and between the goods National Treatment (NT)[18] and transparency and a commitment to the progres-sive liberalization of international trade. The present amendment also expanded the treaty's scope to cover the trade in every related equip-ment and sets out a mechanism for introducing in future a legally–bind-ing standstill or customs duties and charges for energy related imports and exports.[19] Apart from these, Article IV says that the provisions of this treaty is non-derogatory to the GATT 1994 establishing WTO and others related instruments. Article IV reads, 'Nothing in this treaty shall derogate, as between particular contracting parties which are parties to the GATT, from the provisions of GATT and related instrument as they are applied between those contracting parties.'[20]

Part VI of ECT deals with transitional provisions for member coun-tries. Article 29 says that trade in energy material and products between contracting parties, at least one of which is not a party to that GATT or relevant related instruments shall be governed by the provisions of GATT 1994 and related instruments.[21] The 1998 Trade Amendment to the ECT brought it in line with WTO provisions. This amendment became complimentary to the WTO membership for non-members and also a stepping stone towards WTO membership. The agreement was one of energy specific and lack of separate instrument under WTO can be compensated through this instrument. It is a step towards implementation of WTO rules by WTO non-members.

The Organization for Economic Cooperation and Development

The Organization for Economic Co-operation and Development (OECD) was officially born on 30 September 1961, when the Convention entered into force. The mission of the Organization for Economic Co-operation and Development (OECD) is to promote policies that will improve the economic and social well-being of people around the world. The OECD provides a forum in which governments can work together to share experiences and seek solutions to common problems.[22] It also publishes various data of energy conservation and efficiencies and its impact upon the environment, and so on.

WTO activity cooperates with OECD in different ways. The WTO Director General regularly participates in the OECD Ministerial Council Meeting (MCM) and OECD forum. This brings together the ministers from OECD members and other selected invited countries to discuss economic and trade issues.[23] Sustainable economic growth is the ultimate goal of OECD. Apart from this, promoting employment and improving the standards of living, facilitating word trade growth, and marinating financial stability, and so on, are the objectives of the OECD.[24]

The International Energy Agency

The International Energy Agency (IEA) is an autonomous organization founded in 1974, which works to ensure reliable, affordable, and clean energy for its 29 member countries and beyond. The IEA has four main areas of focus—energy security, economic development, environmental awareness, and engagement worldwide.[25] Its four main areas of focus are Energy security—promoting diversity, efficiency, and flexibility within all energy sectors; Economic development—ensuring stable supply of energy to IEA member countries and promoting free markets to foster economic growth and eliminate energy poverty; Environmental awareness—enhancing international knowledge of options for tackling climate change; and, Engagement worldwide—working closely with non-member countries, especially major producers and consumers, and to find solutions to shared energy and environmental concerns.[26]

By providing periodical statistics and reports, it guides the governments on how to take precautions and measures to improve the energy trade.

Organization of the Petroleum Exporting Countries

The Organization of the Petroleum Exporting Countries (OPEC) is a permanent, intergovernmental organization, created at the Baghdad Conference on 10–14 September 1960, by Iran, Iraq, Kuwait, Saudi Arabia, and Venezuela. The five Founding Members were later joined by nine other Members—Qatar Indonesia, Libya, United Arab Emirates, Algeria, Nigeria, Ecuador, Angola, and Gabon.[27] The OPEC mainly regulates crude oil price control at the international level and maintains market stability.

The objectives of OPEC set out in Art II of its statute is as follows:

1. The principal aim of the organization shall be the co-ordination and unification of the petroleum policies of member countries and the determination of the best means for safeguarding their interests, individually and collectively.[28]
2. The organization shall devise ways and means of ensuring the stabilization of prices at international oil markets with a view to eliminates harmful and unnecessary fluctuations.[29]
3. Due regard shall be given at all times to the interests of the producing nations and to the necessity of securing an income to the producing countries an efficient, economic, and regular supply of petroleum to consuming nations and a fair return on their capital to those investing in the petroleum industry.[30]

A country, other than net petroleum exporting country, which does not fundamentally have interest and aims similar to those of member countries cannot be admitted as a member to the OPEC.[31] This organization is limited only to the oil producing countries and the total number of members is 11 at present. WTO is a multilateral organization with more than 150 members and any country can become a member. Recently, more and more petroleum exporting countries are joining the WTO and other related forums.

At the time of Negotiating GATT 1947, none of the petroleum producing countries were involved. The five founding members established OPEC on 14 September 1960 in Baghdad. Later 6 members joined the OPEC which changed the picture of OPEC. This made the GATT Nations (member countries) not to take any special interest in energy trade. However, among the 11 member countries, 6 countries have already become members of the WTO and two are negotiating their terms of accession. Iran and Libya have also applied for accession, but their application remains blocked by the US to this date. The only OPEC country that has never applied for accession to the WTO is Iraq. This development seems to indicate that the acute divergence between these two influential intergovernmental organizations does not make it impossible for one and the same country to satisfy the membership condition of both at one and the same time.[32] But it is very difficult for any country to be compliable with divergent rules and objectives of these organizations. For example, WTO prohibits its members from using export restrictions while OPEC often demands for export restrictions. However, WTO system allows several important exceptions to these rules.[33] For instance, conservation of exhaustible natural resources is most potent here. On the basis of this exception the member countries are able to balance the rules of both organizations.

European Union Energy Policy

The European Union (EU) was established after the Second World War and it consists of 28 countries as its members. The EU is the largest energy importer in the world, importing 53 per cent of the energy it consumes.[34] Still Europe is facing rising energy demand, increased prices, and disruptions to supply. To tackle these problems, EU, has adopted an energy policy which has three main goals, that are, security of supply, competitiveness, and sustainability. A separate energy union which ensures that affordable and climate friendly energy is supplied.[35] The main aim of energy union is to see that energy will flow freely across national borders within EU, to encourage new technologies, energy efficiency measures, and renewed infrastructure to reduce energy consumption.

The Directorate-General for Energy is one of the 33 policy-specific departments in the European Commission. It focuses on developing and implementing the EU's energy policy—secure, sustainable, and competitive energy for Europe.[36] These policies aim at the regulation of energy market by providing for affordable energy prices, promoting sustainable energy production, transport, and consumption and also for enhancing the conditions for safe and secure energy supply between EU Countries.

North American Free Trade Agreement

In 1994, the North American Free Trade Agreement (NAFTA) came into effect, creating one of the world's largest free trade zones and laying the foundations for strong economic growth and rising prosperity for Canada, the US, and Mexico. Since then, NAFTA has demonstrated how free trade increases wealth and competitiveness, delivering real benefits to families, farmers, workers, manufacturers, and consumers.[37] The objective of the agreement was to remove trade barriers between the US, Canada, and Mexico. The NAFTA contains separate Charter on the issues relating to energy trade and basic petrochemicals. It covers various issues relating to energy trade, including investment, cross-border services, measures related to trade in energy services, limitations, and barriers and measures related to investments in the territories of the parties as well as cross-border trade.[38]

Chapter 6 of NAFTA, which addresses energy and basic petrochemicals for the most part, extended to Mexico, the energy trade provisions that were established by the US and Canada in their 1988 Free Trade Agreement.[39] NAFTA liberalized the energy trade much more than energy investment by eliminating tariffs and quantitative restrictions on the trade in energy products; although Mexico was allowed to keep its licensing system with reserves petroleum trade to premix and electricity trade to the Capital Finance International (CFI).[40]

There are many Conventions and Treaties at the international level to protect environment and prevent climate change, through energy issues.[41] From the point of view of international law, all these agreements regulate the international energy market, but the establishment of WTO has created confusion by ignoring the energy trade totally.

Whether Energy Is a Good or a Service?

Traditionally, the classification of energy into goods and services did not necessitate because of less demand for energy and energy trade. But presently due to privatization and liberalization, the trade in energy increased, and the establishment of the WTO resulted in the division of energy into goods and services. Energy trade means trade in natural resources producing energy and also trade in electricity and nuclear resources. Trade in natural resources can be regulated by GATT. Trade in natural resources can be regulated by GATT as it is considered as goods. But electricity whether it is goods or service? The question is rightly taken up by WTO, where all primary energy like fossil fuels, oil, natural gas, and coal, were considered as goods. For WTO, all primary energy like fossil fuels, oil, natural gas, and coal, were considered as goods. The secondary energy which results from the conversion of primary energy, like supply of electricity is generally considered as service. Because, electricity is a physical process which takes place through the cables that carry it and it has to be generated more or less at the same time as it is being used.[42] But the International Convention on the Harmonized Commodity Description and Coding System (HS Nomenclature) declared Electricity energy as a commodity.[43] Now several countries have regulated the international trade of electric energy. Controls and specific customs procedures have been designed to deal with its particularities. For example, due to practical reasons, customs clearance is done after the entry of the goods in the territory and according to measurements at the power plant. Electric energy is thus considered to qualify as goods and is subject to the rules of the World Trade Organization (WTO).[44]

Thus it is very difficult to differentiate between goods and services aspects of trade in energy sector. In the European Union Law, Electricity is considered as goods and not service. Until now the word 'energy' or 'trade' is not used anywhere in the WTO agreement.

Before discussing the WTO provisions relating to energy trade, it is pertinent to know the other international regulations of energy and trade.

The Current Status of Energy in WTO Law

In the recent years, energy topics have reappeared on the negotiations agenda of the WTO. Several energy-exporting countries have recently joined the WTO and others are currently negotiating their accession, hence a substantially larger amount of energy trade is now in the hands of WTO Members trade.[45] Increasing energy needs have led to a growing interest in competition rules and export restriction practices. Progressive privatization of vertically integrated state-owned entities who offer a way for private operators to enter energy markets. The relationship of trade and environment and the debate on sustainable development is strongly bound to energy. The correlations between trade, energy, and climate change and the role of biofuels are also bringing attention to the trade in energy under multilateral trade regulation.[46]

Now the World Trade Organization (WTO) has also no special agreement on trade in energy, and it has not emerged in any of the sectarian agreements that have been drawn up since the Kennedy Round. Yet, since basic WTO rules are applicable to all forms of trade, they also apply to trade in energy goods and services. It is commonly understood that under the WTO rules, production of energy goods comes within the scope of the General Agreement on Tariffs and Trade (GATT), while energy related services, including transmission and distribution, fall under the scope of the General Agreement on Trade in Services (GATS).[47]

The biggest challenge to petroleum-exporting countries in this area probably comes from the United Nations Framework Convention on Climate Change (UNFCCC). A number of studies have been carried out on the impact of emission-reduction policies on the economies of different country-groupings. However, one point of consensus in the international economic community is that countries whose economies are highly dependent upon the export of fossil fuels are likely to suffer the most economic damage from climate change measures. OPEC's research, for example, suggests that OPEC Member countries could collectively suffer losses in revenue flows to the order of US$ 20 billion each year as a result of the proposed mitigation measures being implemented.[48]

A fundamental principle of GATT 1994 is the prohibition of quantitative restrictions on trade, which in principle applies equally to exports and imports (Article XI). It is subject, however, to a number of exceptions to petroleum products.[49]

WTO and Multilateral Agreement

Many multilateral treaties have been concluded along with Marrakesh Agreement establishing the WTO. Even though all agreements are not directly connected to the energy trade, there are some agreements which regulate energy trade. Some of them are explained here.

Agreement on Technical Barriers to Trade and Energy Trade

This Agreement deals with the rules relating to Technical Regulation and product standards. Members are free to adopt only those technical regulations which should not create 'unnecessary obstacles to international trade' and must not be more trade-restriction than necessary to fulfil a legitimate objective. The legitimate objectives include, national security, requirement, prevention of deceptive practices, protection of human health, and safety of animal or plant life, or health, or the environment.[50] These Technical Regulation are relevant to energy trade. For example, a number of WTO members have technical regulations relating to energy efficiency for products like automobiles and appliances.[51]

Now it is common to use energy efficiency labels to the products and provide information on the products energy performance. These labels are affixed to the products and provide information on the products energy consumption of electrical appliances and to show the levels of carbon dioxide emitted by the vehicle, and so on. The International Organization for Standardization (ISO) has developed standards on solar energy, hydrogen, and wind technology, solid and liquid bio-fuels, as well as concerning the calculation of thermal properties of buildings or construction materials.[52]

General Agreement on Tariffs and Service and Energy

Energy-related services include full range of activities from exploration for raw materials to distribution of finished products. The following services can be regarded as 'core' energy services, exploration, drilling, processing and refining, transport, transmission, distribution, waste management, and disposal. Other services include consulting, design and engineering, construction, storage, maintenance of the network, and services related to distribution, such as metering and billing.[53]

Energy related services were usually provided by the government, but privatization and liberalization has opened up many new services which has become a growing part of international trade. Energy services were not taken up by the WTO members and were negotiated as a separate sector during the Uruguay Round,[54] but some members made small commitments among themselves which covered only oil fluid source and mining and created electricity and natural gas. These groups have recently begun the process of bringing energy services fully under GATS discipline. There are hundreds of barriers to continue energy services. For example, the energy services across the borders, is possible only between the neighbouring countries especially in case of gas pipe lines.

The GATS signatories developed the WTO service sector classification list (W/120) which does not include energy service as a separate division. The service incidental to energy distribution, as well as transportation of fuels through pipe lines[55] recognized as part (sub clause) but not under separate rules. The general principles of GATS as applicable to all other services, specially to transport sector is governing these services also. There was a proposal for creating a separate division for energy services. But it has met resistance from member countries, who pointed to the potential for additional time and confusion in gaining acceptance of many WTO members that have not begun to focus on the energy services.[56]

There are four modes of supply covered by the GATS among which three modes of supply relate to energy. Mode I covers service that are supplied cross-border but do not require the physical movement of suppliers.[57] It covers cross-border transit of oil and gas through pipe lines and electronic power transmission. Mode II covers Consumption abroad, refers to situations where a service consumer moves into another Member's territory to obtain a service[58] which has less relevance to the energy service. Mode III covers Commercial presence, that is, a service supplier of one Member establishes a territorial presence, including through ownership or lease of premises, in another Member's territory to provide a service,[59] and Mode IV covers Presence of natural persons, which consists of persons of one Member entering the territory of another Member to supply a service[60] which covers the service of energy service companies. Thus, the GATS rules are applicable to the supply of energy services like all other commercial services.

Agreement on Subsidies and Countervailing Measures

According to this Agreement, providing of export subsidies and also local content or import substitute subsidies are prohibited because it directly affects the interests of the other members. Local content subsidies are prohibited as it violates the principle of National Treatment by encouraging the people to prefer domestic product over imported goods. Now these provisions of Agreement on Subsidies and Countervailing Measures (SCM Agreement) are invariably applicable to the energy trade. The fossil fuels and renewable energy subsidies are recognized by the SCM Agreement. Most Environmentalists have focused their attention on eliminating the subsidies for fossil fuels. The SCM Agreement prohibits providing of export or local content subsidies in all forms of energy resources.

Energy Trade under GATT/WTO

As mentioned earlier, the GATT/WTO has not made any special provision for the regulation of trade in energy sources. But one school of thought believes that international trade in energy is included in and is subject to GATT/WTO disciplines. It holds that international trade in energy is governed by the WTO laws like any other trade in goods and services and that it is not excluded from the coverage of GATT and WTO laws.[61] So the major principles of WTO, like National Treatment, Most Favoured Nations, Transparency, and so on, are applicable to energy trade also. Protection of domestic market and welfare of the nations is the concern of all the countries. So, if there is a threat to the national security or any such situation, then deviation from rules and principles is permitted under article XX of GATT which deals with general exception.

Electric Energy

Under the International Convention on the Harmonized Commodity Description and Coding System (HS Nomenclature), 2716 heading, Electricity energy is already declared as a commodity. Much of Article VII of the General Agreement on Tariffs and Trade 1994 relates to electricity energy. The WTO Customs Valuation Agreement (CVA)[62] controls as well as specific customs procedures that have been designed to deal with its particularities. Among the WTO rules, for implementation is a separate instrument known as the Agreement on Implementation.[63] Historically there were insufficient cross-border interconnections and the lack of large multi-country network in electricity trade. There was limitation to electricity exchange between the governmental bodies, without necessarily relying on GATT/WTO or other trade agreements. Apart from these, electricity is mostly traded on a regional basis, due to the reliance on network as the only means of transportation and the condition of trade liberalizations are often determined by the regional trade agreements rather than WTO. The de-regulation and liberalization of electricity markets and the introduction of competition have more recently led the governments to the recognition of relevance of WTO rules.[64]

Thus, neither the GATT nor the WTO Agreement was considered as having a direct bearing on international energy trade. Trade in hydrocarbons, fissionable materials, and cross-border transmission of electricity largely take place outside the multilateral trading system. The energy trade and GATT/WTO rules seem to largely operate in isolation from one another, since, for the most part, these agreements contain rules of general application. Neither refers to 'energy' or deals specifically with energy matters, although the coverage of the GATS includes market access commitments on various kinds of services equally pertinent to the energy sector.[65]

Protection of Biodiversity

The term biodiversity refers to the variety of life on earth at all its levels, from genes to ecosystems, and the ecological and evolutionary processes that sustain it.[66] It is the primary and paramount duty of every nation to protect the ecosystem. For this purpose the member countries of WTO are free to override the principles of WTO. Article XX of GATT reads as follows:

> Subject to the requirement that such measures are not applied in a manner which would constitute a means of arbitrary or unjustifiable discrimination between countries where the same conditions prevail, or a disguised restriction on international trade, nothing in this Agreement shall be construed to prevent the adoption or enforcement by any contracting party of measures:
>
> 1. Necessary to protect public morals;
> 2. Necessary to protect human, animal, or plant life or health;
> 3. Relating to the importations or exportations of gold or silver;
> 4. Necessary to secure compliance with laws or regulations which are not inconsistent with the provisions of this Agreement, and the protection of patents, trademarks, and copyrights, and the prevention of deceptive practices;
> 5. Relating to the products of prison labour;
> 6. Imposed for the protection of national treasures of artistic, historic, or archaeological value;
> 7. Relating to the conservation of exhaustible natural resources if such measures are made effective in conjunction with restrictions on domestic production or consumption;
> 8. Undertaken in pursuance of obligations under any intergovernmental commodity agreement which conforms to criteria submitted to the contracting parties and not disapproved by them or which is itself so submitted and not so disapproved;
> 9. Involving restrictions on exports of domestic materials necessary to ensure essential quantities of such materials to a domestic processing industry during periods when the domestic price of such materials is held below the world price as part of a governmental stabilization plan;
> 10. Essential to the acquisition or distribution of products in general or local short supply; Provided that any such measures shall be consistent with the principle that all contracting parties are entitled to an equitable share of the international supply of such products, and that any such measures, which are inconsistent with the other provisions of the Agreement shall be discontinued as soon as the conditions giving rise to them have ceased to exist.

Article XX allows exception to the non-discriminatory rules for legitimate environmental and conservation measures applied by the government.[67] Among these, exception measures necessary to protect human, animal, or plant life, or health and measures relating to the conservation of exhaustible natural resources taken in conjunction with domestic restrictions are directly connected to the conservation of bio-diversity. It says that conservation of bio-diversity and protection of environment is superior to the WTO principles and rules.

WTO—Multilateral Agreements Relating to Biodiversity

Agreement on Government Procurement promotes the government purchasing and procurement to achieve domestic policy goals and promoting renewable energy. The TRIPS directly relates to the Industrial polices and it prohibits the patenting of animals and human beings. Even the plant life is non-patentable and protected by the TRIPS. Agreement on Technical Barriers to Trade also supports the trading of energy efficient technical goods. The general exception to the rules of WTO allows the member countries to restrict import or export of energy consuming goods by imposing restrictions on the basis of protecting the environment. WTO promotes the member countries to trade and use the energy efficient technologies. If any member country is able to prove the adverse effect of such technology to the environment or plant or human life, then it is able to impose restrictions upon the international trade. The environmental reason to promote renewable energy may not be sufficient to invoke general exception to trade rules. For example, under GATT Article XX, in order to justify the use of certain subsidies or industrial policies supporting renewable energy technology manufacturing or deployment may require member countries to demonstrate a link between renewable energy technology and health and that such measures are necessary to displace fossil fuels and prevent climate change.[68] Further providing subsidy to the manufacturing of environment friendly and renewable energy sources is disputed under subsidies and countervailing measures agreement.

WTO—Dispute Settlement on Energy Issues

For the purpose of protecting the environment, many member countries of WTO supported the production of sustainable energy. They provided more importance to the renewable sources of energy. But unfortunately in the recent years there is a sharp increase in the dispute over renewable energy subsidies. Since 2010, four disputes have been filed before the WTO-DSB by challenging the government programs supporting renewable energy.[69]

The first ever case concerning the renewable energy subsidies that came before the DSB was between the Japan and Canada.[70] At first the dispute was filed by Japan—which was later joined by EU—they challenged the Feed-In-Programme of the government of Canada where they promoted the use of domestically produced equipment for renewable energy generation. Japan challenged these measures as it violates Art III.4 and Art III.8 (a) of GATT, Art I.1 of SCM Agreement, and also Art II.1 of TRIMS. Both the Panel and Appellate Body found that it did not fall within the scope of derogation of Art III.8(a). The Appellate Body held that the product being procured by the government of Ontario is electricity generated through renewable source, but the foreign product suffering from discrimination was electricity generation equipment, where there is no competitive relationship. However, with regard to Art II.1 of TRIMS and Art III.4, the Appellate Board upheld the Panel conclusion that the minimum requirement of domestic content levels are inconsistent and also the measures at issue constituted the financial contribution in the form of government purchase of goods to the Art III.4 of GATT. After the Canada Renewable energy dispute on 6 February 2013, US requested for the constitution of a Panel with India against the Jawaharlal Nehru National Solar Mission Programme.[71] Similar allegations like Canada Renewable Energy Issues were filed against India. The Government of India filed documents with WTO to explain how a number of State and Local incentive programmes for renewable energy were consistent with the WTO obligations. In another dispute,[72] the US Commerce Department and International Trade Centre imposed Anti-dumping and countervailing duties on Chinese imports of Wind Towers and Silicon Solar Plants. These decisions invariably show that the regulation of energy trade is covered under the various multilateral instruments of GATT/WTO.

Effects of Energy Use on Bio-diversity

Energy is inevitable for human life, use of energy, in any form, causes environment pollution. At the same time the process of energy generation also affects the environment adversely. For example, mining, nuclear energy, fossil fuels, and so on. The process of generation of renewable energy also distracts bio-diversity in certain places, for example, construction of dams for generating hydro-electricity power adversely affects the ecosystem of that area. But compared to the pollution caused by the non-renewable energy, the pollution caused by the renewable energy is very less. The construction of dam for hydro-electricity power affects the biodiversity of the area, but if such energy is not promoted then the biodiversity of the whole world will be affected. Some technology assist directly with energy conservation; while other technologies are emerging that help the environment by reducing the amount of waste produced by human activities.[73] Energy sources such as solar power create fewer problems for the environment than traditional sources of energy like coal and petroleum.[74]

The sustainable development of energy must be promoted to protect the environment. Sustainable development is defined in Brundtland Commission's 1987 report—development that meets the needs of the present without compromising the ability of future generations to meet their own needs.[75] The primary objective of this principle is to meet current and future human needs and aspirations. The future generations have the same rights to develop as we do and preventing such development would be unfair. Apart from these, energy efficiency and energy conservation measures can also be promoted at the international level to protect biodiversity.

***

There is no international energy law as such to regulate energy trade. The WTO rules are applicable to trade in energy goods and services. But there is no separate Agreement under the institutional framework of WTO. The GATT/GATS do not define the energy trade and trade in energy is not expressly included in its purview. There are many conventions and treaties at the international level to protect environment and prevent climate change, through energy issues. From the point view of international trade law, all these agreements regulate the international energy market, and the WTO also indirectly regulates the energy trade. As the population increases, the demand for energy also increases, which results in increasing of energy trade. It affects environment as well as the biodiversity. Like energy trade, even environmental protection is also not considered through separate instruments by the WTO. Each and every Multilateral Instruments of WTO makes provision for environmental protection and conservation of natural resources. As we know, energy transactions cause highest environmental pollution. For the purpose of protecting biodiversity, the WTO trade agreements can accept separate instruments so that along with the regulation of trade, environment pollution can be minimized.

Notes and References

1. UNIDO, 'Energy for Sustainable Development: Policy Options for Africa', available https://www.iaea.org/OurWork/ST/NE/Pess/assets/un-energy_africa_pub.pdf, accessed on 10 December 2015

2. UNIDO, 'Energy for Sustainable Development', p. 10.

3. 'Energy Law', available www.wikipedia.com, accessed on 13 December 2016.

4. The agreement of IMF and IBRD adopted in July 1944 by the United Nations Conference at Bretton Woods, New Hampshire.

5. K. Koul, *The General Agreement on Tariffs and Trade/World Trade Organisation Law, Economics and Politics*, First Edition (New Delhi: Satyam Books Publication, 2005), p. 7.

6. T. Cottier, et al., 'Energy in WTO Law and Policy', WTO available https://www.wto.org/english/res_e/publications_e/wtr10_forum_e/wtr10_7may10_e.pdf, accessed on 13 December 2015.

7. Saudi Arabia, the United Arab Emirates (UAE), China, the Russian Federation, and Venezuela were not original members of GATT 1947.

8. Royal Dutch Shell, Exxon, Gulf, Texaco, BP, Mobil, and Standard Oil of California (today's Chevron).

9. A. Marhold, 'The World Trade Organization and Energy: Fuel for Debate', *European Society of International Law*, 2(8), available http://www.esil-sedi.eu/sites/default/files/Marhold%20-%20ESIL%20Reflections.pdf, accessed on 21 January 2016.

10. UNGA Res. 1803 (XVII) (14 December 1962). General Assembly Resolution 1803 (XVII) of, 'Permanent Sovereignty over Natural Resources', available http://www.ohchr.org/EN/ProfessionalInterest/Pages/NaturalResources.aspx, accessed on 21 January 2016.

11. N.S. Bansal, *WTO After CANCUN*, First Edition (New Delhi: Mittal Publication, 2004), p. 14.

12. Organization of Petroleum Exporting Countries, available http://www.opec.org/opec_web/en/about_us, accessed on 13 December 2015.

13. The Energy Charter Treaty was signed in December 1994 and entered into legal force in April 1998, available http://www.energycharter.org/process/energy, accessed on 12 January 2016.

14. Art 2 of Energy Charter Treaty.

15. The Energy Charter Treaty Art 10–17 deal with investment promotion and protection in cases of loss.

16. The Energy Charter Treaty, Art 19 deals with environmental aspects.

17. Art I of GATT, Art II of GATS, and Art IV of Trade Related Aspects of Intellectual Property Rights (TRIPS).

18. Art III of GATT, Art XVII of GATS, and Art III of TRIPS.

19. Secretary General, 'Energy Charter Treaty and Related Documents: A Legal Framework for International Energy Cooperation', Energy Charter Secretariat, available http://www.ena.lt/pdfai/Treaty.pdf, accessed on 14 January 2016.

20. Art IV of Energy Charter Treaty.

21. Art XXIX, Para 2 (a) of Energy Charter Treaty.

22. International Energy Agency, available http://www.oecd.org/about, accessed on 13 December 2015.

23. The WTO and Organization for Economic Cooperation and Development, available www.wto.org, accessed on 13 December 2015.

24. Organization for Economic Co-operation and Development, available www.oecd.org, accessed on 13 December 2015.

25. International Energy Agency, available http://www.iea.org/aboutus, accessed on 13 December 2015.

26. International Energy Agency, available http://www.iea.org/aboutus, accessed on 13 December 2015.

27. Organization of Petroleum Exporting Countries, available http://www.opec.org/opec_web/en/about_us, accessed on 13 December 2015.

28. Art II (a) of Statutes of Organization of the Petroleum Exporting Countries.

29. Art II (b) of Statutes of Organization of the Petroleum Exporting Countries.

30. Art III (c) of Statutes of Organization of the Petroleum Exporting Countries.

31. Art VII of Statutes of Organization of the Petroleum Exporting Countries.

32. M.G. Desta, 'OPEC and the WTO: Petroleum as a Fuel for Cooperation in International Relation', *Journal of World Trade*, 379(10) (2003), available http://www.mafhoum.com/press7/185E17_fichiers/a47n10d01.htm, accessed on 12 September 2016.

33. Art XX of General Agreement on Tariff and Trade.

34. International Energy Agency, 'Energy Policy of IEA Countries, European Union' (2014), available https://www.iea.org/publications/freepublications/publication/EuropeanUnion_2014.pdf, accessed on 20 December 2015.

35. European Aluminum, available www.atueuropea.eu, accessed on 20 December 2015.

36. Directorate General for Energy, available https://ec.europa.eu/energy/en/about-us, accessed on 20 December 2015.

37. North American Free Trade Agreement (NAFTA), available www.naftanow.org, accessed on 20 December 2015.

38. Chapter VI, Art 601 to 609 of NAFTA.

39. The Institute for International Economics, available www.iie.com, accessed on 13 December 2015.

40. The Institute for International Economics.

41. The United Nations Framework Convention on Climate Change (UNFCCC) 1771 UNTS 107 [1994] ATS 2/31 ILM 849 (1992) and its Kyoto Protocol 2303 UNTS 148/[2008] ATS 2/37 ILM 22 (1998).

42. Energy Quest, available http://www.energyquest.ca.gov, accessed on 12 January 2016.

43. The International Convention on the Harmonized Commodity Description and Coding System (HS Nomenclature) (concluded on 14 June 1983) (UNTS Vol. 1503, 1-25910) from the Customs Co-operation Council (C.C.C.), designated the code (heading) 2716 to electricity energy. The heading is under chapter 27, which comprehends, 'Mineral fuels, mineral oils, and products of their distillation; bituminous substances; mineral waxes'.

44. Cottier, et al., 'Energy in WTO Law and Policy'.

45. Cottier, et al., 'Energy in WTO Law and Policy'.

46. Cottier, et al., 'Energy in WTO Law and Policy'.

47. Cottier, et al., 'Energy in WTO Law and Policy'.

48. UNCTAD, 'Trade Agreements, Petroleum and Energy Policies', United Nation Publication (2000), available http://unctad.org/en/Docs/itcdtsb9_en.pdf, accessed on 12 September 2016.

49. UNCTAD, 'Trade Agreements, Petroleum, and Energy Policies'.

50. Art II, Para 2 of Agreement on Technical Barriers to Trade and Energy Trade.

51. A. Yanovich, 'WTO Rules and the Energy Sector', in Julia Selivanova (ed.), *Regulation of Energy in International Law: WTO; NAFTA and Energy Charter* (Netherland: Wolters Kluwer Publication, 2011).

52. Yanovich, 'WTO Rules and the Energy Sector', p. 11.

53. UNCTAD, 'Trade Agreements, Petroleum, and Energy Policies'.

54. P.C. Evans, 'Strengthening WTO Member Commitments in Energy Services—Problems and Prospects', in Pierre Sauve and Adithya Matoo (eds), *Domestic Regulation and Service Trade Liberalization*, (World Bank and Oxford University Press, 2003), p. 167.

55. Classification of Service Sector—MTN/GNS/W/120 World Trade Organization, available www.wto.org, accessed on 14 January 2016.

56. P.C. Evans, *Liberalizing Global Trade in Energy Services* (Washington: The American Enterprise Institute Press, 2002), p. 37.

57. Art I, Para 2 (a) of GATS.

58. Art I, Para 2 (b) of GATS.

59. Art I, Para 2 (c) of GATS.

60. Art I, Para 2 (d) of GATS.

61. R.L. Arcas, A.F., and E.S.A. Gosh, *International Energy Governance-Selected Legal Issues* (UK: Edward Elgar Publishing Ltd., 2014), p. 90.

62. Art VII of the General Agreement on Tariffs and Trade.

63. World Trade Organization, available www.wto.org, accessed on 30 December 2015.

64. J. Bielecki and M. Geboye Desta (eds), *Electricity Trade in Europe, Review of Economic and Regulatory Changes* (Netherlands: Kluwer Law International Publication, 2004), pp. 122–3.

65. Organization for Economic Co-operation and Development.

66. Centre for Biodiversity and Conservation, American Museum of Natural History, available www.amnh.org, accessed on 29 December 2015.

67. D.R. Downes, *Integrating Implementation of the Convention on Biological Diversity and the Rules of World Trade Organization* (Switzerland and UK: IUCN Publication, 1999), p. 9.

68. J. Lewis, 'The Rise of Renewable Energy Protectionism: Emerging Trade Conflict and Implication for Low Carbon Development', *Journal of Global Environmental Policies*, XIV (November 2014).

69. *Canada v. Japan*, DS412, *China v. USA*, DS419, *Canada v. European Union*, DS426, *European Union v. Argentina*, DS473, available http://.www.wto.org, accessed on 19 January 2016.

70. DSB June 5 2014, Appellate Body Decision, DS 412–26.

71. T. Meyer, 'Energy Subsidies and WTO', *American Society of International Law Journal*, 17 (22) (September 2013), available www.asil.org, accessed on 19 January 2016.

72. DSB, 7 July 2014, Appellate Body Decision, DS 449.

73. P. BalLabh, *International Environmental Law*, First Edition (New Delhi: Cyber Tech Publication, 2011), p. 5.

74. BalLabh, *International Environmental Law*.

75. S. Bell and D.Mc Gillivray, *Environmental Law*, Sixth Edition (Oxford: Oxford University Press), p. 62.

5

NIKITA PATTAJOSHI

AKASH KUMAR

Intellectual Property in the Way of a Clean and Green Environment

Is Licensing the Solution?

Technology in the commercial context would mean 'knowledge of how to make use of the factors of production, to produce goods or services for which there is an economic demand'.[1] Technology innovation both hurts and helps the environment. Therefore, intellectual property, by promoting innovation may be both good and bad for the environment.[2] However, technology, as far as Environmentally Sound Technologies (ESTs)[3] are concerned, works for the benefit of the environment. ESTs can be considered to be those beneficial environmental technology innovations which provide a net environmental benefit compared to existing technologies in terms of resources consumed, wastes produced, and risks to human health and the environment, for example, waste management technologies for solid and hazardous wastes, products and methods for cleaning up pollution, recycling equipment, and processes.[4]

There are national attempts to move towards cleaner sources of energy and away from fossil fuels not just because of the environmental concerns surrounding the use, but also to avoid the supply (of fossil fuels) problem to some extent.[5] Clean, renewable energy is thus not only important from the environmental protection point of view, but also has enormous political and national security implications. Currently, all energy produced by green technologies—wind, solar, hydroelectric—is significantly more expensive than fossil fuels.[6] Considering the high cost involved, individual consumers would not embrace cleaner technologies unless the cost is at par with traditional technologies available to them.[7] For instance, unless the Government comes up with a mandate to run all vehicles on CNG and no other fuel, no users would switch to CNG operated vehicles considering higher cost involved in use of CNG over other fuels. This leaves a heavy burden on the State to create a market for clean and green energy. However, considering the high cost involved, access to such green technologies, for countries which do not have capability to come up with technology of their own, becomes an expensive affair, thereby augmenting the access problem. Promoting greater access to and transfer of these technologies, in particular of Environmentally Sound Technologies (ESTs), was already a central concern at the Rio Earth Summit in 1992 and its important outcome, the Agenda 21 on sustainable development.[8] In this context it is imperative to analyse the existing standards and flexibilities under various legal frameworks in general and Trade Related Aspects of Intellectual Property Rights (TRIPS) in particular.

While discussing the various challenges posed by climate change on the world at large, the role of Intellectual Property (IP) in addressing these challenges cannot be ignored. This is essentially because IP is intricately linked with innovation and is often considered a prerequisite of it and thus any innovation in climate change technologies would include various IP rights in its fold. Transfer and diffusion of environmentally sound technologies (ESTs)[9] or climate change mitigation technologies (MTs), as it may be called, to developing countries is an essential part of any effective international response to global climate change challenge.

International Legal Framework

UNFCCC

The United Nations Framework Convention on Climate Change (UNFCCC) under Article 4 provides for parties, particularly developed countries, to attempt to promote development and diffusion of climate change related technologies.[10] The framework thus requires developed countries to promote the transfer of, or access to, environmentally sound technologies (ESTs).[11] Besides this, the Bali Action Plan[12] calls for 'enhanced action' in area of technology development and 'transfer' and requires 'effective mechanism' for removal of 'obstacles' to transfer of technology to developing country party. Despite these requirements countries have been facing difficulty in coming up with any mechanism that can counteract the obstacles to transfer of technology.

TRIPS

Traditionally, governments of various countries display unwillingness to surrender their sovereignty over specific implementation of Intellectual Property Rights (IPRs). In light of these traditional reservations of sovereign control, the WTO TRIPS agreement is drafted in a manner that would provide significant government flexibility in implementation of various norms.[13]

The TRIPS Agreement provides for minimum standards for the availability, acquisition, and enforcement of intellectual property rights in general for the members of WTO. Thus, the TRIPS framework does not provide for development and diffusion of ESTs or clean and green technologies in specific. However, climate change technologies may come from different fields of innovation and thus different category of IP rights may be relevant for development of different climate change technologies.

For example, patents can be available for a range of climate change technologies including, photovoltaic sector, wind energy, and so on. Article 30 of TRIPS permits members to provide for 'limited exceptions' to patent rights that do not unreasonably conflict with normal exploitation of the patent.[14] Article 31 of TRIPS provides for unauthorized use beyond scope of Article 30, including 'compulsory licenses' and sets out conditions for such licenses. These provisions provide optimum flexibility in the grant of government use licenses, including by eliminating any requirement for prior negotiations with patent holders as a precedent to granting licenses.[15]

Besides, trade Secrets are also common in fields related to climate change technologies.[16] However, it is less likely that TRIPS in this sense would foster diffusion of climate change related technologies as measures that would seek to compel disclosure of trade secrets associated with climate change technologies would not be consistent with the TRIPS level protection.[17]

WIPO

Article 10 of the Agreement between United Nations and the World Intellectual Property Organization[18] stipulates that WIPO must cooperate with other UN bodies in 'promoting and facilitating the transfer of technology to developing countries in such a manner as to assist these countries in attaining their objectives in fields of science and technology and trade and development'.

Further, WIPO has come up with a platform 'WIPO Green' which is a 'match making' ground to foster development and dissemination of green technologies.[19] It is thus a network of non-governmental organizations, intergovernmental organizations, universities, innovators, and governmental agencies from around the world working towards capacity building, knowledge transfer, and licensing of technology to foster development in developing and least developed countries.[20]

IP as a Barrier v. IP as a Facilitator

How IPR affects international transfer of technology has become an increasingly pressing issue for developing countries given the mounting pressure that they face in bringing their IPR protection to the level of that in developed countries.[21] While it can be said that patents, by their very nature pose a substantial barrier to diffusion of technology, certain studies undertaken with regards to specific technologies have shown that patents and other IPRs may not be acting as barriers to market entry.[22]

It is true that intellectual property can be helpful in driving environmental technology innovation just as it drives innovation in other sectors. For example, patenting rates in selected clean energy technologies has increased around 20 per cent per year since 1997.[23] However, it is believed that giving special treatment to environmental technology innovation under patent laws is unlikely to have much positive impact on innovation.[24] This is evident from the fact that despite flexibilities being in place under the TRIPS Agreement and the agreement specifically incorporating technology transfer within the ambit of its provisions, there has not been a significant flow of ESTs (or other technologies for that matter) from developed to developing countries.

Further, it is feared that patents might pose an issue as a broad patent would complicate the development of a major category of new, more efficient or less expensive technology. It is true that while patents are 'the' IPR with respect to a host of environmental technologies, other forms of IP protection too come into picture down the transfer chain— from research and development (R&D), to product design, commercialization, deployment, use, and absorption.[25] Thus, protection may be accorded under technology transfer, trade secrets, and so on. But the problem with technology transfer is that environmental technology transfer requires careful licensing of IP rights as innovators may be reluctant about transferring their technology to a country where it will be difficult or impossible for them to prevent immediate copying and competition. Similarly, it is difficult to foster diffusion of climate change related technologies as measures that would seek to compel disclosure of trade secrets associated with climate change technologies would not be consistent with the TRIPS level protection.[26]

It should also be understood that a successful technology transfer depends on the 'absorptive capacity' of the recipient country. But where there exists a knowledge gap between developed and developing countries and an acute lack of skilled workforce in many of such countries, it may also serve as an impediment against successful transfer.

The proponents of strong IP rights believe that one way how government can promote green technology is through a robust patent system. However, such government sponsored effort to incentivize green energy may not work well with all kind of economies. It is a well-known principle that patent is awarded as a quid pro quo for an inventor's disclosure of the invention in the patent and includes in itself a right to exclude others.[27] This right to exclude first, incentivizes the inventor to invent and secondly, inspires and provides ideas to other inventors. The patent system thus perpetuates more invention in an ongoing cycle of innovation. However, a robust patent regime not only provides incentive to innovate, but also incentive to invest. This is because investors seek help of patent protection to gauge the security of their investment.

Patentable technology which is energy efficient can be vaguely classified into the following types, depending on its market maturity.[28] Venture stage technologies (for example, nanomaterials and wireless power); Emerging market stage technologies (for example, ethanol, biofuels, and smart grid); Mature market stage technologies (for example, solar and wind power, hybrid vehicles, and light-emitting diode (LED) lighting).

It is seen that substantial innovation in green technologies takes place in start-up companies characterized by large intangible assets, technological uncertainty, and low liquidation value.[29] Thus flow of private capital is an essential part of research and development in green technology. Such investment is in turn secured by a strong patent system. Particularly companies with venture stage technology focus on patent filing strategy, building a strong patent portfolio, and using their patent position to reserve access to technology aimed at serving a particular market.[30] Companies with emerging market stage technology focus on value creation, developing market, share and using their patent portfolio to create a network of users of their technology. It is these companies which would like to increasingly license their technology in order to obtain market dominance in a particular technology area and create substantial value for its technology.[31]

Thus, the question whether or not IP is a significant barrier to access to technology will depend on the type of market and type of technology supply.[32] For instance, for export of photovoltaic cells or ethanol, it is unlikely that IP would be a significant barrier as the key concern may be other tariff and similar barriers but for wind engines, it seems that IP may probably pose a little problem.[33] This is because in these sectors developing nations have succeeded in entering the industry.

It is imperative that in case of a transfer of technology with regards to climate change mitigation and adaptation, adequately assessing benefits of the technology that is in public domain and accordingly building capacities on those technologies is essential. However, one of the views that act against smooth transfer of ESTs to developing countries is that very often recipient countries of technology transfer in this field do not have a clear assessment of their technological needs.[34]

Thus, there have been diverse opinions over the issue whether IP rights act as a barrier to dissemination of technology. There is one line of thought which believes that IP rights, especially patents act as a major barrier to diffusion of green technologies and are a clear and present problem for developing countries wishing to access them and hence these countries should use compulsory licensing mechanism to access the required technologies.[35] However, the argument to the contrary that patent does not stand in the way of transfer and diffusion of green technology is based on the finding that there is a dearth of patents on such technologies even in some developed countries and they in turn depend on countries like China and India. Such line of thought is further substantiated by the argument that most fundamental green technologies are off-patent and thus green tech patents, if they exist, rarely confer market power.[36] This view has been countered with help of the argument that many patents cover minor or trivial developments and may be used to block genuine innovation and competition in the field of green technology.[37]

An examination by Dinopoulos and Kottaridi of the growth effects of National Patent Policies led them to the conclusion that harmonization resulting in stronger protection of patent rights has 'an ambiguous effect on the rate of international technology transfer'. [38] For example, the UNDP has pointed out that in some instances, the bilateral investment treaties (BITs) instead of boosting Foreign Direct Investment (FDI) and technology transfer, there is likelihood that such treaties do not translate into affordable technology transfer for developing countries.[39]

On the other hand, there has been evidence to point out that advent of strong patent regime has inhibited technology transfer due to excessive direct and indirect costs associated with restrictive clauses and a decline in bargaining power of the technology buyer.[40] Another factor affecting transfer of know-how has been the 'working of patents'.[41] It is being seen that a large number of patents held by inventors have not been put to work for profits, and that many SMEs have been unable to achieve transformation or upgrade to better technology due to lack of R&D funds and time.

The most prominent issue therefore in this debate is that the debate is not backed by sufficient empirical evidence and reliable objective study or data[42] with regards to innovations in field of ESTs and thus rides on general notions of dichotomy between IP and innovation. Hence, there is an enormous 'grey area' in the debate. In light of this, whether compulsory licensing can actually enhance diffusion of patented technologies is an unanswered question.

Lessons Learnt from the 'Right to Health' Discourse

The perceived tension between relying on strong IP rights to promote clean technology innovations and enabling global access to these technologies can be seen *analogous* to the tussle between strong IP rights to protect pharmaceutical inventions and fight for access to life saving drugs in developing countries. Thus the 'right to environment' discourse, in this sense can be seen running parallel to the 'right to health' discourse and questions of 'access' or dissemination of technology would raise parallels to the debate on TRIPS and public health.

Weak Parallel

However, there exists opinion that despite the similarity of concerns, the parallels are weak.[43] Two main characteristics of ESTs and products that distinguish it from pharmaceutical products are product substitutability and diversity of technological inputs.[44] In the pharmaceutical sector, the element of substitutability of drugs is absent and thus a patent on a specific drug has substantial impact. The patent holder in this sector thus has a very strong market position and can conveniently charge a price well above the production cost. Whereas, in the renewable sectors what is patented is usually specific improvements over existing technologies and thus there is competition between a number of patented products, in addition to the competition with the cheaper traditional sources of energy.[45] As a result of such substitutability and competition, prices can be brought down to a point at which royalties and the price increases available with a monopoly are reduced.[46] In this sense, IP is a much less significant bottleneck in the development and transfer of ESTs and MTs as compared to that of pharmaceutical technologies and products.

Lessons Learnt

International action regarding IPRs and public health has improved the situation of 'right to health' discourse in various developing countries, though there has been no phenomenal success. Thus all attempts to draw a parallel aim at trying to provide a positive roadmap for climate change in lines of the public health developments. Thus, it can be said that the public health discourse has provided us the following lessons: (a) widespread public health negotiations and deliberations suggest that 'soft commitments' in this regard usually do not bear any fruit. Thus, technology transfer commitments from developed countries to developing countries resulting from various climate change negotiations has to be in 'specific' and 'concrete' terms.[47] (b) If the current international IP framework incorporates in it certain flexibilities and exceptions, adequate enough to counteract against all possible hindrances to technology transfer, a balance may be struck between rights of innovators and right to access benefits of ESTs. In this case and in light of a strong analogy drawn above, a declaration comparable to the Doha Declaration on the TRIPS Agreement and Public Health with respect to IPRs and climate change may be useful in the development of law and practice in this regard.[48] The core element of such a Declaration would be to limit the conditions associated with issuing a compulsory license, particularly the limitation in Article 31(f) confining such licenses to the supply of the domestic market in the country authorizing the use.[49] This in effect will create a waiver in favour of ESTs similar to one created for 'public health'. (c) Another step with respect to climate change technologies declaration may be extending the lifting of the restriction under TRIPS for compulsory licensing to these technologies too. This would ensure adequate supply of 'generic' technologies and products to countries that were otherwise deprived of the 'parent' technology.[50] This would extend the 'generic medicine' solution to 'public health' issue to climate change technologies as well.

However, it is to be noted that utility of such a mechanism in the environmental context might not be at par with its utility in the public health context owing to heterogeneity of technologies widely classified as ESTs or MTs. Thus, it is seen that even though a parallel can be drawn between the two fields, policy balances for IP rights in field of pharmaceuticals are very different from that in the environmental technology field and hence any attempt to draw a parallel between the two would prove futile.

Examining Licensing as a Tool

The role played by IP regimes in the process of North–South technology transfer[51] continues to be a highly controversial topic in various

national and international forums. The North–South Technology transfer becomes an important means by which developing countries can
gain access to ESTs that are new to them. The unavailability of proper
R&D facilities and resources renders licensing of foreign technology
rights the only means of obtaining them.

It is a well-known fact that individual owner of IP rights are hostile
to the extensive use of 'exceptions and limitations' provision by the government as it is seen as an intrusion into their exclusive rights. This has
led to increased attempt by the private entities seeking to limit the use
of 'exceptions and limitations' in form of widespread lobbying efforts.[52]
Such lobbying efforts include approaching to legislature, financing of
campaigns, and so on.

It is true that Intellectual Property licensing, particularly patents, can
be used as a mechanism to strike a balance between rigid IP rights and
access to benefits flowing from it.[53] For example, as observed by Barton,
in sectors like export of ethanol or photovoltaic cells or wind engines,
IP has been a less significant barrier and thus developing countries
have been able to enter the market by tool of cross-licenses or product
modifications.[54]

But, licensing mechanism faces challenge in a situation when the
patent owner refuses to license the best technology for a particular
environmental application. It is the role of Courts, regulators, and the
licensing parties in this case to strike a balance between licensor's ability
to profit from his or her IP rights and the public interest in using the
most appropriate technology.[55] Further, patent misuse doctrines and
anti-trust doctrines may further limit the extent to which a patentee
may refuse to grant a license or extend effective term of patent.[56]

A licensing mechanism or arrangement with regards to conferring
or conveying rights in any technology can be of various types. Either
rights may be assigned against a royalty based on sales, or the licensee
be given an exclusive license with an agreement that he would not
license anyone else, or a non-exclusive license in which the licensor is
free to license others to use the technology.[57] Further, depending on the
type of technology, field of innovation, and so on, different mechanisms
may be applied for different ESTs. For example, exclusive license may be
granted for a Nitrogen Oxide removal system technology for fossil fuel
utility applications while a non-exclusive license may be granted for the
process of solid waste incineration.[58]

In a noteworthy attempt by the European Parliament, it has adopted a resolution in November 2007 recommending launching a study on possible amendments to the TRIPS Agreement in order to allow for the compulsory licensing of environmentally necessary technologies.[59] However we have seen that despite the most significant and detailed international norms regarding 'exceptions' being found in the TRIPS Agreement, licensing mechanism under such flexibility has failed to adequately address the access to medicines and public health concern in developing countries. In a joint study conducted by the United Nations Environment Programme (UNEP), European Patent Office (EPO), and International Centre for Trade and Sustainable Development (ICTSD) it was seen that there was a little 'licensing-out' activity towards developing countries, however there was felt a need to improve market conditions and encourage licensing to further improve transfer towards developing countries.[60]

Limitations of Licensing as a Tool in Field of ESTs

The technologies relevant to ESTs and MTs are primarily, though not exclusively held by private firms.[61] And for any activity or research undertaken in the private sector, IP protection has been a time-proven incentive, without which private sector funding would dry up. In this context, no proposal to encourage or mandate technology transfer can merely rely on 'licensing' of government owned patents or other technical data.[62] First, because the private sector is increasingly insensitive to appeals of 'equity' or imbalance and thus in no case would respond to attempts to create balance between countries by means of transfer of technology from one to the other. Further, there continues to exist strong mercantilist and nationalist biases among governments that favour policies designed to strengthen national industries.[63] For example, we would not find OECD governments 'directing' their companies to supply technology to developing countries.[64] In such a situation we see that the framework of licensing would lose its objective. Instead, any proposal for transfer of technology to address climate change should take refuge of other forms of 'private incentive mechanisms'.[65]

Second, the kind of licensing we see in the agricultural and medical or pharmaceutical field can thus be considered as 'humanitarian licensing'[66] and thus would involve asking developed nations to agree to forego their national favouritism in licensing publicly funded inventions, at least with respect to technologies of global environmental importance.[67] It can be said with conviction that any such form of licensing would not stand the test when it comes to environmental technologies.

Third, licensing as a policy balance might not be very attractive if a direct export of a given technology is more profitable than investments in local manufacturing. Thus, licensing as a solution is highly dependent on transaction costs and ability of local capacity building.

Fourth, valuation of a license in the course of licensing negotiations would require looking into investment by the licensor, licensee's potential profits, and prospective benefits from not licensing.[68] For instance, if an industry has found a technology that along with being EST also helps reduce production and compliance cost. Now, by not licensing such technology or transferring it, the industry can continue incurring costs much lower than its competitors and having competitive advantage. Thus the industry would continue to do so despite licensing framework being in place. This is another ground on which the licensing framework needs to be tested.

Therefore, licensing framework, as a policy balance thus needs to be supplemented by other incentives by developed countries for their organizations to transfer technologies to developing countries. This may be in the form of patent pooling or product development partnership between organizations in developing countries and potential sources of technologies.[69]

Other Policy Balances

There is a pressing demand for a 'middle path' between those who defend strong IP rights and those demanding a compulsory licensing regime for ESTs. In view of Professor Abbott, one such path is a possible joint venture between entities in developing and developed countries, and optimizing scope for investment in developing countries.[70] Other such possible solutions are patent pooling[71] or product development partnerships.

More than licensing what would help is negotiations at international level as part of climate change deliberations that would include 'strong' commitment on part of developed nations to make the technology more readily available as a quid pro quo for stronger environmental constraints upon developing nations.[72]

A leading example in this regard can be proposal of Ecuador before the WTO TRIPS Council. In February 2013, Ecuador submitted a proposal to the TRIPS Council, dealing with the 'Contribution of Intellectual Property to Facilitating the Transfer of Environmentally Rational Technology'.[73] The proposal calls for a review of Article 30 and 31 of TRIPS so as to include provisions on the transfer of expertise and know-how to implement compulsory licenses as well as a reduction in the term of protection for a patent of (x) years in order to facilitate free access to specific patented ESTs.[74] The proposal also called for a special provision should be adopted allowing for the exemption from patentability on a case-by-case basis.[75] Thus, the proposal aimed at coming up with a publicly funded mechanism for promotion of open and adaptable technology licensing.

***

While it is true that stronger patent encourages technology transfer in the form of imports, FDI and licensing to developed countries, stronger patents have little effect on technology transfer of ESTs and climate change related technologies to low income countries.[76] Experience across jurisdictions show that it cannot even be said that strong patent positively impacts innovation. Also, various legal uncertainties have continued to undermine possibilities of meaningful balancing action from sides of developed and developing countries.

Further, ESTs and climate change-related technologies encompass a wide variety of technologies, each focusing on a different aspect of environment and climate change. Thus, effectiveness of patent will substantially differ accordingly across different fields of technology. In such a situation it is very difficult to come up with a single, universal mechanism that adequately explains the relationship between IP rights on one hand and diffusion of green technologies within countries on the other. Further, since appropriateness of EST or green technology varies significantly across countries depending on the location, industrial structure, type of economy, and stage of development, it is again unlikely that a single mechanism would adequately explain the relationship between IP rights and diffusion of the technology. Besides, it has been seen that the ability of developing countries to use technology transfer to develop their domestic capabilities and allowing such countries to reap the social and economic benefits out of them has been very mixed.[77]

Thus, there is need to focus on building more and more frameworks for mutually beneficial economic arrangements between developed and developing countries to encourage innovation and transfer of technology to address climate change. The consideration of various measures related to IP should consider all opportunities within climate negotiations and not be limited to TRIPS Agreement. Any suggestion should also be made in light of difficulties and vast political cost of modifying existing regime under TRIPS.

Lastly, at the end of the day we must not lose sight of the fact that IP rights are only one of the various factors affecting flow of technology across jurisdictions. At the international level too, technology transfer has become increasingly drawn into political negotiations between developed and developing countries, particularly those relating to environment related issues.[78] Thus other factors like enabling environment, financing mechanism, adequate incentives, and so on, need equal attention, in the absence of which no amount of flexibilities in the IP regime would help achieve desired levels of technology diffusion and dissemination. Thus, it is the need of the hour that more and more practical initiatives that would enable diffusion of green technologies into developing countries should be focused upon and issues like changing the IP regime should be addressed subsequently. Thus, any suggestion considering only the need for developing flexibilities in the IP regime to facilitate flow of clean energy technologies can be scrapped off.

Thus, the conclusion may be stated in precise term as follows:
First, the question whether the existing IP regime and the flexibilities provided by the legal framework is adequate for healthy development and dissemination of green technologies to developing countries evokes mixed response. Therefore, there is no single policy solution or balancing mechanism to solve the issue of sustainable access to ESTs by developing countries.

Second, in view of the discussion above it may be concluded that the call for developing countries to resort to compulsory licensing mechanism is a laudable step, but may not be a very feasible one as it poses an unattainable standard.[79] Third, we need to look at the debate from a multidisciplinary perspective that would look at not just flexibilities in the law, but also developmental and social politics as well as environmental politics.

Notes and References

1. A. Brown, 'Impact of Patents and Licenses on the Transfer of Technology', in S. Gee (ed.), *Technology Transfer in Industrialized Countries* (New York: John Wiley and Sons, 1979), p. 311.

2. M.A. Gollin, 'Patent Law and the Environment—Technology Paradox', *Environmental Law Reporter*, 20(5) (1990), pp. 101–71.

3. United Nations Framework Convention on Climate Change (UNFCCC) 1771 UNTS 107 [1994] ATS 2 / 31 ILM 849 (1992), Art 4.5. [It is to be noted that UNFCCC does not define ESTs. Thus the term 'ESTs' shall be used to refer to all climate change technologies that are subject of consideration under UNFCCC].

4. M.A. Gollin, *Driving Innovation: IP strategies for Dynamic World* (New York: Cambridge University Press, 2008), p. 334.

5. P. Gattri, 'The Role of Patent Law in Incentivizing Green Technology', *Northwestern Journal of Technology & Intellectual Property*, 11(2) (2013), p. 41.

6. See, P.E. Morthorst, 'Wind Energy the Facts, Costs & Prices', vol. 2, available http://www.ewea.org/fileadmin/ewea_documents/documents/publications/WETF/Facts_Volume_2.pdf, accessed on 22 January 2016) (providing a cost comparison and analysis).

7. For instance, a study on transfer of technologies for substitutes for ozone-damaging chemicals under the Montreal Protocol has given details for some cases in which technology transfer to developing countries' firms was hindered by either high prices or other unacceptable conditions imposed by companies holding patents on the chemical substitutes onto companies in developing countries that wanted a license to manufacture the substitutes. See, 'Some Key Points on Climate Change, Access to Technology and Intellectual Property Rights', *Third World Network* (October 2008), available http://unfccc.int/resource/docs/2008/smsn/ngo/065.pdf, accessed on 20 January 2016.

8. 'Agenda 21, Chapter 34, paragraph 14, 'Transfer of Environmentally Sound Technology, Cooperation and Capacity-building', *UN Documents Cooperation Circles,* available https://sustainabledevelopment.un.org/content/documents/Agenda21.pdf, accessed on 22 January 2016.

9. UNFCCC 1771 UNTS 107, Art 4.5. [It is to be noted that UNFCCC does not define ESTs. Thus the term 'ESTs' shall be used to refer to all climate change technologies that are subject of consideration under UNFCCC].

10. UNFCCC 1771 UNTS 107, Art 4.1(c).

11. UNFCCC 1771 UNTS 107, Art 4.5.

12. Decision of Conference of Parties, (COP-13) 1/C.P.13, 'Bali Action Plan', FCCC/CP/2007/6/Add.1 (14 March 2008).

13. For example, Agreement on Trade-Related Aspects of Intellectual Property Rights, 1869 UNTS 299; 33 ILM 1197 (1994), Art 30 and 31; See, UNCTAD/ICTSD, *Resource Book on TRIPS and Development* (New York: Cambridge University Press, 2005), pp. 25–7.

14. UNCTAD, *Resource Book*, p. 26.

15. TRIPS, 1869 UNTS 299; 33 ILM 1197 (1994), Art 31.

16. Keith Maskus, (2010), 'Differentiated Intellectual Property Regimes for Environmental and Climate Technologies', OECD Environment Working Papers, No. 17, OECD Publishing.

17. J.P. Santamauro, 'Failure is Not an Option: Enhancing Use of Intellectual Property Tools to Secure Wider and More Equitable Access to Climate Change Technologies', in Abbe Brown (ed.), *Environmental Technologies, Intellectual Property and Climate Change: Accessing, Obtaining and Protecting* (Cheltenham: Edward Elgar Publishing, 2013), p. 86.

18. This Agreement entered into effect on 17 December 1974.

19. WIPO Green, available https://www3.wipo.int/wipogreen/en/, accessed on 21 January 2016.

20. WIPO Green.

21. A. Arora, 'Intellectual Property Rights and the International Transfer of Technology: Setting Out an Agenda for Empirical Research in Developing Countries', *WIPO International Round Table on the Economics of Intellectual Property* (Geneva, 26 November 2007), available http://www.wipo.int/export/sites/www/ip-development/en/economics/pdf/wo_1012_e_ch_2.pdf, accessed on 08 September 2016.

22. See, for example, J.H. Barton, 'Intellectual Property and Access to Clean Energy Technologies in Developing Countries: An Analysis of Solar Photovoltaic, Biofuel and Wind Technologies', *ICTSD Trade and Sustainable Energy Series*, Issue 2 (December 2007), p. 16, available http://www.ictsd.org/downloads/2008/11/intellectual-property-and-access-to-clean-energy-technologies-in-developing-countries_barton_ictsd-2007.pdf, accessed on 8 September 2016; C. Hutchison, 'Does TRIPS Facilitate or Impede Climate Change Technology Transfer into Developing Countries?', *University of Ottawa Law & Technology Journal*, 3 (2) (2006), p. 517, available http://www.uoltj.ca/articles/vol3.2/2006.3.2.uoltj.Hutchison.517-537.pdf, accessed on 8 September 2016.

23. United Nations Environment Programme (UNEP), European Patent Office (EPO), and International Centre for Trade and Sustainable Development (ICTSD), Joint Final Report on Patents and Clean Energy: Bridging the Gap between Evidence and Policy (2010), available http://www.unep.ch/etb/

events/UNEP%20EPO%20ICTSD%20Event%2030%20Sept%202010%20 Brussels/Study%20Patents%20and%20clean%20energy_15.9.10.pdf, accessed on 23 January 2016.

24. Gollin, *Driving Innovation*, p. 335.

25. R.L. Okediji, 'Intellectual Property Rights and the Transfer of Environmentally Sound Technologies', *Commonwealth Trade Hot Topics*, Issue 67 (December 2009), p. 4, available http://www.oecd-ilibrary.org/docserver/ download/5k3w8fb9pl32.pdf?expires=1473329792&id=id&accname=guest& checksum=035826D0B9F150F89A1853B63706253A, accessed on 8 September 2016.

26. Santamauro, 'Failure Is Not an Option: Enhancing Use of Intellectual Property Tools to Secure Wider and More Equitable Access to Climate Change Technologies', p. 86.

27. R. Miller, et.al., *Terrell on the Law of Patents*, 17th Edition (London: Sweet and Maxwell, 2011), p. 9.

28. Gattri, 'The Role of Patent Law in Incentivizing Green Technology', p. 41, 43.

29. Gattri, 'The Role of Patent Law in Incentivizing Green Technology', p. 43.

30. Gattri, 'The Role of Patent Law in Incentivizing Green Technology', p. 43.

31. Gattri, 'The Role of Patent Law in Incentivizing Green Technology', p. 44.

32. Barton, 'Intellectual Property and Access to Clean Energy', p. 16. [… for biofuel technologies, IP does not appear to be barring developing countries for accessing the current generation technologies as shown by developments in many countries including Brazil, Malaysia, and South Africa.]

33. Barton, 'Intellectual Property and Access to Clean Energy', p. 16.

34. C. Saez, 'WIPO the Sleeping Beauty of Climate Change Policy, Urged to Awaken' (July 2011), available http://www.ip-watch.org/2011/07/14/wipo-the-sleeping-beauty-of-climate-change-policy-urged-to-awaken/, accessed on 23 January 2016 [quoting Jose Romero, Swiss Federal Office for Environment].

35. E. Lane, 'Going Round on IP and Climate Change' (16 March 2015), available http://www.greenpatentblog.com/2015/03/16/going-round-on-ip-and-climate-change, accessed on 23 January 2016 [quoting C. M. Correa, *The Burden of Intellectual Property Rights* (February 2015)].

36. Lane, 'Going Round on IP and Climate Change' [quoting F. M. Abbott].

37. Lane, 'Going Round on IP and Climate Change' [quoting Carlos M. Correa, *The Burden of Intellectual Property Rights* (February 2015)]

38. E. Dinopoulos and C. Kottaridi, 'The Growth Effects of National Patent Policies', *Review of International Economics*, 16(3) (2008), p. 499.

39. UNDP, *Towards a Balanced Sui Generis Plant Variety Regime: Guidelines to Establish a PVP Law* (New York: UNDP, 2008), p. 10, available http://www.undp.org/content/dam/aplaws/publication/en/publications/poverty-reduction/poverty-website/toward-a-balanced-sui-generis-plant-variety-regime/TowardaBalancedSuiGenerisPlantVarietyRegime.pdf, accessed on 23 January 2016; R. Ramachandran, *International Intellectual Property Law and Human Security* (New York: Springer, 2013), p. 25.

40. A. Arora, 'Intellectual Property Rights and the International Transfer of Technology: Setting Out an Agenda for Empirical Research in Developing Countries'; L. Brantetter, et al., 'Do Stronger IPR Increase Technology Transfer? Empirical Evidence From US Firm Level Panel Data', *Quar. Jour. of Eco*, 121(1) (2006), pp. 321–49; E. Bascavusoglu, and M.P. Zuniga, *Foreign Patent Rights, Technology and Disembodied Knowledge Transfer Cross Borders: An Empirical Application* (France: Unpublished Manuscript, University of Paris, 2002).

41. Ramachandran, *International Intellectual Property Law and Human Security.*

42. N. Fink, Report on the Contribution of IP to Facilitating Transfer of Environmentally Sound Technologies (Friedrich Ebert Stiftung, Geneva May 2015), p. 3, available http://www.fesglobalization.org/geneva/documents/2015/2015_05_12_IPR%20and%20EST.pdf, accessed on 23 January 2016.

43. See for example, F.M. Abbott, 'Innovation and Technology Transfer to Address Climate Change: Lessons from the Global Debate on Intellectual Property and Public Health', *ICTSD Programme on IPRs and Sustainable Development*, Issue Paper no. 24 (Switzerland: ICTSD Geneva, 2009).

44. C. Ebinger and G. Avasarala, 'Transferring Environmentally Sound Technologies in an Intellectual Property Framework' (November 2009), available https://www.brookings.edu/research/transferring-environmentally-sound-technologies-in-an-intellectual-property-friendly-framework/, accessed on 23 January 2016.

45. Ebinger and Avasarala, 'Transferring Environmentally Sound Technologies in an Intellectual Property Framework'.

46. Barton, 'Intellectual Property and Access to Clean Energy'.

47. Abbott, 'Innovation and Technology Transfer to Address Climate Change'.

48. 'Some Key Points on Climate Change, Access to Technology and Intellectual Property Rights', *Third World Network* (October 2008); M. Littleton, 'The TRIPS Agreement and Transfer of Climate-Change-Related Technologies to Developing Countries' (Unpublished Working Paper No. 71, UN/DESA, ST/ESA/2008/DWP/71, October 2008); Okediji, 'Intellectual Property Rights and the Transfer of Environmentally Sound Technologies', p. 4.

49. Okediji, 'Intellectual Property Rights and the Transfer of Environmentally Sound Technologies'.

50. See 'Some Key Points on Climate Change, Access to Technology and Intellectual Property Rights'.

51. R. Mittal, *Licensing Intellectual Property: Law and Management* (New Delhi: Satyam Law International, 2011), p. 420 [North–South technology transfer is a genre of international technology transfer where technology is transferred from the developed countries of the North to the developing countries of the South.]

52. See generally, Ellen F.M.'t Hoen, *The Global Politics of Pharmaceutical Monopoly Power: Drug Patents, Access, Innovation and the Application of the WTO Doha Declaration on TRIPS and Public Health* (AMB Publishers, 2009), p. 58.

53. For instance, in US we have the Clean Air Act, which provides for compulsory licensing in cases of necessity to comply with emission requirements and non-availability of reasonable alternatives.

54. Barton, 'Intellectual Property and Access to Clean Energy', p. 18.

55. M.A. Gollin, 'Using Intellectual Property to Improve Environmental Protection', *Harvard Journal of Law and Technology*, 4 (Spring 1991), pp. 194–217.

56. Burchfiel, 'Patent Misuse and Antitrust Reform: Blessed be the Tie?', *Harvard Journal of Law and Technology*, 4 (Spring Issue 1991), p. 1.

57. See generally, Mittal, *Licensing Intellectual Property: Law and Management*, p. 420.

58. Gollin, 'Using Intellectual Property to Improve Environmental Protection', pp. 194, 218.

59. European Parliament Resolution on Trade and Climate Change (November 2007) 2007/2003/INI, available http://www.europarl.europa.eu/sides/getDoc.do?type=TA&reference=P6-TA-2007-0576&language=EN, accessed on 19 January 2016.

60. United Nations Environment Programme (UNEP), European Patent Office (EPO), and International Centre for Trade and Sustainable Development (ICTSD), Joint Final Report on Patents and Clean Energy: Bridging the Gap between Evidence and Policy.

61. Abbott, 'Innovation and Technology Transfer to Address Climate Change', p. 20.

62. Lane, 'Going Round on IP and Climate Change' [quoting Frederick M. Abbott].

63. Abbott, 'Innovation and Technology Transfer to Address Climate Change', p. 21.

64. Abbott, 'Innovation and Technology Transfer to Address Climate Change', p. 23.

65. Abbott, 'Innovation and Technology Transfer to Address Climate Change', p. 22.

66. See, for example, A. Brewster, 'A. Chapman and S. Hansen, Facilitating Humanitarian Access to Pharmaceutical and Agricultural Innovation', *Innovation Strategy Today*, 1(3) (2005), pp. 205, 216, available www.biodevelopments.org/innovation/index.htm, accessed on 10 January 2016.

67. Barton, 'Intellectual Property and Access to Clean Energy', p. 18.

68. See, generally, Mittal, *Licensing Intellectual Property: Law and Management*, p. 425; *Geaorgia–Pacific Corporation v. United States Plywood Corp.*, 318 F Supp. 1116 (1970) [Provides useful checklist of 15 factors to be checked in case of a patent license, in the context of determining a reasonable royalty as damages for infringement].

69. P.S. Subbarao, 'International Technology Transfer to India an Impedimenta & Impetuous', Unpublished W.P. No. 2008-01-07, IIM, Ahmedabad (January 2008), p. 13.

70. Lane, 'Going Round on IP and Climate Change' [Quoting Frederick M. Abbott].

71. WIPO Secretariat, 'Patent Pools and Anti-trust: A Comparative Analysis' (March 2014), available http://www.wipo.int/export/sites/www/ip-competition/en/studies/patent_pools_report.pdf, accessed on 20 January 2016.

72. Barton, 'Intellectual Property and Access to Clean Energy', p. 20.

73. WTO Doc. No. IP/C/W/585, TRIPS Council Meeting, (June 2013).

74. WTO Doc. No. IP/C/W/585.

75. WTO Doc. No. IP/C/W/585.

76. B. Hall, et al., 'The Role of Patent Protection in Clean/Green Technology Transfer', *Santa Clara Computer and High Technology Law Journal*, 26(4) (2010), pp. 487, 521.

77. Subbarao, 'International Technology Transfer to India an Impedimenta & Impetuous', p. 14.

78. Subbarao, 'International Technology Transfer to India an Impedimenta & Impetuous', p. 13.

79. See Lane, 'Going Round on IP and Climate Change'.

NATIONAL PERSPECTIVES

6

ARMIN ROSENCRANZ
RAJNISH WADEHRA
NEELAKSHI BHADAURIA
PRANAY CHITALE

Clean Energy in India

Supply and Prospects

E lectricity is vital for human well-being and betterment in any developing nation. The benefits of electricity are far reaching and go beyond economic considerations. India's electricity sector requires immediate attention. An average Indian rural household receives only a few hours of electricity per day. This gives an entire family a limited space to do their basic household chores. The uses of electric heaters or gas stoves are alien to many rural households. Even in urban households, electricity is not subsidized, leaving a marginal migrant worker with skyrocketing electricity bills. India is a densely populated country, comprising 17.74 per cent of the world's total population.[1] This figure is projected to rise to 18.90 per cent by the end of 2022.[2] In relation to its population, India's electricity production is negligible.

The lack of electricity corresponds to the failure on the part of the government to provide electricity in an efficient and cost-effective manner. India has an enormous renewable and non-renewable electricity pool. India's generation capacity is 303 gigawatts (GW) of which about 212 GWs is thermal or fossil fuel–based generation capacity, about

46 GWs is hydroelectric capacity, about 43 GWs is renewables, and about 5 GWs is nuclear power.[3]

The present scenario is undergoing a shift. India is gearing to transform its electricity sector towards a greener and more reliable side, and is revamping its energy sector by utilizing its renewable stocks. India is striving to achieve a target of producing 175 GW of electricity from renewable sources by 2022.[4]

This ambitious target coupled with a rapid growth in population and industrialization is a challenging task for all participants, public and private alike. There is significant criticism[5] and doubt as to how India is to achieve this target unilaterally set by Narendra Modi, Prime Minister of India.

According to the Central Electricity Authority (CEA), the per capita electricity consumption reached 1075 Kwh in 2014–15.[6] Despite being an increase from the preceding years, India is lagging far behind developed countries, which have average per capita electricity consumption at around 15,000 kWh.[7] India is consuming about a quarter of what the average Chinese person consumes and 8.5 per cent of what an average American consumes (Bhaskar 2015). India's per capita emissions are among the lowest in the world in light of its high population.

There is a pressing need to increase India's per capita consumption in order to achieve growth and economic development. But the targets set in India's Intended Nationally Determined Contributions (INDCs) presented at the 2015 Paris climate meeting means that India has committed, in effect, that its economic development agenda would not be at the cost of degrading the environment. It has a responsibility to the environment and the world to sustainably generate renewable electricity.

Right to Electricity and the Electricity Act, 2003

To provide a fundamental right to electricity is no longer a distant dream in India. The government is on its path to electrifying the remaining 18,452 villages.[8] However, unrealistic electricity prices are a cause of concern for the government. Electricity utilities have failed in managing generation, transmission, and distribution companies (gencoms, transcoms, and discoms) in an effective manner. Even factories bear the price of paying unrealistic prices. This, as a result, affects the consumers in the form of increased prices of all household goods produced by such factories. Thus, when electricity is made available at very high costs, the burden is entirely on the consumers. To facilitate the development and to enhance the nation's GDP, cheap power must be provided directly to the consumers and small and medium enterprises. A rights based approach would enable a consumer to ensure electricity at a reasonable price mandated by law.

The Electricity Act of 2003 (Act) has taken some bold steps to transform power generation in India. The erstwhile electricity boards have been dismantled under this Act to form gencoms, transcoms, and discoms. However, the three wings have not been made efficient or privatized due to political clashes. Regulatory bodies have strictly implemented the fixed return stipulated by the Act. This helps utilities load a fixed return on all their investments and the burden is passed on to the consumer. There is no focus on efficiency, and in fact over 23 per cent of India's power is lost in Aggregate Technical and Commercial Losses (AT&C Losses), which are an unexplained camouflage for theft and inefficient working of the discoms and transcoms.[9]

In December 2015, COP 21, under the United Nations Framework Convention on Climate Change (UNFCCC), set up strong goals to change the global climate policy. The treaty did away with the differentiation between developed and developing countries, thereby making all countries accountable to achieve their set targets. The treaty put an obligation on every country to report their emissions and implement their efforts. This would in turn be subject to international review.[10]

Environmentalists and climate advocates have looked down upon the Paris Agreement with a certain level of uncertainty and scepticism. The reduction of greenhouse emissions by reducing carbon dioxide emissions was appreciated. This would in turn facilitate in achieving the difficult goal of keeping temperatures below two (2) degrees Celsius. The 'Ratchet mechanism' for 2015 commitments, also known as 'ambition mechanism', was approved. This mechanism allows for commitments to climate change to become more progressive and ambitious with time and allows for review of progress every five years. [11]

However, environmentalists and climate advocates deplored the non-focus on sea level rise; the lack of any mechanism to hold countries accountable for any commitment made in Paris; the absence of any pledge to leave fossil fuels in the ground; the absence of any plan to upgrade electricity grids or to develop adequate storage for excess wind and solar energy; no mention of clean coal technologies such as IGCC, coal gasification, supercritical, carbon capture, and geological storage; no incentives to the private sector to develop and market low-carbon energy technologies; no coverage of aviation or shipping; and no firm promises on climate finance or technology transfer.

India's Electricity Scenario

Overview

Despite failing to directly address certain factors, the Paris Agreement aimed to be extremely ambitious. India, like other nations, has devised its own contributions (INDCs) to achieve targets. The main argument put forth in this chapter is that India's set target is overly ambitious and is unlikely to meet any of its renewable energy commitments in Paris. India's government has declared that it will restrict energy provided by fossil fuels to 40 per cent of its total energy needs by 2035, reduce greenhouse gas emissions intensity of its GDP by 33 per cent and create a carbon sink of 2.5 billion tonnes of CO_2 equivalent through re-forestation.[12]

There are four major factors governing India's energy policy challenge today:

1. There is a requirement to move to renewable power. To achieve this, India needs to tap as much power from coal, gas, diesel, hydro, and nuclear to reach its promised energy potential. Therefore, India struggles to increase all—traditional fossil fuels, nuclear, as well as renewables.
2. There is inefficient management of electricity, resulting in enormous AT&C Losses. Such losses need to be reduced to ensure efficiency.
3. There is huge public resistance faced by state government-owned distribution companies (discoms) to realistically price the cost of electricity to consumers. This has led to huge accumulated losses, which are putting pressure on state treasuries and banks. These losses are backed by government guarantees, but their sheer size— Rs 3.80 lakh crores or about $60 billion and growing every year, is worrying.[13]

4. India desires to present itself as a good global citizen. It is the third highest greenhouse-gas-emitting nation in the world after the US and China.[14] Before Paris, India had already promised low-carbon energy improvements to the US under the Energy Cooperation Agreement of January 2015.[15]

The fine balance that India manages to achieve in this situation is key to India's future growth and development, as well as its global standing. The Modi government's will and resourcefulness is being tested under this stress. India's bold solar, wind, and biofuel initiatives have been showcased, though not a part of its INDCs, as a bold move engineered to draw admiration. These have managed to get India off the hook from promising any absolute reductions in either its carbon emissions or its coal production.

India has also retreated from its prior position of leading the G77 countries to protest and insist on the 'polluter pays' principle of getting developed nations to pay them for switching to renewables. The West has promised a climate change fund of $100 billion in 2020 whereas countries in Asia Pacific (specifically India) would now require an investment of $2.5 Trillion.[16] In the current slowdown, it cannot be expected that India, even though it carries weight in being a $2 Trillion economy, its G77 friends cannot hope to raise funds anywhere near the huge amount that would be required for a successful and time bound makeover.

Renewable Energy in India

India is in the midst of operating the world's largest capacity development agenda, with a tense deadline of 2020. In June 2016, India recorded about 44 GW of total electricity produced from its renewable resources.[17] This figure is a long way from its target of 175 GW. Presently, India has four years to increase its solar generated electricity from 7 GW to 100 GW, wind generated electricity from 27 GW to 60 GW and biofuels from 5 to 10 GW.[18] (See below). Generation from renewable sources will need to increase dramatically in the next four years.

Even though the share of renewable energy in the total power generation has increased from 4.97 per cent in 2012 to 5.7 per cent in

2016, the percentage of growth of renewable energy requires a major enhancement to achieve the targets set in the INDCs.[19] India is unlikely to achieve this target owing to the fact that the growth of renewable energy in India is slow and time consuming, as discussed below. It requires a major boost in infrastructure, funds, and policies. A number of pressing issues in the present established electricity generation sector require immediate attention and action.

The Ministry of New and Renewable Energy (MNRE) has prepared a state-wise blue print for tapping renewable energy from state stocks. The question arises as to whether India will stick by the targets set, as these top down driven targets seem to be passing the burden to the states to meet the central government's ambitious goals. Also, in the process, these goals are now being listed as a target for 2022 instead of 2020. The amount of generation intended in each region prescribed is: Northern region 46319 MW (further divided among solar, 31120 MW, 8600 MW Wind, 2450 SHP, and 4149 MW Biomass), Western region 54010 MW (28410 MW solar, 22600 MW wind, 125 MW Small Hydro Power (SHP), and 2875 MW Biomass), Southern Region 56650 (26531 MW solar, 28200 MW wind, 1675 MW SHP, and 2612 MW biomass), Eastern Region 12616 MW (12237, 135, and 244 MW Solar, SHP, and Biomass respectively) and North Eastern Region 1820 MW (1205 and 615 MW Solar and SHP).[20]

Renewable Energy Target of 175 GW

India has set the goal to increase its renewable energy capacity (excluding hydro power) to 175 GW by 2022. Out of this ambitious goal, Solar PV is to be 100 GW (of which 40 GW is expected to come from rooftop solar installations),[21] 60 GW wind, and 10 GW biofuel-based power. The 175 GW renewable energy target is further complicated by India's commitment at the latest Paris climate conference to reduce carbon emissions to 30–5 per cent and increase renewables to 40 per cent of the energy mix by 2030.[22]

In order to scale up renewable energy output as planned, India will need an investment of $ 140 billion over the next six years.[23] Most of this is being sought from a hesitant private sector, driven by the state government auctions for setting up solar farms with certain subsidies provided by the state and central governments. But the industrial climate in the country is poor, and returns on investments in solar are falling rapidly with competitive bidding in the tenders so far called by some states. With high NPA (Non-Performing Assets) levels of Indian public sector banks, some of these bids could result in poor returns, thereby crippling the chances for setting up these renewable energy capacities.[24]

In a rush to showcase efforts towards achievement, however, Minister Piyush Goyal went on record to state that the target of 100 GW of installed capacity of solar energy can be achieved by the end of 2017 itself, which seemed outrageously ambitious.[25]

Even though the government has talked up the usage of renewable energy in order to meet its set target of 175 GW, it seems highly unlikely. In particular, to meet India's solar target of 100GW by 2022 would imply average annual solar capacity additions of over 16 GW; a similar level has so far been observed only in one year in one country (China).[26] To achieve this target would mean attracting large capital funds, taking steps to ensure that projects are financially robust, that land is available, and that regulatory approvals are granted rapidly.[27] This target therefore seems preposterous. In order to achieve this target, government will have to aim towards large public-private partnerships and foreign investments; it will have to persuade, induce, and incentivize Indians to get on board and contribute for the good of the country.

The Indian government has been unable to allocate the substantial capital needed to increase India's 2016 budget aimed towards 100 per cent electrification in India by 2018.[28] The huge financial debts faced by State utilities (Discoms) and the huge NPAs faced by its banks are areas of concern that the government must tackle effectively while raising funds for renewable energy targets. Several billion dollars of debt on state power utilities has prevented the government from measurably increasing renewable energy despite consistent ambitious targets.[29] This fact has only been strengthened by the failure of the so-called Renewable Energy Certificate scheme, and low compliance with the renewable purchase obligation (Mittal 2015).

Sources of Renewable Energy

Solar Energy

Solar energy development is highly feasible as India receives 300 days of natural sunlight over an extensive landmass of nearly three million square kilometers.[30] In the 2015 budget and INDCs, the government decided to drastically enhance its solar energy target from 20 GW to 100 GW.[31] The five-fold increase is met with extensive speculation and criticism, as measures taken by the government are minimal as compared to the ambitious targets.

To achieve the target, incentives and schemes established by the government ought to have been working in full swing, aiming to rapidly increase solar energy production. However, facts indicate that there are a number of constraints, even at the grass roots level, making the target unachievable. As of May 2016, there are only 10 GW of solar projects under development and 8.4 GW that are anticipated to be auctioned in the second half of 2016–17.[32]

However, the government has stepped up its fiscal and promotional incentives to increase investments in the solar energy sector and thereby increase production. It has made provisions for ensuring capital subsidies, tax holidays on earnings for 10 years, generation based incentives, accelerated depreciation, and viability gap funding (VGF) to ensure a greater inflow of capital from private players.[33] The government has further made facilities for financing solar rooftop systems as part of home loans,[34] concessional excise and custom duties, a preferential tariff for power generation from renewables, and foreign direct investment up to 100 per cent under an automatic route (no requirement of government approval).[35] Further, the government has introduced renewable purchase obligations set at 8–10 per cent, whereby non-renewable energy suppliers are mandatorily required to buy certain solar energy.[36]

Despite establishing incentives and schemes, there are a number of concerns pertaining to the solar industry:

1. The renewable purchase obligations have implementation issues. The obligation is not implemented by the government and consequently, not adhered to by the utilities.
2. The solar electricity tariffs have reached an unprecedented low, posing a serious issue in the market. Potential investors bid extremely low tariff rates to win a tender, creating an aggressive bidding trend.[37] As a result, investors may not be able to sustain their business at such low rates, creating an unstable market at a very early stage and making it extremely difficult to earn a return on their investments.[38] Solar tariffs have fallen to an unprecedented low of

Rs 4.34/kWh (a drop of about six per cent between April and July 2016).[39] Domestic banks have raised concerns about low tariff rates and are reluctant to fund projects below Rs 5/kWh.[40] However, developers hope that in the time period between bidding and procurement, module and balance of system (BOS) costs will continue to drop along with interest rates to make these projects feasible, as most such projects are slated for commissioning by 2017.[41]

3. In addition to low tariffs, there is already a scarcity of funds in India's renewable energy sector. Even though the government has approved a number of major projects in India, there limited funds to finance solar investment projects.[42] Requisite funds for provision of the VGF support will be made available to MNRE from the National Clean Energy Fund (NCEF) operated by the Ministry of Finance.[43] However, the funds earmarked are insufficient to make the scheme successful.

4. The international market for solar energy is on a downfall. Sun Edison Inc., a US based company, which was due to set up a number of solar power plans in India, filed for bankruptcy on 16 April 2016 in the New York Federal Court.[44] This has caused a regression in the market, as almost 1.3 GW of clean energy projects are up for sale.[45] This implies that even sanctioned and planned projects could be scrapped in the near future if investors are not ready to take up Sun Edison's pending projects. In light of the Sun Edison crisis, it is imperative for government to relax nearing deadlines set by power purchase agreements for investors to take up left-over projects. The government needs to take measures to prevent the Indian clean energy market from falling apart.

5. The lack of available technology for solar storage spells trouble for the Indian market.[46] Solar storage could be used to reduce energy fluctuations, thereby facilitating inter-state energy transfer.[47] This would also allow solar power plants to produce more electricity and transfer energy in the absence of sunlight.[48] Solar storage would increase the solar tariff by a considerable amount.[49] Such technology has not been developed and its costs have not been taken into consideration when investors bid for tariffs below Rs 5/kWh. Adding storage costs could increase the cost of solar energy to a whopping Rs 14/kWh.[50] The government might provide subsidies and take in such storage costs. However, as stated above, the govern-

ment has an immense deficit of funds to even finance projects as low as 5 GW. Storage and warehousing are important concerns if the government has any intention of achieving its 2022 targets.

6. 100 GW of solar energy is likely to produce only 30 GW to 40 GW of electricity.[51] Taking factors such as night, dust, rain, cloud cover, or snow would reduce electricity generation. Most solar arrays being set up today are not equipped to manoeuver their direction according to the changing course of sunlight (Rosencranz 2016). A majority of PV solar panels are fixed in position owing to cost reduction considerations (Rosencranz 2016). Therefore, the output of electricity generated from sunlight reduces considerably depending on the location, the weather, and daylight (Rosencranz 2016). Thus, even if these targets are achieved in 2022, the total power generated will be much lower. That would be the best we can hope for. If these targets slip and are not realized on time, as the thermal power capacity is slated to grow substantially, the composition of solar power generation would prove to be even smaller.[52]

To achieve the target, the government ought to address the above issues with a high degree of urgency. The MNRE should establish a fixed and decisive deadline to achieve its targets in a timely and efficacious manner.

Wind Energy

The government proposed to achieve 60 GW of wind-installed capacity in the 2015 budget and its INDCs. As of May 2016, the cumulative capacity of all wind farms was 26.8 GW, making India the 5th largest wind power producer in the world.[53] Wind energy has been the predominant contributor of renewable energy in India, accounting for over 65 per cent of installed renewable capacity.[54] With a potential of more than 100 GW, India's aim is to achieve 60 GW of installed wind power capacity by 2022.[55] The potential has been worked out on the basis of availability of areas with high wind speeds. But these too vary substantially over seasons and over years. Wind energy was, until a couple of years ago, the fastest growing renewable energy sector in the country. It seems wise to put a focus on achieving as much as can be achieved by setting up wind farms.

To achieve growth in the wind energy sector, the government aims to incentivize investors by providing a tax depreciation allowance.[56] An amount equivalent to 80 per cent of tax depreciation is granted during the first year of installation.[57] This essentially implies that 80 per cent of the installation cost can be added to the income generated by selling wind power.

A major step towards facilitating the growth of wind energy is the approval of the National Offshore Wind Energy Policy, which aims to tap and establish wind farms within India's Exclusive Economic Zone (EEZ).[58] There is enormous potential for development of offshore wind energy along the 7600 km Indian coastline, thereby replicating the onshore wind power development offshore.[59] Potential developers will be allocated blocks through a bidding process.[60] Seabed lease agreements will be executed between the potential developers and the government.[61] The National Institute of Wind Energy (NIWE) or State-owned off-taking companies will then execute power purchase agreements with developers in accordance with the regulations of the Central Electricity Regulatory Commission.[62]

There are a number of fiscal incentives for offshore wind power projects. Project developers will be entitled to a ten-year tax holiday under the Income Tax Act 1960. There will be exceptions on custom and excise duties for manufacturing and importing equipment. Projects will also be exempt from service tax on third party services relating to resource assessment, environmental impact assessment and oceanographic studies, and the use of survey and installation vessels.[63] However, this seems to be an uncertain and expensive proposition. It would be extremely difficult for NIWE, as the principal agency for the development of offshore wind power projects, to attract investment.

Despite this major move towards development of offshore wind power projects, there are issues pertaining to the weak national grid transmission structure. In 2014, the government announced its plans to make a separate corridor for all renewable energy resources, but with the many constraints already discussed, all this seems unlikely to materialize.

Further, wind turbines and mills currently in use have reached the end of their useful life and need replacement.[64] The old technology has undergone significant improvement through the use of taller towers and lighter blades. These improvements could increase wind energy capacity, but replacement is essential—and, of course, costly.

Wind energy development is further constrained by local challenges such as land acquisition from local landowners, delay in approvals and processes, and disputed power purchase distribution agreements. There is immense competition from solar energy, which is fast growing.[65] Despite proliferating strongly, installed wind capacity grows at less than half the pace of solar PV, in part due to the wide gap in the cost of solar compared with wind. The costs of solar panels have fallen steeply and there is no such fall in the costs of installing wind turbines. This imposes the largest constraint on investment for wind power generation. The returns are far too small and slow. Despite the huge incentives put forth by the government, investment is likely to falter. Accordingly, India is very likely to witness less than the targeted growth in wind energy.

Bio Fuels

Biomass is the most widely used energy source in India. Seventy per cent of India's population is reliant on biomass—wood and dung—for cooking in rural households, at a great cost to public health.[66] In the regulated sector, bio-energy constitutes only 4.8 GW, which is less than 1.3 per cent of the total primary energy capacity in the country. Of this, bio-energy from waste (0.118 MW) is a small and insignificant fraction.[67] India's internal target is to establish an installed capacity of 10 GW of biofuels by 2022.[68] India's potential to generate energy from biofuels, is not even beginning to be tapped with these meagre targets.

The National Policy on Biofuels has envisaged a target blending of diesel and petrol with up to 20 per cent biofuels (bio-diesel and bio-ethanol). But to achieve this aim and expand the use of biofuels, India is moving slowly. The government plans on blending only 5 per cent of biofuels in diesels that would be consumed by bulk users such as the railways and defence establishments.[69] Despite being economical for expanding biomass, a major concern is with respect to the implementation of blending.

The government proposes to encourage farmers and landless labourers to plant non-edible oil seeds and plants like jatropha to boost the production of bio-diesel and bio-ethanol. The agricultural produce would be procured by public or private processing entities through the government's Minimum Support Price Mechanism.

To expedite the proliferation of bio-diesel and bio-ethanol across the country, the government will enhance the incentives for processing and production activities.[70] Foreign investment in the sector would also be encouraged.[71]

The Motors Vehicles Act already allows 'conversion of an existing engine of a vehicle to use biofuels'.[72] Engine manufacturers will be required to make the necessary changes to the engines to ensure compatibility with biofuels.[73]

Very little growth, however, is expected to come from bio-waste. India generates sufficient nutrient-rich sewerage and solid wastes which should be used to generate power.[74] The few pilot plants that have come up are mired in controversy and large scale generation of power from bio fuels is not now envisaged. The technology exists and the raw material of waste is available in abundance; what remains is to put together the right policies and to create the space for key investors to come in and fuel its growth.[75]

Government Measures to Achieve Targets

Progress in 2016

India is close to achieving its solar capacity additions for 2016. The MNRE has been proactive by approving 15 GW of new solar projects, of which 12 GW of projects commenced operation in 2017.[76]

India is taking steps to deal with the problem of its electricity retailers and an insufficient grid. For instance, an initiative that will be executed by the Powergrid Corporation of India Limited and other state transmission utilities is the Green Energy Corridor (GEC).[77] Under this initiative, additional transmission system will be set up, thereby allowing renewable energy to be transmitted on the grid. This would also alleviate the issue of power deficit. Many states such as Andhra Pradesh, Rajasthan, and Tamil Nadu have already begun the initiative. Another initiative, the Central Electricity Regulatory Commission ('CERC') has been introduced to prescribe regulations that support and increase transmission capacity.[78] It also supports more renewable energy on the grid.

The government's UDAY (Ujjwal Discom Yojana) scheme was launched in 2015, which provides lifelines to state-based power distribution companies.[79] These state government-owned distribution utilities will be improving their operational efficiencies and at the same time reducing their debt as well as their interest thereon. This scheme will particularly benefit those power distribution companies lacking financial aid due to power theft being committed on a large scale, high costs of power, inadequate operations, and inaccurate billing practices.[80] However, not all States have signed up for it.

The Solar Energy Corporation of India (SECI) aims to cover matters of different kinds of renewable energy.[81] The intention was to increase direct investment from the public sector, so that more projects could be run by the public sector or in collaboration with private parties. So far the push has been only on private investment, but the country's economy seems to show the right mix between the public and private sectors. The PPP models for developing solar power have not yet been formulated. It is expected that SPCI would bring up innovative routes for promotion of renewable energy.

A new policy for distributed generation has been announced recently by the central government.[82] This would enable the operation of micro grids, small storage neighbourhoods that can store extra power, and share in limited areas to benefit local users. The policy has just been revealed and the ministries are formulating rules for its implementation.

Other Important Measures

This section highlights the measures taken by the government to achieve its target of 175 GW by 2022. It argues that the measures adopted are insufficient and inefficient, comprising many loopholes. These are as follows:

Coal Cess

The increased coal cess (clean energy cess) is seen as one measure to provide funds in the renewable energy sector. The coal cess, initially introduced at Rs 50 (US 8 cents) per tonne of coal, has increased to Rs 400 per metric tonne.[83] The cess, forming the corpus of the National Clean Environment Fund, is being used for financing clean energy technologies and related projects.[84] The total collection of Rs 170.84 billion

($2.7 billion) till 2014–15 is being used for 46 clean energy projects worth Rs 165.11 billion ($2.6 billion).[85]

Despite the substantial increase, the cess will make solar power marginally more competitive by increasing the cost of coal. The increased costs will eventually be passed on to consumers in the form of higher electricity bills.[86] Disappointingly, the funds are not being utilized for its renewable energy purpose. Much of the cess collected has been given to the Ministry of Water Resources for the Ganga rejuvenation project. We, therefore, cannot assume that the entire amount collected from the new coal tax will go towards renewable energy.[87] In any case, the increased cess is still not sufficient to achieve the ambitious goals set by government.

Financial Package for State Debts

The MNRE specifically called upon states to make a significant contribution to meet the steep renewable energy installation.[88] The government had announced a comprehensive financial package to restructure the debt on state power utilities.[89] The debt on these utilities will be taken over and guaranteed by the respective states.[90] These utilities will not be given any additional credit from banks and will have to gradually increase their tariffs and reduce losses.[91] The government failed to consider the massive amount of state-utility debt involved before making its renewable energy commitments.

Role of Private Investors

State-owned institutions such as Power Finance Corp. Ltd and Rural Electrification Corp. Ltd have committed a total of $300 million to the fund for renewable energy.[92] However, no progress has taken place to work towards the set target.

Public-Private Partnerships (PPP)

PPPs bring in the technology, expertise, and capital needed to tap into the unrealized potential of renewable energy in India.[93] The International Finance Corporation ('IFC'), part of the World Bank, advises government on structuring PPP transactions. It is currently helping the Indian government establish partnerships to support the quest for a sustainable future (for example, the Gujarat rooftop solar project).

The IFC advised the government to set up a major project by structuring a PPP. Two private sector companies won a 25-year concession for a 2.5 MW solar rooftop project each in Gandhinagar, capital of Gujarat.[94] Under the agreement, the companies installed solar panels on the rooftops of public buildings and private residences and connected them to the grid, and many people are expected to benefit from increased access to power.[95] The IFC is advising on the replication of this successful pilot in five other cities in Gujarat.[96] This method is also going to generate renewable energy over a larger period of time and not immediately.

Slow Auctions

Private parties play an important role in setting up solar and wind projects. The government awards tenders primarily through a reverse auction mechanism.[97] However, such auctions were announced as recently as in 2016, leaving no time to actually allow parties to start their projects given the oncoming deadline.[98]

Two Major Renewable-Energy Related Policies

The Strategic Plan for New and Renewable Energy, which provides a broad framework, and the National Solar Mission, which sets capacity targets on solar renewables.[99] The government has auctioned many projects under the 'National Solar Mission'.[100] The government now plans to raise the capital needed for renewable energy through auctioning several solar projects ('ultra-mega solar projects'), which are to be set out across States. The government has increased the solar ambition of its National Solar Mission from 20 to 100 GW installed capacity by 2022, a five-time increase and over 17 times more solar than it currently has installed.[101] To this end, the government also announced its intention to bring solar power to every home by 2019 and intends to invest in 25 solar parks, which have the potential to increase India's total installed solar capacity almost tenfold.[102]

Solar Mission

The original Solar Mission 2008 targets for 2017 were 27.3 GW wind, 4 GW solar, 5 GW biomass, and 5 GW other renewables. For 2022, the targets were to increase to 20 GW solar, 7.3 GW biomass, and 6.6 GW other renewables.[103] Without the aggressive participation of the private sector, any rise of India's renewable energy market will not be possible.[104] This aggressive march towards grid parity has been possible only because of large-scale participation of national and international private sector companies.[105] Some of the companies leading the Indian solar power market include, First Solar, SkyPower Global, ReNew Power, Azure Power, and several others.[106] Previously, only Indian companies were investing in these projects. However, as the targets and volume for projects increased, it has attracted foreign investors as well.

State Auctions

Several States have set up plans and targets with special emphasis given to solar energy. Auctions of State solar projects have also taken place.[107] Delhi is one such State where the activities with regard to solar power are going to increase. Delhi has extreme temperatures and this incentivizes the development and adoption of renewable energy. The weather in Delhi also makes it suitable for generating electricity through solar panels. Delhi's peak power demand is 6.5 GW a day.[108] Delhi Chief Minister Arvind Kejriwal has stated that, 'Making Delhi a solar city is on our 70-point agenda'.[109] For this purpose, the target set by the State is higher than any other State. It has set a target of generating 1 GWHr of solar power a day through rooftop installations by 2020 and 2 GWHr by 2025.[110]

International Relations with US

One more measure that was taken to help achieve India's targets was the signing of the Partnership to Advance Clean Energy (PACE) in 2009 between India and US. US promised to support India's goals of shifting to renewable energy completely. The main goal of the policy is to accelerate inclusive, low carbon growth by supporting research and deployment of clean energy technologies.[111] This was further strengthened by joint agreement between President Obama and Prime Minister Narendra Modi in January 2015,[112] but has not yet been supported by any visible action. There has not been any sufficient assistance to achieve the 2022 ambitious goal, or any noticeable increase in US funding for PACE.

Policies for the High Investment

The ambitious goal of achieving 100 GW that the Indian Government has set would require a high investment of at least Rs 6 Lakh Crore ($100 billion).[113] Energy Minister Piyush Goyal has said, 'we are working on a policy to promote large-scale domestic manufacturing of solar equipment for making it more competitive'.[114] Minister Goyal further added that, 'We are trying to bring in a policy wherein we are thinking what support we can [obtain] for large-scale production of equipment like silicon wafers. A policy in this regard is being considered which will be put up for Cabinet approval soon for a quantum jump in domestic production of solar equipment.'[115]

While the government is moving in the right direction by identifying the needs and the investment required, it is being unusually aggressive and perhaps delusional by stating that policies and targets have almost been achieved. As noted earlier, Minister Goyal stated that the 100 GW target of installed capacity of solar energy could be achieved by 2017 itself.[116] He also said that India has the potential to have 750 GW of solar power generation capacity; domestic as well foreign players would have ample opportunities.[117] We are not even close to the target as of now. Government's ambitious goals may attract some investors in the near-term, but we may lose out on future investments. The minister stated that the Department of Industrial Policy and Promotion has laid down certain policies and they are being evaluated for the future.

Adjusting the Tariffs

The government also has to adjust the tariff when foreign investors come in. The Central Electricity Regulatory Commission (CERC) promotes foreign investment by increasing tariffs. They need an open, transparent, and effective regulation system and determination of tariffs (which is critical, as this would directly impact the revenues of investors).[118] Accordingly, CERC has announced certain guidelines and key steps to be taken to help achieve the foreign investment target.

Government's Viewpoint

The government has continued to assert at all international meetings that the 175 GW target is achievable. Recently, even at the Solar Thermal

Technology and Solar Cooker Excellence awards, Goyal stated that the targets are realistic. He said that India exceeded its 2016 fiscal year (FY) solar targets by 116 per cent, and that the country 'has signaled to the world that we're ready to lead'.[119] India will make new plans yearly to achieve its goals.

India may have a suitable climate for the development of solar energy, but the government continuously fails to take into consideration the huge investments required. As per the yearly targets of the government, India aims to add 12 GW of new solar power capacity this fiscal year, and add 32 GW and 48 GW of new solar capacity in FY2018 and FY2019, respectively.[120] The current targets are not close to being achieved as it has reached only 5.8 GW which is not even half of the intended. Certain areas in India do not have sufficient amount of sunlight needed to set up solar panels; this will force the government to shift to other modes of renewable energy to achieve its goals. Careful planning and investment is required to give India a hope of leading in renewable energy.

The targets for 2016 have still not been achieved; to keep up this trend the state must put in greater efforts in the years to come. This will impact national energy climate change commitments as well as solar commitments.

The Indian subcontinent has been actively promoting the transmission of solar energy onto the grid. Nonetheless, it now needs to pay heed to the ever-increasing demand for solar power.[121] This can be achieved by taking steps to ensure the private sector gains access to affordable solar energy. Storage technologies and distributed and off-grid generation systems are essential to achieve India's future solar commitments.[122]

Renewable Targets Being Too Ambitious: Not Achievable

As already indicated, the renewable targets seem too ambitious to actually be achieved. Solar, for example, which is one of the most promising among the renewable technologies, is beset with unrealistically high targets. This is evidenced by the fact that the government raised its solar target from 20 GW (pre-Paris) to 100 GW by 2022. Given that India's output levels are between 3–4 GW, meeting the 100 GW target will require more than a 50 per cent compound annual growth rate for the next few years.[123] This leap seems even more fictional and far-fetched in the context of the fact that Germany, which is the world leader in installed solar, has only targeted 66 GW by 2030.[124] If the German statistical data is to be believed, India's 100 GW goal seems like a figment of her imagination. The fact that India has the advantage of being closer to the equator thereby making it more potent to tap solar resources does not do enough to justify such a huge difference between the two countries.

There are several other factors which go into making a particular country a favourable destination for production of solar energy, financing capital projects being a major one. Evidently, India has suffered in this field. In spite of foreign companies pledging over double the 100 GW goal, foreign investors may still be reluctant to proceed with financing these projects given the high rate of stalled infrastructure projects. Risks including but not limited to market, credit, and counterparty risk result in increased costs and uncertainty, thereby making it difficult for solar to compete with coal, even with declining solar costs. Having said this, India will surely increase its renewable energy capacity by 2022; but the targets of 100 GW of solar and 175 GW of total renewables in a bid to lead the world are unrealistic.[125]

Other countries have gone ahead and set realistic goals, which they can achieve within their stipulated time. India should attract foreign investment by setting ambitious goals and seeming to be determined about achieving them. A competitive cycle could emerge where foreign players compete for a piece of the Indian renewable sector.[126] However, if the financing climate does not improve and/or adequate policy support does not materialize, India will be left with massive unfinished projects and a further tarnished foreign investment reputation.[127]

The investment for renewable energy should be simultaneously achieved while producing renewable energy. However, if the amount being invested by companies does not yield expected results, it will cause more harm to future investments. The government has to ensure that it brings about proper legislation that will govern all aspects of renewable energy. It further needs to adapt to other renewable energy, like wind, that is easily available and does not require such high investments. Further, it needs to generate small capital investments for developing renewable projects. The easiest way to meet a big target is to work with a few developers on immense projects—the financing is easier, and permitting and land acquisition are more tractable for a single project rather than for a multitude.[128]

Further, India's INDCs do not reflect the fact that India plans to triple its usage of coal for thermal power generation, as Coal India Ltd has planned to ramp up its coal output from the current level of about 550 million tonnes to 908 million tonnes by 2020.[129] Additional end user coal production of nearly 500 million tonnes is planned through recently allotted coal blocks. This obviously leads to the conclusion that India has plans to triple its thermal power capacity. We could expect this to rise so much that it overshadows the enhanced renewable capacity. Further, coal being produced in India has high ash content and the efficiency is extremely low. More coal use is, of course, at odds with the government's renewable energy targets. It remains to be explained by the government how any sizeable impact might be expected on its overall emissions from the push it claims to be making to enhance renewables, when thermal power is already slated to be enhanced to such a large level.

Urgent Measures to be taken by the Government

Implementation of Draft Renewable Energy Act

The draft National Renewable Energy Bill, 2015 ('Bill' or 'Proposed Act'), released on 14 July 2015, is aimed to regulate and promote the use of renewable energy in India.[130] The Act only covers some aspects of renewable energy in India. To expand its reach, the Act is under amendment and it is hopeful that the Proposed Act will cover all aspects of renewable energy in India. In addition, the Bill aims to address issues that are not adequately covered in the Act, such as principles of grid planning and operation.[131]

The Proposed Act aims at producing decentralized energy. This will cut down the cost of transmission to a large extent and also prevent any loss of energy due to transmission. Under the Proposed Act, there will be a committee set up which will be called the 'National Renewable Energy Committee' ('RE Committee') and will implement the National Renewable Energy Policy and National Renewable Energy Plan.[132] The RE Committee would also enable inter-ministerial coordination and coordinate matters on the integration of renewable energy into the electricity grid.

The Act has also simplified the process for corporations to get a license to supply energy by doing away with the requirement of obtaining a license for renewable energy. The MNRE will be responsible for setting up an accreditation programme for renewable energy manufacturers, system integrators, and others.[133] There will be a National Renewable Energy Fund and State Green Fund, which will be run by the Centre and the State respectively to meet the expenses of setting up renewable energy.[134]

The Proposed Act displays a significant change in policy.[135] It is an extremely important piece of legislation, intended to create an institutional structure with the objective of promoting renewable energy in the country.[136] It seeks to create a National Renewable Energy Policy to focus on research and development. However, the Act is yet to be passed and still requires changes and amendments.

Use of Geothermal Energy and Concentrated Solar Energy

Geothermal energy is a site specific, green, and reliable source of energy that extracts the steam heat in geologic rock. Unlike solar or wind energy, geothermal energy is not season or weather specific. India has a potential of producing 100 GW of geothermal energy.[137]

In light of the growing need to shift towards green energy, the MNRE has proposed forming the Indian Geothermal Energy Development Framework (Draft) for harnessing geothermal energy in India.[138] The Draft aims to facilitate development, technological advances, and research in geothermal energy. It aims to develop 1 GW of energy by 2022 and 10 GW by 2030.[139] However, presently, the Draft has yet to be implemented. A number of Indian Fortune 500 companies have attempted international collaboration to establish geothermal energy in India. However, there has been no actual development of any such projects.

This sector of renewable energy has been largely neglected owing primarily to the cheap availability of coal and lack of experience in the field, making foreign investors reluctant to enter the Indian market.[140] Geothermal energy is expensive. To make it economical, the cost and quality of drills are required to be lowered. One way to address such issues is through international collaborations with various countries such as US that have established expertise in geothermal energy.

Another technology that needs to be fostered quickly is the emerging new area of concentrated heat and power (CHP). Solar panels in parabolic formations concentrate solar just like a magnifying glass does. This is beamed to a central tower in the CHP Array, which uses its heat as well as power.[141] Large establishments of CHP have already been pioneered in desert locations in the US. India could take this technology forward by indigenizing it to suit its own needs. A focus area of research in such technologies needs to be created so that India comes up with its own novel methods, innovations, and inventions. Schemes should be created to develop these alternatives for the future.

Involvement of Indian People

One of the reasons the shift to renewables is not taking place at an accelerated pace is that Indian people are not involved; it seems like someone else's target. Rooftop solar photovoltaic has been recently enabled by the regulatory bodies and this can take off when people want to invest a bit of time, money, and energy to install solar panels on their rooftops. Solar water heaters are already widespread and work well on Indian roofs. Awareness of the new technology, benefits offered by the local governments as incentives with a drive to use natural energy, reduce pollution, achieve self-sufficiency, and conserve fossil fuels can be achieved just the way awareness of a clean India is being achieved through Swachh Bharat.

To augment India's funding needs for this switch, inviting foreign investments is already being practiced, but innovative schemes need to be created to inspire higher domestic investments.

India has an exceptionally high rate of savings. If the nation were to call upon people to invest in new government energy schemes, buy government energy bonds, as well as take part in crowd funding, and if the right awareness were to be created of the nation's polluted and health-damaging environment, the response might be surprising.

***

The choices faced by India need to be judiciously exercised and the nation needs to arrive at its real prerogatives. It appears that populist sentiment plus pressures in negotiations with the US might have nudged

India into declaring its INDCs, the most critical parts of which India has chosen to offer in relative terms. There is no specific quantified carbon capping, halting, or reduction committed.

It will take India a massive dose of investment in transmission and distribution infrastructure, apart from just making arrangements to generate more renewable energy. There is no evidence so far of a plan for what the nation needs to do to achieve these commitments. These should have obviously been planned before committing the INDCs.

India's bold internal goals of taking renewables up to 175 GW by 2020 are quantified but these do not contain any greenhouse gas reduction or low carbon commitment. Further, these goals are already slipping to 2022 and are likely to slip further.

Our analysis above indicates that coal output, on the other hand, is planned to triple from the present 550 million tonnes to about 1.5 billion, to feed the planned growth of energy demand in the country.[142] There is no license required to setup thermal or any other power generation plant.[143] Import of coal continues to be duty free and over 140 million tonnes were imported last year. Being cheap, due to a depressed international coal market, the import of coal is likely to grow, despite substantial growth in internal output. Compressed natural gas (CNG) terminals are in place with a substantial network of pipelines to import more gas. India has begun importing large amounts of CNG.

There is no evidence of financial means to achieve the INDCs, as against the need of trillions of dollars of investments required to achieve these. All that can be seen is a small loan of $1 billion recently committed by the World Bank. This is too little to achieve these grand goals. India has watered down its earlier stand of insisting on western funding for the switch to renewables.

Concern about breathing polluted air is missing in popular Indian sentiment. There is little consciousness in the people of this nation to urge the decision makers to make their lives healthier. An implicit internal consensus appears to be in place which supports increasing generation of non-renewable energy. The government therefore seems to be playing to appease international sentiment with a grand show of its much hyped internal and INDC goals, while going ahead behind the scenes with its real plans to generate more from fossil fuels. Unless public opinion in the country becomes alarmed at the threats this poses to the health of its citizens, India might not manage to swing out of its present energy trajectory.

Notes and References

1. 'India Population', *Worldometers*, available http://www.worldometers. info/world-population/india-population/, accessed on 17 July 2016.

2. 'India Population', *Worldometers*.

3. Installed Capacity, Central Electricity Authority, available http://www. cea.nic.in/reports/monthly/installedcapacity/2016/installed_capacity-06.pdf, accessed on 15 July 2016.

4. India's Renewable Energy Targets, Ministry of New and Renewable Energy, Govt. of India, available http://mnre.gov.in/file-manager/UserFiles/ Tentative-State-wise-break-up-of-Renewable-Power-by-2022.pdf, accessed on 06 September 2016

5. K. Ross, 'India Charts a Roadmap to Achieve Ambitious Solar Targets', *World Resources Institute* (31 May 2016), available http://www.wri.org/ blog/2016/05/india-charts-roadmap-achieve-ambitious-solar-targets, accessed on 18 September 2016.

6. *Government of India, Ministry of Power, Central Electricity Authority, Power Sector*, Central Electricity Authority, Ministry of Power, 16 May, available http://www.cea.nic.in/reports/monthly/executivesummary/2016/exe_summary-05.pdf, accessed on 17 July 2016.

7. U. Bhaskar, 'India's Per Capita Electricity Consumption Touches 1010 kWh', *Livemint* (20 July 2015), available http://www.livemint.com/Industry/ jqvJpYRpSNyldcuUlZrqQM/Indias-per-capita-electricity-consumpion-touches-1010-kWh.html, accessed on 18 July 2016.

8. 'Work in 1/3rd of Un-electrified 18,000 Villages Over: Piyush Goyal', *The Economic Times* (5 February 2016), available http://articles.economictimes. indiatimes.com/2016-02-05/news/70373436_1_power-minister-piyush-goyal-electrification-work-villages, accessed on 18 July 2016.

9. M.S. Bhalla, 'Transmission and Distribution Losses (Power)', available http://www.teriin.org/upfiles/pub/papers/ft33.pdf, accessed on 18 July 2016.

10. 'Outcomes of the U.N. Climate Change Conference in Paris', Centre for Climate & Energy Solutions, available http://www.c2es.org/international/ negotiations/cop21-paris/summary, accessed on 12 July 2016.

11. S. Yeo, 'Explainer: the 'Ratchet Mechanism' within the Paris Climate Deal', *Carbon Brief* (3 December 2015), available https://www.carbonbrief.org/ explainer-the-ratchet-mechanism-within-the-paris-climate-deal, accessed on 7 September 2016.

12. Press Information Bureau, Ministry of Environment and Forests, Government of India, 'India's Intended Nationally Determined Contribution is Balanced and Comprehensive: Environment Minister', 2 October 2015, available http://pib.nic.in/newsite/PrintRelease.aspx?relid=128403, accessed on 6 September 2016.

13. Ministry of Power, Coal and New & Renewable Energy, 'Towards Ujwal Bharat UDAY: The Story of Reforms' (9 November 2015), available http://ujwalbharat.gov.in/sites/default/files/Towards_Ujwal_Bharat.pdf, accessed on 7 September 2016.

14. International Energy Agency, 'CO2 Emissions from Fuel Combustion' (2015), see particularly Figure 9 'Top Ten Emitting Countries in 2013', available https://www.iea.org/publications/freepublications/publication/CO2EmissionsFromFuelCombustionHighlights2015.pdf, accessed on 6 September 2016.

15. Press Secretary 'Fact Sheet: US and India Climate and Clean Energy Cooperation', The White House Press Office (25 January 2015), available https://www.whitehouse.gov/the-press-office/2015/01/25/fact-sheet-us-and-india-climate-and-clean-energy-cooperation, accessed on 7 September 2016.

16. Noemi Glickman, '$ 2.5 Trillion to be Invested in Renewables in Asia-Pacific to Build the Power Capacity Needed in 2030', Bloomburg New Energy Finance (1 July 2014), available https://about.bnef.com/press-releases/2-5-trillion-invested-renewables-asia-pacific-build-power-capacity-needed-2030/, accessed on 7 September 2016.

17. Ministry of New and Renewable Energy, 'Physical Progress (Achievements')', available http://mnre.gov.in/mission-and-vision-2/achievements/, accessed on 25 August 2016.

18. Government of India, Ministry of Power Central Electricity Authority Power Sector, Central Electricity Authority, Ministry of Power, (16 May 2016), available http://www.cea.nic.in/reports/monthly/executivesummary/2016/exe_summary-05.pdf, accessed on 17 July 2016.

19. Press Bureau of India, Government of India, 'International Solar Alliance Cell and World Bank Signs Declaration for Promoting Solar Energy' (30 June 2016), available http://pib.nic.in/newsite/pmreleases.aspx?mincode=28, accessed on 18 July 2016.

20. Ministry of New & Renewable Energy, 'Tentative State-Wise Break-up of Renewable Power Target to be Achieved by the Year 2022 So that Cumulative Achievement Is 1,75,000 MW', available http://mnre.gov.in/file-manager/UserFiles/Tentative-State-wise-break-up-of-Renewable-Power-by-2022.pdf, accessed on 18 July 2016.

21. Solar Energy Research Institute for India and the United States 'US India Partnership to Advance Clean Energy (PACE)', June 2013, available http://

www.seriius.org/pdfs/062013_indo_us_pace_report.pdf, accessed on 16 June 2016.

22. Press Information Bureau, 'India's INDC Balanced'.

23. Press Information Bureau, 'India's INDC Balanced'.

24. At Rs 4 Lakh Crore, Bad Loans Exceed Market Value of PSU Banks, *The Economic Times* (21 February 2016), available http://economictimes.indiatimes.com/industry/banking/finance/banking/at-rs-4-lakh-crore-bad-loans-exceed-market-value-of-psu-banks/articleshow/51078318.cms, accessed on 16 June 2016.

25. Piyush Goyal, '100 GW Solar Capacity by 2017-end Likely', *The Economic Times* (2 April 2016), available http://articles.economictimes.indiatimes.com/2016-04-02/news/71995348_1_power-minister-piyush-goyal-installed-capacity-gw, accessed on 16 June 2016.

26. Energy and Climate Change, World Energy Outlook Special Report 2015, International Energy Agency, at 33, available https://www.iea.org/publications/freepublications/publication/WEO2015SpecialReporton EnergyandClimateChangeExecutiveSummaryUKversionWEB.PDF, accessed on 26 August 2016.

27. Energy and Climate Change, World Energy Outlook Special Report 2015.

28. India Budget, 'Key Features of Budget 2016-2017', available http://indiabudget.nic.in/ub2016-17/bh/bh1.pdf, accessed on 18 June 2016.

29. S. Mittal, 'India Aims to Achieve Colossal Renewable Energy Targets 2 Years in Advance', *CleanTechnia* (9 November 2015), available http://cleantechnica.com/2015/11/09/india-aims-achieve-colossal-renewable-energy-targets-2-years-advance/, accessed on 18 June 2016.

30. World by Map, 'Land Area', available http://world.bymap.org/LandArea.html, accessed on 18 June 2016.

31. A. Jaiswal, 'Rapid Growth in India's Solar Energy Market: Stronger Policies to Achieve 100 GW by 2020', Natural Resources Defense Council (30 June 2015), available https://www.nrdc.org/experts/anjali-jaiswal/rapid-growth-indias-solar-energy-market-stronger-policies-achieve-100-gw-2020, accessed on 25 August 2016.

32. Mercom Capital Group, 'Solar Installations in India to Double in 2016 with More than 4 GW', available http://mercomcapital.com/solar-installations-in-india-to-double-in-2016-with-more-than-4-gw-reports-mercom-capital-group#sthash.gLRLLnmA.dpuf, accessed on 25 August 2016.

33. Financial Support, Renewable Energy, Make in India, available http://www.makeinindia.com/sector/renewable-energy, accessed on 25 August 2016.

34. G. Balachandar, 'PSBs to Accept Solar Rooftop Cost as Part of Home Loan Proposals', *The Hindu* (5 January 2015), available http://www.thehindu.

com/business/Industry/psbs-to-accept-solar-rooftop-cost-as-part-of-home-loan-proposals/article6757354.ece, accessed on 25 August 2016.

35. Financial Support, Renewable Energy, Make in India.

36. T. Kenning, 'India Releases State Specific Renewable Purchase Obligation Targets' (22 February 2016), available http://www.pv-tech.org/news/india-releases-state-specific-renewable-purchase-obligation-targets, accessed on 25 August 2016.

37. Mercom Capital Group, 'Solar Installations in India to Double'.

38. Mercom Capital Group, 'Solar Installations in India to Double'.

39. Press Information Bureau, Ministry of New and Renewable Energy, 'Solar Tariff at a New Low of Rs 4.34 Per Unit' (19 January 2016), available http://pib.nic.in/newsite/PrintRelease.aspx?relid=134602, accessed on 25 August 2016.

40. Mercom Capital Group, 'Solar Installations in India to Double'.

41. Mercom Capital Group, 'Solar Installations in India to Double'.

42. Abhishek Pratap, 'Greening industry with Greenpeace', *The Economic Times*, available http://economictimes.indiatimes.com/Interviews/shellarticleshow/12378831.cms, accessed on 7 September 2016.

43. G.C. Prasad, 'Govt Setting up $1.25 bn Renewable Energy Fund, *Livemint* (5 February 2016), available http://www.livemint.com/Industry/yYAQPlNQt0URb4jKvIYWeJ/Govt-setting-up-125bn-renewable-energy-fund.html, accessed on 7 September 2016.

44. E. Wesoff, 'The End of SunEdison: Developer Now Looking Into Liquidating Its Assets', *Green Tech India* (18 May 2016), available http://www.greentechmedia.com/articles/read/The-End-of-SunEdison-Developer-Now-Looking-Into-Liquidating-Its-Assets, accessed on 25 August 2016.

45. Priyanka, 'Sun Edison to Sell India Projects, Hires Rothschild to Run Sale Process', *VC Circle* (17 June 2016), available http://www.vccircle.com/news/cleantech/2016/06/17/sunedison-sell-india-projects-hires-rothschild-run-sale-process, accessed on 25 August 2016.

46. ET Bureau, 'India Readies Plan to Improve Renewable Power Storage', *The Economic Times* (22 August 2016), available http://economictimes.indiatimes.com/industry/energy/power/india-readies-plan-to-improve-renewable-power-storage/articleshow/53802021.cms, accessed on 25 August 2016.

47. Centre for Sustainable Energy, 'Optimizing the Use of Solar Power with Energy Storage', available https://energycenter.org/article/optimizing-use-solar-power-energy-storage, accessed on 25 August 2016.

48. M. Kohli, 'Two Revolutions Set to Transform India's Energy Sector', *The Economic Times* (13 June 2016), available http://blogs.economictimes.indiatimes.com/et-commentary/two-revolutions-set-to-transform-indias-energy-sector/, accessed on 25 August 2016.

49. Kohli, 'Two Revolutions Set to Transform India's Energy Sector'.

50. A. Upadhyay, 'India Opens Market for Solar Battery Makers Such as Tesla', *Bloomberg* (15 March 2016), available http://www.bloomberg.com/news/articles/2016-03-15/india-solar-tender-opens-market-for-battery-makers-such-as-tesla, accessed on 18 July 2016.

51. A. Rosencranz, 'Prospects for Clean Energy in India', *Law and Policy Brief*, vol. II (5 May 2016).

52. This is explained by the simple fact that solar panels can generate power only when they receive sunshine. The night takes away about half of the capacity and, further, rains, cloudy days, storms, dust, and climate variations hamper full production. Thus, solar power capacity is not able to generate more than about 25 to 30 or 35 per cent at best, or in other words, that these plants run at a plant load factor of 25 to 35 per cent of their installed capacity. Whereas comparable capacity set up in thermal or nuclear power would generate steady flow of power 24 hours a day. So an investment in 1 GW of solar power should in fact be compared with an investment of only a quarter of this capacity in thermal or nuclear power. This is further complicated by the fact that there is no technology available to store power generated in large volume or at a suitable cost.

53. D. Sengupta, 'Headwinds in MP and Maha to Affect Wind Power Capacity Addition', *The Economic Times* (30 May 2016), available http://energy.economictimes.indiatimes.com/news/renewable/headwinds-in-mp-and-maha-to-affect-wind-power-capacity-addition/52505169, accessed on 18 July 2016.

54. M. Chadha, 'India's Wind Energy Capacity Could Top 65 GW by 2020', *Clean Technia* (14 April 2011), available http://cleantechnica.com/2011/04/14/indias-wind-energy-capacity-could-top-65-gw-by-2020/, accessed on 18 July 2016.

55. India's INDCs.

56. S. Jai, 'Budget 2016: Wind Sector to Take a Hit as Accelerated Depreciation Tax Benefit Capped at 40%', *Business Standard* (2 March 2016), available http://www.business-standard.com/budget/article/budget-2016-wind-sector-to-take-a-hit-as-accelerated-depreciation-tax-benefit-capped-at-40-116022900591_1.html, accessed on 25 August 2016.

57. Jai, 'Budget 2016'.

58. Press Information Bureau, Government of India, 'Approval of National Offshore Wind Energy Policy' (9 September 2015), available http://pib.nic.in/newsite/PrintRelease.aspx?relid=126754, accessed on 25 August 2016.

59. PIB, 'Approval of National Offshore Wind Energy Policy'.

60. DSK Legal, 'Overview of the National Offshore Wind Energy Policy, 2015', as Approved by the Union Cabinet to Facilitate Offshore Wind Farming

in the Territorial Waters of India (6 November 2015), available http://www.mondaq.com/india/x/441280/Renewables/National+Offshore+Wind+Energy+Policy+2015, accessed on 25 August 2016.

61. DSK Legal, Overview of the National Offshore Wind Energy Policy, 2015.

62. PIB, 'Approval of National Offshore Wind Energy Policy'.

63. Trilegal, 'Cabinet Approves National Offshore Wind Energy Policy' (22 October 2015), available http://www.mondaq.com/india/x/437044/Renewables/Cabinet+Approves+National+Offshore+Wind+Energy+Policy, accessed on 18 July 2016.

64. Global Wind Energy Council, 'India Wind Energy Outlook 2012' (November 2012), p. 10, available http://www.gwec.net/wp-content/uploads/2012/11/India-Wind-Energy-Outlook-2012.pdf, accessed on 25 August 2016.

65. World Energy Outlook, 'India Energy Outlook', International Energy Agency, available http://www.worldenergyoutlook.org/media/weowebsite/2015/IndiaEnergyOutlook_WEO2015.pdf, accessed on 18 July 2016.

66. European Business and Technology Centre, 'Biofuels and Bio-Energy in India', p. 2, available http://ebtc.eu/pdf/111031_SNA_Snapshot_Biofuels-and-Bio-energy-in-India.pdf, accessed on 25 August 2016.

67. Central Electricity Authority, available www.cea.nic.in, accessed on 15 July 2016.

68. The United Nations Framework Convention on Climate Change, 'India's Intended Nationally Determined Contribution', available http://www4.unfccc.int/submissions/INDC/Published%20Documents/India/1/INDIA%20INDC%20TO%20UNFCCC.pdf, accessed on 17 July 2016.

69. The United Nations Framework Convention on Climate Change, 'India's Intended Nationally Determined Contribution'.

70. A. Aradhey, 'Biofuels India 2016', USDA Foreign Agricultural Service, available http://gain.fas.usda.gov/Recent%20GAIN%20Publications/Biofuels%20Annual_New%20Delhi_India_6-24-2016.pdf, accessed on 24 June 2016.

71. Aradhey, Biofuels India 2016.

72. Biofuels International, 'India Plans to Boost Biofuels to Reduce Pollution' (26 August 2015), available http://biofuelsnews.com/display_news/9556/india_plans_to_boost_biofuels_to_reduce_pollution/, accessed on 25 August 2016.

73. Biofuels International, 'India Plans to Boost Biofuels to Reduce Pollution'.

74. India Waste to Energy—Concepts, Energy Alternatives India, available http://www.eai.in/ref/ae/wte/concepts.html, accessed on 7 September 2016.

75. India Waste to Energy—Concepts, Energy Alternatives India.

76. Ross, 'India Charts a Roadmap to Achieve Ambitious Solar Targets'. Also see, 'India's Solar Capacity Crosses 12 GW', *The Economic Times*, April 10, 2017 available at https://economictimes.indiatimes.com/industry/energy/power/indias-solar-power-capacity-crosses-12-gw/articleshow/58113364.cms.

77. Ross, 'India Charts a Roadmap to Achieve Ambitious Solar Targets'.

78. Ross, 'India Charts a Roadmap to Achieve Ambitious Solar Targets'.

79. Ross, 'India Charts a Roadmap to Achieve Ambitious Solar Targets'.

80. Ross, 'India Charts a Roadmap to Achieve Ambitious Solar Targets'.

81. Solar Energy Corporation of India 'Objectives', available http://www.seci.gov.in/content/innerpage/objectives.php, accessed on 7 September 2016.

82. Ministry of New and Renewable Energy, 'Draft National Policy on RE based Mini/Micro Grids' (1 June 2016), available http://mnre.gov.in/file-manager/UserFiles/draft-national-Mini_Micro-Grid-Policy.pdf, accessed on 7 September 2016.

83. Press Trust of India, 'Govt. Doubles Clean Energy Cess on Coal to Rs 400 per tonne', *Business Standard* (29 February 2016), available http://www.business-standard.com/article/pti-stories/govt-doubles-clean-energy-cess-on-coal-to-rs-400-per-tonne-116022900394_1.html, accessed on 17 July 2016.

84. International Energy Agency, Energy and Climate Change, World Energy Outlook Special Report 2015.

85. The United Nations Framework Convention on Climate Change, 'India's Intended Nationally Determined Contribution'.

86. Mercom Capital Group, 'Solar Installations in India to Double'.

87. Mercom Capital Group, 'Solar Installations in India to Double'.

88. Mittal, 'India Aims to Achieve Colossal Renewable Energy Targets 2 Years in Advance'.

89. R. Nair, 'New Bailout Plan for State Power Utilities', *Live Mint* (6 November 2015), available http://www.livemint.com/Politics/ZXS4Wqpr Rv2VlDOz48onkI/New-bailout-plan-for-state-power-utilities.html, accessed on 19 July 2016.

90. By T. Wilkes and D. Tripathy, 'In India, Utility Debts Threaten Modi's Power-For-All Drive', *Reuters* (12 September 2015), available http://www.reuters.com/article/india-power-debt-idUSL5N11D05H20150913, accessed on 19 July 2016.

91. Mercom Capital Group, 'Solar Installations in India to Double'.

92. Gireesh Chandra Prasad, 'Govt Setting up $1.25 bn Renewable Energy Fund', *Live Mint* (05 February 2016), available http://www.livemint.com/Industry/yYAQPlNQt0URb4jKvIYWeJ/Govt-setting-up-125bn-renewable-energy-fund.html, accessed on 18 July 2016.

93. Mittal, 'India Aims to Achieve Colossal Renewable Energy Targets 2 Years in Advance'.

94. International Finance Corporation, 'Renewable Energy and Energy Efficiency' (August 2013), available http://www.ifc.org/wps/wcm/connect/41f4e300407f54ed851595cdd0ee9c33/SectorSheets_Renewables.pdf?MOD=AJPERES, accessed on 18 July 2016.

95. International Finance Corporation, 'Renewable Energy and Energy Efficiency'.

96. International Finance Corporation, 'Renewable Energy and Energy Efficiency'.

97. S. Mittal, 'Charting India's Renewable Energy Future through Public-Private Partnership', *Clean Technia* (03 January 2016), available http://cleantechnica.com/2016/01/03/charting-indias-renewable-energy-future-public-private-partnership/, accessed on 18 July 2016.

98. Mittal, 'Charting India's Renewable Energy Future through Public-Private Partnership'.

99. Centre for Climate & Energy Solutions, 'India's Climate and Energy Policies', available http://www.c2es.org/international/key-country-policies/india, accessed on 18 July 2016.

100. Mittal, 'Charting India's Renewable Energy Future through Public-Private Partnership'.

101. Mittal, 'Charting India's Renewable Energy Future through Public-Private Partnership'.

102. Mittal, 'Charting India's Renewable Energy Future through Public-Private Partnership'.

103. Mittal, 'Charting India's Renewable Energy Future through Public-Private Partnership'.

104. Mittal, 'Charting India's Renewable Energy Future through Public-Private Partnership'.

105. Mittal, 'Charting India's Renewable Energy Future through Public-Private Partnership'.

106. Mittal, 'Charting India's Renewable Energy Future through Public-Private Partnership'.

107. J. Chaudhary, 'Big Delhi Push to Rooftop Solar', *India Climate Dialogue* (10 June 2016), available http://indiaclimatedialogue.net/2016/06/10/big-delhi-push-rooftop-solar/, accessed on 19 July 2016

108. Chaudhary, 'Big Delhi Push to Rooftop Solar'.

109. Chaudhary, 'Big Delhi Push to Rooftop Solar'.

110. Chaudhary, 'Big Delhi Push to Rooftop Solar'.

111. Office of International Affairs, 'U.S.-India Energy Cooperation', available http://energy.gov/ia/initiatives/us-india-energy-cooperation, accessed on 18 July 2016.

112. Office of International Affairs, 'U.S.-India Energy Cooperation'.

113. Office of International Affairs, 'U.S.-India Energy Cooperation'.

114. 'Govt's Policy Soon for Large-scale Solar Manufacturing: Piyush Goyal', *DNA* (5 Mar 2016), available http://www.dnaindia.com/money/report-govt-s-policy-soon-for-large-scale-solar-manufacturing-piyush-goyal-2185844, accessed on 18 July 2016 and 'Govt to Achieve 100 GW Solar Power Generation Target by 2017-end: Goyal', *DNA* (3 April 2016), available http://www.dnaindia.com/money/report-govt-looks-to-pull-off-100-gw-solar-capacity-by-2017-end-2197557, accessed on 18 July 2016.

115. 'Govt's policy soon for large-scale solar manufacturing: Piyush Goyal'.

116. '100 GW Solar Capacity by 2017-end Likely: Piyush Goyal'.

117. 'Govt's Policy Soon for Large-Scale solar Manufacturing: Piyush Goyal'.

118. 'India Is Pursuing an Ambitious Renewable Energy (RE) Target of 175 GW by 2022. Examine the Role of the Central Electricity Regulatory Commission (CERC) in Helping India Achieve This Target', *Insight India* (22 February 2016), available http://www.insightsonindia.com/2016/02/22/7-india-pursuing-ambitious-renewable-energy-re-target-175-gw-2022-examine-role-central-electricity-regulatory-commission-cerc-helping-india-achieve-target/, accessed on 18 July 2016.

119. Ross, 'India Charts a Roadmap to Achieve Ambitious Solar Targets'.

120. Ross, 'India Charts a Roadmap to Achieve Ambitious Solar Targets'.

121. Ross, 'India Charts a Roadmap to Achieve Ambitious Solar Targets'.

122. Ross, 'India Charts a Roadmap to Achieve Ambitious Solar Targets'.

123. V. Sivaram, 'Do India's Renewable Energy Targets Make Sense?', *Council on Foreign Relations* (11 March 2015), available http://blogs.cfr.org/levi/2015/03/11/do-indias-renewable-energy-targets-make-sense/, accessed on 18 July 2016.

124. Sivaram, 'Do India's Renewable Energy Targets Make Sense?'.

125. Sivaram, 'Do India's Renewable Energy Targets Make Sense?'.

126. Sivaram, 'Do India's Renewable Energy Targets Make Sense?'.

127. Sivaram, 'Do India's Renewable Energy Targets Make Sense?'.

128. Sivaram, 'Do India's Renewable Energy Targets Make Sense?'.

129. 'Coal India Will Invest Rs 13,900 Crore to Raise Output to 908 MT by 2020', *The Economic Times* (06 August 2015), available http://articles.economictimes.indiatimes.com/2015-08-06/news/65280800_1_production-target-output-target-coal-india-ltd, accessed on 7 September 2016.

130. Ministry of New and Renewable Energy, 'Request for Comments/Observations/Feed-back on the Draft Renewable Energy Act 2015' (October 2014), available http://mnre.gov.in/file-manager/UserFiles/dra2015-comments.html accessed on 7 September 2016.

131. Ministry of New and Renewable Energy, 'Request for Comments'.

132. Ministry of New and Renewable Energy, 'Request for Comments'.

133. Ministry of New and Renewable Energy, 'Request for Comments'.

134. Ministry of New and Renewable Energy, 'Request for Comments'.

135. A. Rosencranz, 'Prospects for Clean Energy India', *Law & Policy Brief*, Jindal Global Law School, II(5) (May 2016).

136. Rosencranz, 'Prospects for Clean Energy India'.

137. Energy Alternatives India, 'India Geo-thermal Energy' [http://www. eai.in/ref/ae/geo/geo.html], accessed on 7 September 2016.

138. Ministry of New & Renewable Energy, 'Indian Geothermal Energy Development Framework', available http://mnre.gov.in/file-manager/ UserFiles/Draft-Geothermal-frame-work-for-comments.pdf, accessed on 18 July 2016.

139. Ministry of New & Renewable Energy, 'Indian Geothermal Energy Development Framework'.

140. Energy Alternatives India, 'India Geothermal Energy', available http://www.eai.in/ref/ae/geo/geo.html, accessed on 18 July 2016.

141. Concentrating Solar Power (CSP) Technologies, Solar Energy Development Programmatic EIS, available http://solareis.anl.gov/guide/ solar/csp/, accessed on 7 September 2016.

142. IANS, 'Confident of Meeting 1.5 Billion Tonne Coal Target by 2020: Union Coal Secretary', *The New Indian Express* (5 April 2015), available http://www.newindianexpress.com/business/news/Confident-of-Meeting-1.5-Billion-Tonne-Coal-Target-by-2020-Union-Coal-Secretary/2015/04/05/ article2749086.ece, accessed on 7 September 2016.

143. Power, Department of Industrial Policy & Promotion, Ministry of Commerce & Industry, p. 9, available http://dipp.nic.in/English/Investor/ Investers_Gudlines/Power.pdf, accessed on 7 September 2016.

7

SANJAY UPADHYAY

Renewable Energy Development in India

The Need for a Robust Legal Framework

India has come a long way since the Department of Non-Conventional Energy Resources was setup in 1982. The rapid industrialization and urbanization, the fast-depleting conventional energy sources, the concerns on environment and clean technology, and most recently, the visible impacts of climate change makes a strong case for providing a robust legal framework for promoting renewable energy (RE). Clearly, the need for developing RE to compliment conventional energy sources cannot be overemphasized in this rapid changing global climate. The advantages of RE sources as indigenous, non-polluting and virtually inexhaustible resources, especially in this uncertainty of global climate change make a fit case for promoting RE. It also provides national energy security at a time when decreasing global reserves of fossil fuels threatens the long-term sustainability of the Indian economy. The energy security is an issue not only at the national level, but also at the local level. A major part of rural India still suffers from inadequate supply of electricity in particular and other energy sources for other basic needs. India has a vast supply of RE resources, and thus is also a major energy producer and consumer. It is the eleventh largest economy in the world and is poised to make tremendous economic strides over the next ten years, with significant development already in the planning stages.[1] The need for energy thus would only increase and dependence on fossils alone can be disastrous in the long run.

Indeed, it is the only country in the world to have an exclusive ministry for RE development, the Ministry of New and Renewable Energy Sources (MNRE). Since its formation, the Ministry has launched various programmes and schemes on RE. Acceleration of renewable development throughout the nation is required both to meet the underserved needs of millions of rural residents and the growing demand of an energy hungry economy. The development and deployment of RE sources are driven by the needs to, decrease dependence on energy imports, sustain accelerated deployment of RE system and devices, expand cost-effective energy supply, augment energy supply to remote and deficient areas to provide normative consumption levels to all section of the population across the country, and switch fuels through new and RE system/device deployment.

For viable and sustainable development of RE in the country it is thus important to have certainty, stability, coherence, and uniformity in the legal and policy mandate across the country. Moreover, the current legal backup through the Electricity Act and Tariff Policy is limited (explained later) and thus has a transitional role. So although the government is committed to promoting the use of RE sources, the commitment is yet to be backed by legislation.

The need for a law on RE also becomes evident when one considers the requirement of regulating various projects and programmes initiated and sponsored by the MNRE, coordinating the work of different nodal agencies at the central and state levels, and more importantly, for facilitating rapid economic and commercial investment and subsidies, which need to be promoted but also made legally accountable for the security of the environment and equitable development in the new liberalized market. Thus, the primary targets in development of RE sector are, (a) an integrated energy planning, (b) coordinated energy efficiency policy and regulatory framework, (c) focused power sector reforms, (d) accelerating demand side management, (e) fresh approach on past schemes and programmes on energy conservation and energy efficiency, (f) Integrating initiatives by international and national organizations and funding agencies, (g) Introducing innovative measures for enhancing energy efficiency and learning from successful initiatives, and last but not the least, (h) increased exposure to global trade in energy efficient technologies.

It is now well recognized that the exploitation of RE resources can not only supplement the generation of power but also be a viable economic option if the competitive field for all players in the arena is levelled out by putting all the economic cards on the table. The hidden cards in the conventional energy resources are the various subsidies and natural resources cost to the Nation which are not accounted for. It is therefore indeed a matter of necessity and not just what is desirable to firm up the legal frame on RE.

Existing Legal Framework on Renewable Energy

In the past, few regulatory and legislative measures such as the enactment of Energy Conservation Act (ECA), 2001, the Electricity Act 2003 (EA), and Integrated Energy Policy (IEP) 2006 have been undertaken for improving energy efficiency and energy conservation. Recent initiatives include the National Mission on Enhanced Energy Efficiency (NMEEE) under the National Action Plan on Climate Change (NAPCC), 2008. However, notwithstanding weaknesses within the regulatory and legal framework itself, the maintenance and sustainability of energy efficiency services remain the key challenge to be addressed. Further, the regulatory framework providing for energy efficiency measures has been by and large a failure in India. Causes for such failure are to be seen within the policy and legal framework and outside. Thus, there is a need to understand the existing framework so as to suggest measures that could facilitate efforts for energy efficiency. Also, such measures must include reviewing of not only implementation strategies and mechanisms but an assessment and review of national policies, laws, regulations, and schemes concerning energy conservation and energy efficiency (on supply and demand side) so as to identify policy and legal barriers and gaps in the way of enhancing energy efficiency measures. Given the above background, it is pertinent to describe in some detail the current legal framework in order to appreciate necessity for developing a new RE law.

Constitutional Provisions

RE is not a specific item in the Constitution's legislative lists. Although Electricity is a concurrent subject[2] on which both the centre and the state governments are empowered to legislate. It is pertinent to mention that electricity includes power generated from conventional as well as non-conventional energy sources. The subject of non-conventional energy sources or RE has been a subject matter of the Panchayat and thus it plays a key role in RE development! The constitutional validity of such legislation also needs to be examined at the outset.

Constitutional Validity of Renewable Energy Law

The Constitution of India lists the subject matter of laws which can be made by Parliament and by the Legislatures of States under Article 246. Under this Article the Parliament has plenary powers to make any law including the law to amend the Constitution subject to the limitations laid down therein.[3] Further, the power to legislate on a topic of legislation carries with it the power to legislate on an ancillary matter which can be said to be reasonably included in the power given.[4]

The State under Article 48-A read with 51-A (g) 14 and 21 of the Constitution, must endeavour to protect and improve the environment. Promoting the use of non-conventional energy sources and reducing the pressure on the non-renewable sources, to augment the power capacity of the State would be a right step in this direction.

Additionally, under Article 39(b) the State has a duty to direct its policy towards ensuring that the ownership and control of the material resources of the community are so distributed as best to sub-serve the common good. Clearly, a law of this nature would promote this cause.

Further, the legislative competence of the Parliament on matters relating to Electricity is under Entry 38 of the Concurrent List or the List III. This implies that both central and state governments are competent to legislate on aspects of electricity. One of the key results of RE development would certainly be electricity.

Entry 33 of List III relates to trade and commerce in and the production, supply, and distribution of the products of any industry where the control of such industry by the Union is declared by Parliament by law to be expedient in public interest and supply and distribution includes the power to control the price of the commodities coming under this Entry.[5] Also, the State Legislature is competent to provide for levy of surcharge on electricity so long as the relevant provisions do not conflict with any provision in the Central Act (which is the EA).

Further and significantly, the RE sources such as solar or wind do not come under any specific entries of the Seventh Schedule of the Constitution and hence can be legislated upon by Parliament vide Entry 97, List I read with Article 248, giving residuary powers to the Parliament.

Last but not the least, the proposed Renewable Energy Act is in furtherance of the constitutional mandate under the Constitution (Seventy Third amendment) Act, 1992, which received the assent of the President on 20 April 1993. Part IX which deals with the Panchayats has been incorporated under the said Amendment. Under Article 243-G, of the same Part, the powers, authority, and responsibilities of Panchayats have been illustrated. The Eleventh Schedule which has also been included vide the said Amendment Act, specifically includes, inter alia, Rural Electrification including distribution of electricity (Item 14), Non-Conventional energy Sources (Item 15), and Minor Forest Produce (Item 7). The Article clearly provides that the state government may by law endow Panchayats with such power and authority with respect to the implementation of schemes for economic development and social justice and this includes those in relation to the matters listed in the Eleventh Schedule. Note that this power however is with the state governments.

Clearly, the scope and objective of the Renewable Energy Act is within the legislative competence of the Parliament and its effective implementation would further the cause of decentralized development and fulfil the constitutional obligation.

National Framework on Renewable Energy

Having understood that there are no serious constitutional impediments to enacting a RE Law, it is important to understand the current framework that addresses RE in India. The EA and ECA provide the statutory mandate dealing with renewable form of energy. It is important to map the provisions dealing with RE and analyse them in the context of development of a new RE framework.

The Electricity Act, 2003

BACKGROUND TO THE ACT   The policy of encouraging private sector participation in generation, transmission, and distribution and the objective of distancing the regulatory responsibilities from the government to the regulatory commissions, the need for harmonizing and rationalizing the provisions in the Indian Electricity Act, 1910, the Electricity (Supply) Act, 1948, and the Electricity Regulatory Commission Act, 1998, in a new self-contained comprehensive legislation arose, culminated in formulation of EA.[6] The focus of the EA was to bring in power sector reforms and need to provide for newer concepts such as power trading and open access. The Act seeks to encourage competition with appropriate regulatory intervention. Competition is expected to yield efficiency gains and in turn result in availability of quality supply of electricity to consumers at competitive rates. Salient features of the Act are as follows:

1. The central government to prepare a National Electricity Policy in consultation with state government (Section 3).
2. Thrust to complete the rural electrification and provide for management of rural distribution by Panchayats, Cooperative Societies, non-Government organizations, franchisees, and so on. (Sections 4, 5, and 6).
3. Provision for license free generation and distribution in rural areas (Section 14); Generation being delicensed and captive generation being freely permitted.
4. Hydro projects would however need clearance from the Central Electricity Authority (CEA) (Sections 7, 8, and 9).
5. Transmission Utility at the central as well as state level, to be a Government company—with responsibility for planned and coordinated development of transmission network (Sections 38 and 39).
6. Provision for private licenses in transmission and entry in distribution through an independent network (Section 14).
7. Open access in transmission from the outset (Sections 38–40).
8. Distribution licensees would be free to undertake generation and generating companies would be free to take up distribution businesses (Sections 7 and 12).
9. Trading, a distinct activity is being recognized with the safeguard of Regulatory Commissions being authorized to fix ceilings on trading margins, if necessary (Section 12, 79, and 86).

10. Provisions safeguarding consumer interests (Sections 57–9, 166).
11. Ombudsman scheme for consumers' grievance redressal (Section 42).

Renewable Sources of Energy: Statutory Provisions

The EA provides for a formulation of National Electricity Policy and Tariff Policy[7] by central government, in consultation with the state governments and authority for development of the power system based on optimal utilization of resources such as coal, natural gas, nuclear substances, or material, hydro and renewable sources of energy. In keeping with this mandate the National Electricity Policy, 2005 was formulated with the following stated objectives.

1. Access to Electricity—7,108 un-electrified villages electrified in 2015–16 which is 37 per cent higher than previous three years.[8]
2. Availability of Power-Demand to be fully met by 2012.
3. Energy and peaking shortages to be overcome and adequate spinning reserve to be available. Overall generation in the country has increased from 1048.673 during 2014–15 to 1107.386 BU* during the year 2015–16.[9]
4. Supply of Reliable and Quality Power of specified standards in an efficient manner and at reasonable rates. Per capita availability of electricity to be increased to over 1000 units by 2012. The country's per capita electricity consumption has reached 1010 kilowatt-hour (kWh) in 2014–15, compared with 957 kWh in 2013–14 and 914.41 kWh in 2012–13, according to the CEA, India's apex power sector planning body.[10]
5. Minimum lifeline consumption of 1 unit/household/day as a merit good by year 2012. This scheme has been merged into the Rajiv Gandhi Grameen Vidyutikaran Yojana but the target is yet to be met.[11]
6. Financial Turnaround and Commercial Viability of Electricity Sector; Protection of consumers' interests.

As can be seen, promotion of RE is not one of the main objectives of the National Electricity Policy 2005, though one of the issues dealt in the policy relate to RE. Excerpts from National Electricity Policy on RE is given below:

5.12.1 Non-conventional sources of energy being the most environment friendly there is an urgent need to promote generation of electricity based on such sources of energy. For this purpose, efforts need to be made to reduce the capital cost of projects based on non-conventional and renewable sources of energy. Cost of energy can also be reduced by promoting competition within such projects. At the same time, adequate promotional measures would also have to be taken for development of technologies and a sustained growth of these sources.

5.12.2 The Electricity Act 2003 provides that co-generation and generation of electricity from non-conventional sources would be promoted by the State Electricity Regulatory Commissions (SERCs) by providing suitable measures for connectivity with grid and sale of electricity to any person and also by specifying, for purchase of electricity from such sources, a percentage of the total consumption of electricity in the area of a distribution licensee. Such percentage for purchase of power from non-conventional sources should be made applicable for the tariffs to be determined by the SERCs at the earliest. Progressively the share of electricity from non-conventional sources would need to be increased as prescribed by State Electricity Regulatory Commissions. Such purchase by distribution companies shall be through competitive bidding process. Considering the fact that it will take some time before non-conventional technologies compete, in terms of cost, with conventional sources, the Commission may determine an appropriate differential in prices to promote these technologies.

FORMULATION OF POLICY ON RENEWABLE ENERGY AND STANDALONE SYSTEMS[12]

Mandate of the provision: The central government has to formulate after consultation with the state governments, a national policy, permitting stand-alone systems (including those based on renewable sources of energy and other non-conventional sources of energy) for rural areas.

Analysis of the provision: These sections as such do not create any substantive legal mandate apart from bringing RE also as a player in to the centre-stage of policy formulation. To give real boost to this sector it is important to have a clear legal mandate to attract investors. In spite of passage of six years, an RE policy has not been framed under these provisions.

TARIFF REGULATIONS AND DETERMINATION[13]

Mandate of the provision: The appropriate commission (state or central) has to specify the terms and conditions for determination of tariff, and in doing so, is to be guided by the principle for the promotion of cogeneration and generation of electricity from renewable sources of energy; and the National Electricity Policy and Tariff Policy.[14]

Analysis of the provision: Tariff fixation is also the responsibility of the State Electricity Regulatory Commissions (SERCs) and tariffs have to be fixed separately in different category of RE, that is, solar, biomass, wind, hydro, and other sources. Further, within these categories there are a lot of differentiation based on technology used and the size of the project. Ten states namely, Chhattisgarh, Haryana, Kerala, Orissa, Punjab, Rajasthan, Tamil Nadu, Uttar Pradesh, Uttarakhand, and West Bengal have fixed tariff for solar energy and for biomass; fourteen states namely, Andhra Pradesh, Chhattisgarh, Gujarat, Haryana, Karnataka, Kerala, Maharashtra, Madhya Pradesh, Punjab, Rajasthan, Tamil Nadu, Uttar Pradesh, Uttarakhand, and West Bengal have fixed tariff for electricity generated by specified methods. Nine states, that is, Tamil Nadu, Maharashtra, Rajasthan, Karnataka, Gujarat, Madhya Pradesh, Andhra Pradesh, Kerala, and West Bengal have fixed tariff for the wind generated electricity.[15]

Emergent Issues include the timeline within which the tariff has to be fixed. (Tariff can also be fixed on an application made by the developer); there is no guarantee of payment of fixed tariff for fixed number of years which is decided primarily keeping in mind the life of a project and the profitability of the project among other factors. (In Germany for example, it extends up to 20 years for wind and solar based projects).[16] The tariff fixation does not take into account the actual gradual technology advancement and market development in consequence of technology becoming cheaper in determining the tariff for the later years of project life; there should be a specific provision which takes into account the gradual technological advancements as well as market development; tariff fixation is of critical importance for any RE project and is a highly technical and nuanced process which has to take into account a multiplicity of factors—technical, financial, and market related. The location of the project and the quality of the resource also plays a very important role. One set of tariff for wind energy cannot fit all the projects of wind energy. This section has given all the powers to the SERC. There should be a provision in the Act itself about the manner in which this process is carried out.

STATE REGULATORY COMMISSION: ITS FUNCTIONS[17]

Mandate of the provision: The SERCs are empowered to promote cogeneration and generation of electricity from renewable sources of energy by providing, suitable measures for connectivity with the grid and sale of electricity to any person, and also specify, for purchase of electricity from such sources, a percentage of the total consumption of electricity in the area of a distribution license. This essentially means that they are mandated to provide Renewable Power Purchase Obligations (RPPO). This is the first step in creating conducive legal environment for the development of RE in any state. This RPPO generates confidence among the developers that the distribution licensee shall have to purchase the electricity they produce using RE sources.

Analysis of the provision: However, from a look at the number of states in which the SERCs have actually fixed the RPPOs[18] and the percentage they have fixed, we can gauge the reality. This very potent legal tool has not been utilized to its full extent partially because of the mandate in the EA which is not comprehensive.

The mandate of EA does not fix any time line within which this RPPO has to be fixed; EA also does not venture into the minimum percentage of RE that the SERC have to fix; EA is also silent on how a gradual increase in percentage of RPPO can be ensured by the SERCs; there is no mandate to provide a National RPPO; the orders/regulations specifying the percentage of renewable power procurement lack the monitoring and enforcement mechanism at present. Except the Maharashtra SERC, no other Commission has imposed any penalty for non-compliance with the order/regulation; the probable reason for these issues could be a direct fallout of the fact that the SERCs have no control over the RE industry development, availability of RE equipment, availability of and access to latest technology, availability of tailor made or suitable finances to ensure that generation of electricity from RE increase over a period of time. These are some of the critical areas without which RE generation in the country cannot take off and these are also the areas which have not been dealt with in the EA.

Energy Conservation Act, 2001

The legal and institutional framework to address energy conservation in India consists of Bureau of Energy Efficiency (BEE) and designated state agencies of the Ministry of Power (MoP). Owing to phenomenal economic growth, energy consumptions rates both in the industry and domestic use have risen to unsustainable limits. Energy efficiency measures are a proven means to reduce dependence on traditional energy resources by using them more efficiently.

BACKGROUND AND OBJECTIVE OF ENERGY CONSERVATION ACT   ECA is an Act to provide for efficient use of energy and its conservation. Energy in this statute includes RE and as such its importance in development of RE is required to be mapped.

Application of Energy Conservation Act to Renewable Energy Generation and Its Shortcomings

As per the definition of 'energy', electricity from RE sources are included, however other applications of RE that should come within the ambit of 'energy' are not discussed. If the definition of energy[19] is expanded in the context of RE and is made to include any form of energy derived from renewable sources the ambit of the scope of powers and functions of the BEE can be increased vis-à-vis RE in the country. Then it would be under the jurisdiction of the BEE to take steps to increase energy efficiency of the methods of energy consumption in the rural areas of India. India's primary energy balance shows that RE accounts for 32 per cent of primary energy consumption in the year 2003–4.[20] Of this the major contributor is traditional biomass mainly used in cooking followed by electricity generation from large hydro dams. This use of biomass for cooking in rural India can certainly be made more energy efficient and environmental friendly. Increasing the energy efficiency of these methods would also reduce the pressure on the biomass available and help in its sustainable use in the rural parts of India. Keeping in mind the huge proportion of this consumption it should be imperative that we target this sector for energy efficiency enhancement and its use in more user friendly and environmentally friendly manner.

The definition section can also have the word 'conservation' and it should be defined in such a manner so as to bring within its ambit any form of RE generation for its usage at the point of generation so as to decrease the overall energy consumption from the grid and in case where grid connected electricity is not there such generation would have replaced such requirement for the time being. To illustrate this point further, the 'energy conservation building codes' is defined in the ECA and means the norms and standards of energy consumption expressed in terms of per square meter of the area wherein energy is used and includes the location of the building. This energy conservation building code should also include mandatory use of solar water heating or energy conserving building techniques, or insertion of solar panels within the façade of the building or mounted on the roof of the building to generate as much RE as possible for its own consumption and to reduce consumption of energy using latest green or energy saving building technology. For government and public buildings these provisions related to RE generation and green building technology should be made mandatory and for private construction the incentives should prove to be an allurement to adopt these measures.

The ECA has not been able to keep pace with the rapidly rising energy consumption and energy losses, due to its inherent weaknesses. The BEE has also not been effective in implementing its own energy conservations schemes, such as reduction in transmission and distribution of losses, and failure to implement Energy Conservation Building Code are specific examples. Thus, there is a need to strengthen the ECA and to alternatively make all these innovative provisions as part of the new RE law.

Issues Not Addressed by Existing Legal Mandate on Renewable Energy

Having examined relevant provisions from EA and ECA concerning RE, it becomes clear that the following issues are yet to be addressed, in order to improve the energy efficiency measures:

1. The statement of objects and reasons of EA elucidates 13 main features of the Act and not one focuses on development of RE, except allowing for setting up of standalone systems for generation and distribution in rural and remote areas. The focus on development and promotion of RE is missing in EA. Even the tinkering of EA would not address the issues concerning RE development, as EA deals with electricity generation alone and development of RE is not only from the perspective of electricity generation but also its utilization in other energy applications.

2. Development of RE in the country needs a holistic approach to the whole issue of RE sources, their various applications and their efficient use for electricity generation, heat generation, cogeneration, or in the rural context where biomass is the main source of energy for meeting their primary energy demands. This has not been addressed by the current legal framework.

3. Resource mapping is an important aspect in RE development which is not addressed by legal framework and it cannot be left to administrative enabling orders.

4. Technology development in the field of RE is happening in a piecemeal manner and a concerted focus on the same is required. It requires to be legally mandated so that it is not made too discretionary.

5. Development and promotion of RE can be ensured if there are legally mandated provisions for making use of RE resources, for example, solar water heating in at least public, government, and commercial buildings. The EA has no such provision or legal space to provide for such a provision as it is beyond electricity generation.

6. Provision for incentives to people at large for generating their own electricity by putting solar panels or solar water heaters.

7. The existing RE legal mandate does not include issues such as Biofuels. What is needed is a full economic chain from farming, harvesting, and extraction to esterification, blending, and marketing.

8. Fuel cell development and incorporation of hybrid cars, as means of lessening the burden on conventional fuel also needs to be addressed by legal framework.

Why a Separate Renewable Energy law?

A law has a more certain status than a collection of policies and is less likely to be changed in case of a shift in the political environment. Law would largely overcome the issue of inconsistent State regimes (for example, differences in RE targets, enforcement regimes, definitions of RE, taxation benefits (sales tax). A national law would also be welcomed by foreign investors as it would make the regulatory environment clearer and more consistent. A specific RE law allows for several support mechanisms to be addressed in one document, providing a central reference for interested parties and making it simpler to identify any inconsistencies or inefficiencies between concurrent programmes. Market mechanisms for promoting RE are required and need to be mandated by law. One such market mechanism can be Renewable Energy Certificate (REC) that can be traded in the market. Law in general is a very effective means to bring about change and to create conducive environment. The provisions dealing with RE are enabling provisions and not mandatory in present EA. The time has come for having mandatory provisions for development and promotion of RE in the country. The enforcement has not been effective enough. Second, EA speaks only of generation, transmission, and distribution of electricity but does not holistically address energy production from renewables, even extending to transport fuels or biofuels. What needs to be provided is a comprehensive legislative framework for development of all types of RE technologies and usage and promotion thereof that goes beyond electricity, for example, solar thermal systems, bio-fuels, and so on. The law will facilitate enforcement of measures for further promotion of RE in the country.

To ensure RE development, a legal instrument seems to be a more prevalent and successful option being preferred by the countries world over. It is not disputed even in India and that is why RE for the first time was dealt within the framework of EA. Legislation in order to be effective should be comprehensive, adequate, and not piece meal. A single statute should preferably contain all the legal mandate or legal provisions that are required for the development of a particular sector or industry. In India as of today as is evident from the discussion above the current legal/regulatory regime is inadequate and fragmented. In the sections above we have discussed the issues that the current regime does not touch upon. Even if we tinker with the existing legal framework which is limited to EA, ECA, Tariff Policy, Electricity Policy, Rural Electrification Policy, we would not be able to touch upon all the issues that are relevant for the development of this sector in the country. If we make an attempt to regulate all the issues necessary to ensure development of RE then the legal and regulatory framework would become very fragmented. A fragmented system is never effective and to ensure enforceability multitude of authorities are involved.[21]

The objective to raise the share of RE in our energy mix and to encourage other RE applications are gigantic in proportion. With a concerted push and a fortyfold increase in their contribution to the primary energy, renewables may account for only 5–6 per cent of India's energy mix by 2031–2.[22] While this figure appears small, the distributed nature of renewables can provide many socio economic benefits. The MNRE renewable electricity targets of under 43 GW in April 2016 to 175 GW by the year 2022, which includes 100 GW from solar power, 60 GW from wind power, 10 GW from bio power, and 5 GW from small hydro power,[23] are unlikely to be achieved unless legislative measures at the national level are not immediately put in place to attract private investments from within the country and abroad.

If India is serious about achieving this target, a concrete, comprehensive legislation needs to be in place. A single authority in the country needs to be made responsible for the development of the sector and an integrated approach with coordination at the very top is required. Gokak committee in its report[24] has very clearly brought out the issues being faced by the sector in the country and this is despite the efforts being made by MNRE and legal space created by the EA. Following are some of the issues which emerge from the constraints to RE development mentioned in this report.

A vigorous and robust research and development (R&D) in the sector which focuses not only on innovation but also on indigenization of existing technologies has to be ensured; Resource mapping needs to be done comprehensively; the issue of land requirement for RE projects need to be dealt with; and, the distortions in energy market on account of subsidized conventional electricity have to be rectified.

As is also evident from the Gokak Committee Report, a compartmentalized approach is seen in the sector wherein R&D is entrusted to one organization; schemes and programmes are run by the ministry without any long-term vision; law is enforced by another ministry (MOP); and RE industry development and fiscal and financial incentives are decided by multitude of authorities. It is therefore essential that a unitary authority should be responsible for the overall development of the RE sector. Standing committee on energy in the 14th Lok Sabha has so far submitted 26 reports. Standing committee in its 26th report has also recommended formulation of RE law. The RE sectors in India needs a decisive, enforceable, and binding language of the law. It would be interesting to look at what UNEP refers to[25] as *a legislation that 'pushes' the market* towards more RE!

New Changes in the Legal Framework Surrounding Renewable Energy

Keeping all the above-said considerations in mind, certain changes have been proposed in the legal framework surrounding RE. For instance, amendments have been proposed in the EA which is supposed to revolutionize the electricity sector of entire country along with setting new inroads into development of RE.

Amendments vide the Electricity (Amendment) Act, 2014

One of the main amendments is the inclusion of definition on 'decentralised distributed generation' meaning electricity generation from wind, small hydro, solar, biomass, biogas, bio-fuel generation from any kind of waste including municipal and solid waste, geothermal, hybrid power system, or such other sources as may be notified by the central government for end-use at or near the place of generation.[26] In a similar manner, definition for 'renewable energy' is provided to include small hydro, wind, solar, including its integration, with combined cycle, biomass, bio-fuel, co-generation, urban or municipal waste, and other sources approved by the central government in consultation with MNRE.[27] Furthermore, it includes other RE friendly provisions such as ten per cent RE generation capacity to be established in coal and lignite based thermal generating stations;[28] provision for National RE Policy in the country;[29] no requirement for person intending to generate and supply electricity from RE sources, to acquire any license for such generation and supply of electricity[30] In case of non-compliance of directions at the regional level by a company generating RE, the penalty shall be one crore rupees fine;[31] and at the state level, it would be fifty lakh rupees.[32] In case of non-compliance of directions by Appropriate Commission, the penalty for RE generating company would be ten lakh rupees and in case of continuing failure, an additional penalty of 10,000 rupees for every day.[33] In case of non-compliance of orders or directions under this Act, by a company generating RE, the penalty would be imprisonment for a term which may extend to three months or fine of ten lakh rupees or both and in case of continuing failure, ten thousand rupees for every day of non-compliance.[34] The sale of electricity generated from RE sources is exempted from cross subsidy and open access charge for a period as prescribed.[35]

In addition, in October 2014, the MNRE constituted an Expert Committee to provide a draft Renewable Energy Act. The Committee met on 7 November 2014, and decided to constitute a sub-group to 'collate the suggestions and propose a draft Act'. The sub-group prepared a draft Renewable Energy Bill, 2015 hereinafter referred to as 'the Bill'. It broadly proposed provisions for institutional structure, supportive ecosystem, economic and financial framework, constitution, and operation of national and state level funds to support achieving of the objectives of the Act, and RE applications including distributed and grid connected renewable electricity.[36]

National Renewable Energy Draft Bill 2015[37]

The Bill is a landmark legislation that seeks to promote the production of energy through the use of RE sources and contribute to ensuring fulfilment of national and international obligations on climate change. The draft proposes a model constitution of decision-making and advisory bodies in the government, development of a conducive ecosystem that promotes the utilization of RE sources and permits investments,[38] constitution, and operation of national and state level funds and RE applications through measures such as grid connected renewable electricity.

The objective outlined in the Bill is to promote RE and reduce dependence on energy from conventional (fossil fuel) sources, in response to (macro) economic, environmental, and social concerns. RE here includes wind, solar radiation, mini hydro, biomass, biofuels, landfill, and sewage gas, municipal solid waste, industrial waste, geothermal energy, ocean energy, any other energy source, as may be notified by the Ministry and hybrids of above sources.[39]

It provides for an institutional structure at central and state levels. The central government is empowered with planning, advising, monitoring, executing, and reviewing powers with respect to RE policy, national programmes, R&D, utilization of funds, standards/norms for resource assessment, issuance of guidelines for development of state policies and plans, and so on.[40] State governments are empowered to manage and finance their respective projects in compliance with

National policies, plans, and guidelines.[41] A National Renewable Energy Committee (NREC) is to be constituted to facilitate inter-ministerial cooperation for the implementation of the Bill.[42] A National Renewable Energy Advisory Group (NREAG) shall be established to advise the government on such matters and make annual reports on the same.[43] A Renewable Energy Corporation[44] is empowered to act as the national RE procurement entity and to support RE investment zones across the country.

Functions of the NREC include identification of measures for development of indigenous technology, manufacturing base, capacity development, skill development, export of technologies, and establishment and coordination of technology missions created under the Act. The NREC also has the discretionary power to consult the NREAG and other stakeholders. The NREAG is supposed to identify measures required to create awareness and educate the citizens for adoption of RE technologies and promote private sector and community participation. Further, it also has the power to undertake consultation processes with various stakeholders, as deemed necessary and expedient.[45]

A comprehensive National RE policy has to be formulated within six months of commencement of the Act.[46] National RE fund is to be managed by the central government and state green funds are to be managed by state governments.[47] The bill provides for promotion of decentralized and standalone RE applications in rural and urban areas, with designated nodal entities assigned the task of assessing areas where such application is feasible.[48] A list of villages and hamlets where grid-extension is unfeasible within five years is to be published within six months. No license is required for suppliers of RE; only compliance of conditions specified by the CEA under Section 53 and 73 of the EA is required. This law is yet to be passed into an Act.

Apart from bringing clarity to the concept of RE, these legislative changes also seeks to incentivize generation of RE through open access and increased efficiency in the power grid with increased share of renewables in India's energy production. Stricter penalties for non-compliance in the EA as well as the Renewable Energy Act, when they become operational, lays a greater burden on the regulatory institutions as well as power generating companies to meet the targets. The Bill also takes into consideration other legislations such as EA, the Environment Protection Act, 1986 (EPA) and Land Acquisition, Rehabilitation, and

Resettlement Act, 2013 (LAA). Linkages with existing programmes on climate change such as NAPCC and its missions such as NMEEE, National Solar Mission and other missions like National Electric Mobility Mission, National Wind Energy Mission, and Waste to Energy Mission have been promoted within the Bill. Other key linkages have been established with National Manufacturing Policy and National Skill Development Programme as well.[49]

Institutional Concerns

As of today there are multitudes of agencies which are entrusted with the above named tasks. Ministry of Finance decides about the fiscal and financial incentives to RE industry, availability and access to technology is not being specifically looked at and the technology development and demonstration programmes are not adequate and lack desired amount of focus and finances. Multiplicity of agencies and organizations handling different aspects of RE, capacity, lack of coordinated framework, state variance in financial and fiscal incentives necessitates a clear, decisive and well-coordinated legislative framework that sees RE sector as one integral whole.

The role of Indian Renewable Energy Development Agency Limited (IREDA), needs to be reassessed and potential of it as a national refinancing institution on the lines of National Bank For Agriculture and Rural Development / National Housing bank for the RE sector may be explored. IREDA's own equity base can be expanded by the financial institutions of the country instead of continuing with the current system of Government of India support.

State nodal agencies (SNAs) are supposed to play the role of single window clearance agencies for renewable energy projects, thus facilitating the approvals required from different line departments such as pollution control, fisheries, mining, forests, and so on; however, the practice in most states is that the onus is on developers to get the clearances required for their projects.[50] MNRE also needs to have a regulatory and facilitative role along with the SNAs. Currently there is no one to regulate various projects and programmes initiated and sponsored by the MNRE through the instrument of law, to facilitate rapid economic and commercial investment and subsidies in the areas of RE, which need to be promoted as well as being made legally accountable.

R&D needs to be incorporated as a legal mandate. This will further facilitate the promotion of RE. The amendments within EA and the Bill mandates the need for such an R&D by the central government and authorities sought to be established by these legislations.

Implication of Other Laws

Implementation of RE resources, programmes, and projects and the functioning of the agencies involved can attract the provisions of several laws. All these need to be integrated in one legal instrument to facilitate RE sector as a whole. For example, the relevant provisions of following legislations can be codified in one code. Forest Conservation Act, LAA, EPA, and Notifications on Environment Impact Assessment (EIA), Coastal Regulatory Zones (CRZ), other Ecologically Sensitive Areas (ESAs), issued under it, Water Act, Panchayat Acts, business laws, such as Bank and Tax laws, Company, and Industrial Development Acts. Similarly, Municipal Solid Wastes (Management and Handling) Rules, 2000, Hazardous Wastes (Management and Handling) Rules, 1989, Bio-Medical Waste (Management and Handling) Rules, 1990, Batteries (Management and Handling) Rules, 2001, and so on.

Development of Renewable Energy: International Experience

Various countries around the world have recognized the growing importance of RE and have passed specific laws, directives, ordinances, or provisions for RE in order to ensure growth of the sector. These directives usually specify some long-term RE target. This is endorsed by the UNEP in its energy handbook that to ensure generation of electricity from RE sources in many cases, both statutory and regulatory action will be necessary, with legislation required in the first instance to establish the requirement or programme, and regulation then required to fill in the details and implement the legislation.

European Union (EU)

The Directive 2001/77/EC issued by the European Parliament and Council on 27 September 2001 on the promotion of electricity produced from RE sources in the internal electricity market prompted many member states of EC to have national legislation to promote RE. The German RE law is a prime example of what properly designed, stable energy statute can do to bolster the growth of RE.

Germany

In Germany, the share of electricity from RE has increased from 6.3 per cent in 2000 to over 14 per cent in 2007 and this phenomenal increase in Germany is attributed to the Renewable Energy Sources Act. The law was passed in the German Bundestag on 26 February 2000 and was implemented from 1 April 2000. Its tariff fixation system is very elaborate and differentiated and it can be a good guiding principle. Germany's RE sector is among the most innovative and successful worldwide. Net-generation from RE sources in the German electricity sector has increased from 6.3 per cent in 2000 to about 30 per cent in 2014.[51] For the first time ever, wind, biogas, and solar combined accounted for a larger portion of net electricity production than brown coal in the first half of 2014.[52] On Sunday, 15 May 2016 at 14:00 hours, renewables supplied nearly all of domestic electricity demand.[53]

China

The scope of RE law is very comprehensive however the institutional framework and the nature of the content are very general. China has more than doubled its total installed capacity by installing 3,449 MW of wind energy in 2007, a 256 per cent increase from previous year's figure.[54] It has become the strongest market in Asia, even though India remains the country with the largest installed capacity in Asia with 8000 MW, while China reached 6,050 MW. The Chinese market was boosted by the country's new RE Law, which entered into force on 1 January 2006. The goal for wind power in China by the end of 2010 is 5,000 MW, which has already been reached well ahead of time.[55]

Australia

Australia has implemented the Renewable Energy (Electricity) Act 2000 that has set a mandatory national RE target. The framework set to achieve this target, is through the Mandatory Renewable Energy Target Scheme (MRET). It is a trading system based on the creation, trade, and surrender of RECs, each REC corresponding to one megawatt hour of electricity generated from renewable resources.

Austria

Green Electricity Act, 2003 (amended 2006) was established to enact new provisions related to renewable electricity generation and combined heat and power (CHP). This Act regulates various renewable electricity-related matters, including the guarantees of origin of electricity produced from RE sources; the obligations to purchase and pay for electricity; the preconditions for, and the promotion of, electricity produced from RE sources; and, the nation-wide equal sharing of costs associated with the promotion of electricity produced from RE sources and from CHP plants.

United Kingdom

There are essentially three acts that deal with RE.

Sustainable Energy Act, 2003

This Act deals with the provisions for the development and promotion of a sustainable energy policy. The Act makes it mandatory for the Secretary of State to publish annually a 'sustainable energy report' to indicate progress made towards cutting the UK's carbon emissions, maintaining the reliability of energy supplies. The Act also requires specifying targets for electricity production from CHP plants.

Energy Act, 2004

The Secretary of State is required to publish a strategy for promotion of micro-generation after considering its potential for implementing 'sustainable energy report' and enhancing the availability of electricity and heat for consumers in Great Britain. The sources of energy and technologies that are permitted under the micro-generation initiative are biomass, biofuels, fuel cells, photovoltaic, water (including waves and tides), wind, solar power, geothermal sources, and CHP systems.

Climate Change and Sustainable Energy Act, 2006

This Act complements the Sustainable Energy Act, 2003 and the Energy Act, 2004. The purpose of the act is,

> to make provision about the reduction of emissions of greenhouse gases, the alleviation of fuel poverty, the promotion of micro generation and the use of heat produced from renewable sources, compliance with building regulations relating to emissions of greenhouse gases and the use of fuel and power, the renewables obligation relating to the generation and supply of electricity and the adjustment of transmission charges for electricity; and for connected purposes.

Philippines

The Philippines' Biofuels Act of 2006 came into effect in January 2007. The Act makes the limited use of Biofuels (biodiesel and bioethanol) mandatory. The objective being to substitute a fraction of the costly imported crude oil with indigenously made renewable liquid fuel.

***

It is clear from the above that globally wherever there has been a systematic development of RE, a legislative framework has played an integral part. Since the legislative mandate in India has been fragmented for long, it is time to forge the development of RE through a comprehensive legislation that looks at RE in a holistic manner and not just from the perspective of electricity generation. The international experience points to the fact that a comprehensive statutory framework is required to provide clarity and stability for promotion of RE and for inclusion of private sector in this endeavour. It is hoped and believed that the Draft Renewable Energy Act of 2015 will go through a more rigorous consultative process especially incorporating past experiences and efforts and then passed at the earliest, so as to legitimize and incentivize the production and consumption of RE to meet the sustainable developmental goals of India and contribute to the world at large.

Notes and References

1. N. Lotia, R.L. Shrivastava, and M.N. Nasim, 'Development of Framework for the Evolution of Alternative Energies Supply and Demand: A Review', *IOSR Journal of Mechanical and Civil Engineering (IOSR-JMCE)*, pp. 54–9, available http://iosrjournals.org/iosr-jmce/papers/ICAET-2014/me/volume-6/13.pdf?id=7622, accessed on 24 August 2016.

2. Schedule VII, Concurrent list, Entry 38, The Constitution of India.

3. AIR 1967 SC 1643 (1658).

4. AIR 1959 SC 544 (566): 1959 Cri. L.J. 660.

5. *Jain v. Union of India* (1970) 2 S.C.W.R 59.

6. Statement of Objects and Reasons Electricity Act, 2003.

7. Sections 3(1) and 3(2) Electricity Act, 2003.

8. 'Two Years Achievements and Initiatives', Ministry of Power, Coal and New & Renewable Energy, available http://powermin.nic.in/sites/default/files/uploads/Ujwal_Bharat_Brochure_English.pdf, accessed on 12 September 2016.

9. Overview, available http://powermin.nic.in/en/content/overview, accessed on 12 September 2016.

10. U. Bhaskar, 'India's per Capita Electricity Consumption Touches 1010 kWh', *Livemint* (20 July 2015), available http://www.livemint.com/Industry/jqvJpYRpSNyldcuUlZrqQM/Indias-per-capita-electricity-consumption-touches-1010-kWh.html, accessed on 7 June 2016.

11. Shirish S. Garud, 'Rural Electrification: Challenges and the Way Ahead', available http://www.teriin.org/index.php?option=com_featurearticle&task=details&sid=919&Itemid=157, accessed on 5 May 2016.

12. Section 4, Electricity Act 2003.

13. Section 61, Electricity Act 2003.

14. Section 61, 61(h) and 61(i) Electricity Act, 2003.

15. 'Distributed Renewable Energy Projects in Rural Areas of India: A Guide to Implementation', ELDF, 2008, available https://in.boell.org/2009/12/10/distributed-renewable-energy-projects-rural-areas-india-guide-implementation, accessed on 8 September 2016.

16. 'Feed-In Tariffs in Germany', available https://en.wikipedia.org/wiki/Feed-in_tariffs_in_Germany, accessed on 23 August 2016.

17. Section 86 (1), Electricity Act, 2003.

18. 15 States Have Fixed the RPPO. As per annexure 1 to Draft report of working group on policies on renewables.

19. Section 2(h) ECA 'energy' means any form of energy derived from fossil fuels, nuclear substances or materials, hydro-electricity and includes electrical energy or electricity generated from renewable sources of energy or bio-mass connected to the grid.

20. 'Integrated Energy Policy Report of the Expert Committee', Planning Commission, New Delhi (August 2006): Chapter VII, available http://planningcommission.nic.in/reports/genrep/rep_intengy.pdf, accessed on 12 September 2016.

21. Biodiesel; Development in the country is an example in case.

22. 'Integrated Energy Policy Report of the Expert Committee', Planning Commission, New Delhi (August 2006), p. xxii, available http://planningcommission.nic.in/reports/genrep/rep_intengy.pdf, accessed on 12 September 2016.

23. Ministry of New and Renewable Energy, Annual Report, 2015–16, available http://mnre.gov.in/file-manager/annual-report/2015-2016/EN/Chapter %201/chapter_1.htm, accessed on 23 August 2016.

24. Ministry of Power, 'Report of the Gokak Committee on Distributed Generation' (2002), available http://111.93.33.222/RRCD/oDoc/33_gokak.pdf, accessed on 23 August 2016.

25. UNEP, 'UNEP Handbook for Drafting Laws on Energy Efficiency and Renewable Energy Resources' (2007), available http://www.unep.org/delc/Portals/119/publications/UNEP_Energy_Handbook.pdf, accessed on 23 August 2016.

26. Section 2 (15A), Electricity (Amendment) Bill, 2014.

27. Section 2 (57A), Electricity (Amendment) Bill, 2014.

28. Section 7, Electricity (Amendment) Bill, 2014.

29. Section 3, Electricity (Amendment) Bill, 2014.

30. Section 14, Electricity (Amendment) Bill, 2014.

31. Proviso to Section 29 (6), Electricity (Amendment) Bill, 2014.

32. Proviso to Section 33 (5), Electricity (Amendment) Bill, 2014.

33. Section 142, Electricity (Amendment) Bill, 2014.

34. Section 146, Electricity (Amendment) Bill, 2014.

35. Section 42 (4), Electricity (Amendment) Bill, 2014.

36. Ministry of New and Renewable Energy, Request for Comments/Observations/Feed-back on the Draft Renewable Energy Act 2015, available http://mnre.gov.in/file-manager/UserFiles/dra2015-comments.html, accessed on 2 June 2016. Note that this is not the first time that an effort to draft RE law has been done. Way back in 1995, the then MNES had made an effort to draft the RE law through WWF's Centre for Environmental Law where the author was a research associate along with late Dr Chhatrapati Singh. Thereafter, another Non- Government Agency namely, World Institute of Sustainable Energy carried out an exercise under the aegis of the MNRE to develop a draft on RE law. Thereafter, an Environmental Law Firm, Enviro Legal Defence Firm (of which the author is the Managing Partner) was contracted to draft the RE law in 2009.

37. 'National Renewable Energy Act, 2015', available http://mnre.gov.in/file-manager/UserFiles/draft-rea-2015.pdf, accessed on 2 June 2016.

38. This includes, RE Policy and Plan, resource Assessment, policies on testing, monitoring and verification, and indigenous manufacturing of components.

39. Renewable Energy Act, 2015, p. 6.

40. Renewable Energy Act, 2015, p. 7.

41. Renewable Energy Act, 2015, pp. 7–8.

42. Renewable Energy Act, 2015, p. 9.

43. Renewable Energy Act, 2015, p. 10.

44. Renewable Energy Act, 2015, p. 12.

45. Study Report on 'Mapping the Legal Potential of Community Based Biofuel Production in India—Special Focus on the States of Andhra Pradesh, Uttar Pradesh and Maharashtra', ELDF (2015). [Unpublished]

46. Renewable Energy Act, 2015, p. 12.

47. Renewable Energy Act, 2015, p. 16.

48. Renewable Energy Act, 2015, pp. 17–18.

49. Renewable Energy Act, 2015, p. 4.

50. S. Mahajan and P.R. Krithika, 'Governance of Renewable Energy in India: Issues and Challenges', TERI-NFA Working Paper Series, No. 14 (March 2014), available http://www.teriin.org/projects/nfa/pdf/working-paper-14-Governance-of-renewable-energy-in-India-Issues-challenges.pdf, accessed on 23 August 2016.

51. W. Caroline, 'Germany Reaches New Levels of Greendom, Gets 31 Percent of Its Electricity From Renewables', available http://www.bloomberg.com/news/articles/2014-08-14/germany-reaches-new-levels-of-greendom-gets-31-percent-of-its-electricity-from-renewables, accessed on 29 May 2016.

52. Caroline, 'Germany Reaches New Levels of Greendom'.

53. J. Shankleman, 'Germany Achieves Milestone—Renewables Supply Nearly 100 Percent Energy for a Day' (16 May 2016), available http://www.renewableenergyworld.com/articles/2016/05/germany-achieves-milestone-renewables-supply-nearly-100-percent-energy-for-a-day.html, accessed on 29 May 2016.

54. IPCC, 'IPCC Scoping Meeting on Renewable Energy Sources', available http://www.ipcc.ch/pdf/special-reports/srren.pdf, accessed on 29 May 2016.

55. IPCC, 'IPCC Scoping Meeting on Renewable Energy Sources'.

8

ERIMMA GLORIA ORIE

Examining the Legal Impediments in the Development of Renewable Energy in Nigeria

There is a renewed consciousness in the twenty-first century of the essential role that the energy sector plays in the economy of a country ranging from future economic development, employment, and the personal health of a nation's citizens. In particular, energy development has been elevated high up the political agenda with the advent of climate change and policies concerning energy security. This importance of the energy sector was clearly demonstrated, for instance, by the impacts of Russia's ability to affect gas prices in the majority of the European Union (EU) at both continental and member state levels.[1] At the national level, energy propels economic development by serving as the launch pad for industrial growth and, via transport and communications, providing access to international markets and trade. Reliable, efficient, and competitively priced energy supplies also attract foreign direct investment—a very important factor in boosting economic growth in recent times.

By the same token, the need for sustainable energy is rapidly increasing in the world. The global search for 'green', sustainable energy sources has been influenced largely by the increasing need for energy for industrial, manufacturing, and domestic purposes; exhaustive nature of fossil fuel sources and the global need for transition from a fossil economy to one driven by an increasing share of Renewable Energy (RE). The resources are generally well distributed all over the world, even though wide spatial and temporal variations occur. Thus, all regions of the world have reasonable access to one or more forms of renewable energy supply. Renewable energy has an important role to play in meeting the future energy needs in both rural and urban areas. A widespread use of RE is important for achieving sustainability in the energy sectors in both developing and industrialized countries.

China, Brazil, and India are the fastest growing RE countries among the developing countries of the world.[2] Of these three, China is the largest producer and consumer of energy in the world[3] and thus is a country to be reckoned with, if the world is to achieve a low-carbon emission scheme. Consequently, increasing the share of RE in its energy mix has become one of the key pillars of China's low-carbon development strategy. Initially, the country exploited RE in order to resolve rural energy shortage, and issued policies in this respect as far back as the 1970s.[4] Today, China has earned a leading position in the exploitation of RE, coupled with the promulgation of the necessary legal regimes to attain its RE targets. The same cannot be said of many developing economies including Nigeria where developing RE alternatives has posed a daunting and intractable phenomenon.

The Nigerian energy situation is a paradox of abundance and lack as well as misplaced priority in the long and complex effort to increase electricity supply by more than fivefold by 2020, from the current level of about 3,400 MW to 20,000 MW. According to the Department for Petroleum Resources (DPR),[5] hydrocarbons reserves have traditionally attracted the vast majority of domestic and foreign investment in Nigeria and more than 78 per cent of total energy consumption is derived from petroleum supply. The electric power capacity demand by projection in Nigeria would be approximately 3.5-fold between 2010 and 2020 and 7.5-fold between 2010 and 2030, at a growth rate of 7 per cent while the projected supply by fuel mix shows a similar trend with the demand at both growth rates of 7 per cent and 13 per cent respectively. Nevertheless, the government is falling short of its targets to boost electricity to revamp its ailing economy. This was succinctly captured by the Council for Renewable Energy, Nigeria (CREN) which estimates that power outages brought about a loss of 126 Billion Naira ($984.38 million) annually. Overall, the supply of electricity, the country's most used energy resource, has been erratic.[6] In the present predicament as a nation, depending mainly on fossil fuel (petroleum) is not enough to meet the energy needs of the country. The country is in dire need to up its game in the development of RE.

In addition, the abundant natural gas extracted in oil wells is continuously flared at a rate of approximately 70 million m^3 per day. This gas flaring by oil companies is a major means through which greenhouse gases are released into the atmosphere. Some 45.8 billion kW of heat are discharged into the atmosphere of the Niger Delta from flaring 1.8 billion ft^3 of gas every day. Gas flaring has raised temperatures and rendered large areas uninhabitable. Between 1970 and 1986, a total of about 125.5 million m^3 of gas was produced in the Niger Delta region, about 102.3 (81.7 per cent) million m^3 were flared, while only 2.6 million m^3 were used as fuel by oil-producing companies and about 14.6 million m^3 were sold to other consumers.[7] Obviously, the use of RE sources will reduce the over dependence on the burning of fossil fuel and help to reduce the effect of climate change on the environment.

Despite its enormous and abundantly available RE resources like hydropower, biomass, wind power, biogas, solar energy, and geothermal power, Nigeria is grappling to produce adequate energy for its economy. For instance, the identified hydroelectricity sites have an estimated capacity of about 14,250 MW but very little is being realized.[8] Similarly, Nigeria has significant biomass resources to meet both traditional and modern energy uses, including electricity generation. Nigeria's production and usage of hydroelectric energy, geothermal, and solar energy is still trifling due to misplaced priority, uneven development of these resources, and unregulated activities in the energy sector. The result is a wide disparity in the energy demand to supply ratio in Nigeria at the present and in the future. There has also been a supply and demand gap of the conventional electricity from fossil fuel. This necessitates an urgent need for alternative energy sources and efficient energy usage in order to avert worsening energy crises.[9]

The Nigerian government is mindful of the advantages of RE. Because RE are constantly being replenished from natural resources, they have security of supply, unlike fossil fuels, which are negotiated on the international market and subject to international competition, sometimes even resulting in wars and shortages. Their rate of use does not affect their availability in the future; thus, they are inexhaustible. They are clean and pollution-free and are therefore a sustainable natural form of energy. They can be cheaply and continuously harvested and are therefore a sustainable source of energy. Perhaps, these notions propelled the government to formulate some policy initiatives including RE master plan to develop the RE sector. Regrettably, most of the policies are out of tune with current global realities on the subject matter and lack legal frameworks as they are still undergoing legislative procedures at the National Assembly for many years, thereby slowing down the process of enactment into law. The consequences are a poorly developed RE sector arising from misplaced priorities, inadequate legal, institutional, and regulatory framework, as well as implementation lapses.

The word 'policy' generally stands for a set of principles and long-term goals that constitute the substratum for making rules and guidelines, and that gives overall direction to planning and development of the organization or country as in this case.[10] In the context of this discussion it would therefore mean the very values or fundamental objectives of a country that guides such country not only in the assessment of its short term goals but also in the pursuit of its targets. Usually articulated over a period through extensive consultation, a policy framework therefore gives focus and direction to a country, ensuring that it does not derail from its chosen path. At this juncture, it is pertinent to state that the scope of this work covers the period between 2005 when Nigeria adopted the first RE master plan till date.

The chapter starts with conceptual clarifications of the key words to present a proper understanding before reviewing the legal and regulatory frameworks for RE in China to enable relevant lessons to be drawn. This is followed by RE potentials of Nigeria leading to the essence of treating the legal framework for RE. Next, are sections on the challenges militating against the development of RE in Nigeria and strategies to resolve the issues. The work ends with a conclusion and a couple of relevant recommendations.

Conceptual Clarification

The concept of energy is one that is difficult to define in a simple and accurate way.[11] According to physicists, energy is the ability to do work, and work is moving something against a force, like gravity. There are different kinds of energy in the universe, and energy can do different things.[12] For example, food is the main source of energy that fuels our bodies, helping us to grow, run, talk, and think; Electrical energy is used to power most things in our modern society including computers, televisions, fridges, and freezers; Fossil fuel energy powers our cars, buses, trucks, boats, and airplanes. Generally, energy is classified into two basic types, namely, RE and non-renewable energy.

As W. Mingyuan[13] noted there is no consensus of opinion on the definition of RE. However, there appears to be a consensus in describing it as energy sources that are inexhaustible, regenerative, and continuously replenished by the natural world.[14] RE is energy from limit-less natural resources, such as, biomass, geothermal, sunlight, wind, water, and tide/wave. This form of energy like the conventional type of energy is fundamental for socioeconomic development and poverty eradication. On the other hand, Non-renewable energy is energy produced by burning fossil fuels such as coal. They are non-renewable because there are finite resources of fossil fuels on the planet. If they are continually used, one day they will run out.

Furthermore, the word 'development' is derived from the verb 'to develop', which means 'to bring out the capabilities or possibilities of, to bring to a more advanced or effective state'.[15] Development thus means to make something better than it was, to improve, development does not mean growth.[16] However, as a *complex* issue, with many different and sometimes contentious definitions, there is a basic perspective that equates development to economic growth.[17] The term 'development' in international parlance therefore encompasses the need and the means by which to provide better lives for people in poor countries. It includes not only economic growth, although that is crucial, but also human development—providing for health, nutrition, education, and a clean environment.[18] It is in this latter context that the term development shall be employed in this paper.

Renewable Energy Policy Framework for RE in China

The year 2015 was notable as the first in which investment in renewables excluding large hydro in developing countries outweighed that in developed economies. The entire developing world including China, India, and Brazil committed a total of $156 billion, up 19 per cent in 2014, and a remarkable 17 times the equivalent figure for 2004, while developed countries invested $130 billion, down 8 per cent and their lowest tally since 2009. Apart from these three countries, other developing countries equally lifted their investment by 30 per cent last year to an all-time high of $36 billion, some 12 times their figure for 2004. Among those 'other developing' economies, those putting the largest sums into clean power were South Africa, up 329 per cent at $4.5 billion as a wave of projects winning contracts in its auction programme reached financial close; Mexico, 105 per cent higher at $4 billion, helped by funding from development bank Nafin for nine wind projects; and Chile, 151 per cent higher at $3.4 billion, on the back of a jump in solar project financings. Morocco, Turkey, and Uruguay also saw investment beat the $1 billion barrier in 2015. However, the single largest element entity in this turn-around for renewables excluding large hydro was China, which boosted its investment by 17 per cent to $102.9 billion, or 36 per cent of the world's total.[19] Although US was first among the developing countries, however, it came a distant second overall (developed and developing countries), with $44.1 billion, up 19 per cent. Japan was a clear third in the ranks of investing nations, its $36.2 billion, level with 2014, followed at a distance by the UK with $22.2 billion, up 25 per cent.

China is within the same global development strata with Nigeria and thus provides a good basis for comparison. The country is also the largest producer and consumer of energy in the world.[20] The histrionic economic development of China in the past few decades has increased demands for natural resources within and beyond the country itself in ways that are unprecedented in human history. As would be expected, this has created rising environmental challenges linked to the extracting, processing, and use of those natural resources in areas from construction to power generation. However, a different way to look at this development trajectory is the fact that in comparison with the global and regional picture, China's track record in improving resource efficiency has been in some cases among the best internationally. For example, according to a report,[21] China's energy efficiency, improved over the 1970 to 2009 period at an annual compound growth rate of over 3.9 per cent, exceeding the global performance of just under 0.7 per cent and that for the Asia Pacific region as a whole which was 0.13 per cent.[22] Similarly, China's per capita energy consumption, as measured by total primary energy supply (TPES), increased from 31 per cent of the world average levels in 1970 to over 74 per cent of the world average in 2005 and 95 per cent in 2009.[23] Thus expanding the share of RE in its energy mix became one of the key pillars of China's low-carbon development strategy.[24]

Various Policy Measures

In the past few years China has introduced considerable number of policies in areas from renewable energies to vehicle emissions standards that are contributing to boosting resource efficiency and assisting towards a transition to green economy and an ecological civilization. As a major step towards achieving the target, China introduced some policies as listed below:[25]

1. China began to implement cleaner production (CP) in the early 1990s as a way of confronting the country's serious environmental problems. A network of national and local CP policies incorporated CP activities such as demonstration projects, training and promotion centres, and the creation of the National Cleaner Production Centre (CNCPC).[26]

2. National RE Target: The 2005 Renewable Energy Law (REL)[27] in its articles 7 and 8 mandated the State Council's energy department[28] to create mid-term and long-term national targets for RE production, and a national RE development and utilization plan.

3. Mandatory Connection and Purchase Policy: This principle obviously embodies the 'feed-in- tariff' system. Given that unsubsidized RE is currently more expensive than the conventional fossil fuel alternative, feed-in tariffs guarantee investors that a RE project will be able to sell the electricity it generates to the grid at price above the market rate for electricity from conventional sources of energy.[29] In line with the RE law, the State Electricity Regulatory Commission (SERC) issued the 'Measures on Grid Company Full Purchase of Electricity from Renewable Energy' (Full Purchase Measures) in 2007, which specified requirements for connection and purchase of renewable power. The SERC also provides sanctions for erring grid companies. It has been submitted that the non-compliance with these obligations by grid companies coupled with the fact that there were no reported cases of SERC exercising its discretion to impose

penalties for non-compliance led to the amendment of the law in 2009 to improve implementation. Feed-in Tariff System: Chinese Feed-in- tariff system appears unique as it guarantees the price at which power will be purchased on a project-by-project basis through competitive bidding. Thus, China practices the general feed-in-tariff under the mandatory connection and purchase policy guarantee.

4. Cost-sharing Mechanism (including a special fund for RE development): The mechanism ensures that the costs associated with feed-in tariffs and the reasonable costs associated with connecting renewable generators to the grid are to be shared nationwide through a surcharge on end-users of electricity. In order to compensate grid companies for the additional cost of purchasing RE, there is a nationwide renewable surcharge levied on electricity users at a uniform rate based on the users' consumption of electricity.[30]

5. Medium and Long-Term Development Plan for Renewable Energy in China 2007 is that by 2010, the share of RE in total primary energy consumption will be raised to 10 per cent, and by 2020 to 15 per cent.[31] Renewable Energy Law 2006:[32] This promotes the development of RE, provides a framework for pricing, special funding, special import facilities for equipment, and provisions for grid management. Medium and Long-Term Plan for Energy Conservation 2004: This sets out specific energy conservation targets for industrial, transportation, and building sectors and calls for a revision of existing energy policies and recognizes the importance of economic incentives.

6. Construction of RE generation systems: This funding supports the construction and maintenance of household biogas digesters, and some of it will go to the construction of large-scale biogas utilization sites. Subsidies for rural end-users: In early 2009, China launched a subsidy programme for the rural use of household electrical appliances, including SWH. Production subsidies: To encourage the collection and processing of a wide variety of wastes for more-efficient energy use, a subsidy of RMB130–150 (USD19–22) is provided for every ton of biomass pellets made from agricultural or forestry residues.

7. Tax Policies: Tax incentives for renewable energy development have been limited in China. National policies include a reduced 6 per cent VAT collection from small hydropower projects, a 50 per cent tax reduction for wind power projects, and a tax reduction or exemption

for imported renewable energy equipment that China is not able to produce domestically. Investment Policy: In addition to the financial support from the Chinese government for R&D, demonstration projects, and key equipment manufacturing, China's Treasury Bond also supports RE. Industry Support: The Chinese government has provided various financial supports, such as tax breaks, investment subsides, and bonuses, to facilitate the commercialization of renewable energy. Voluntary initiatives with industry are also part of China's energy conservation efforts. Under the Top-1000 Energy-Consuming Enterprises Programme (NDRC, 2008) participating enterprises sign energy conservation agreements with local governments, and are expected to formulate energy conservation plans and efficiency goals, establish reporting and audit systems, and conduct training. The programme was aimed to save 100 million tonnes of coal equivalent between 2005 and 2010.

In China, environmental objectives are integrated into several national policies and regulations, including the *Circular Economy Promotion Law* and the *Cleaner Production Promotion Law*. Sustainable consumption and production (SCP) principles are also integrated into China's Five-Year Plans (FYPs) for Social and Economic Development. The FYPs form the basis for coordinating Chinese national public policy priorities. They are developed by the National Development and Reform Commission (NDRC) and approved by the National People's Congress. SCP principles are integrated through quantified pollution emission targets as well as quantified resource efficiency targets.

In sum, China has achieved impressive results in its energy efficiency programmes due to factors such as government leadership, focused and well-targeted measures to achieve energy efficiency, and climate resilient environment as priority, comprehensive system design, expansion programme, enforcement, continuous consultative and participatory process, expert design, and research teams.

Legal Framework for RE in China

China's first law for RE regulation came into force in 2006 (amended in December 2009). The law established some strategies for RE development such as the target system of gross amount of RE sources, approval of RE generation through synchronization, a full acquisition system, on-grid price, financial incentives, and cost allocation system of RE sources.

There is also the Law on Energy Conservation 2007 which codifies the major elements of the Medium and Long-Term Plan for Energy Conservation and places great importance on the behaviour and performance of the government itself with regards to energy conservation. The revised *Energy Conservation Law* (2007) equally promotes the use of clean, alternative fuels and provides incentives for the development and use of high-efficiency vehicles, including alcohol-fuelled, hybrid, electrical, and compressed natural gas vehicles.[33] Mandatory fuel economy standards have been instituted to achieve emission reductions in private vehicle use.

The Cleaner Production Promotion Law (2003) governs the implementation of all CP activities in the country. It seeks to promote cleaner production, increase resource efficiency, and reduce and avoid the generation of pollutants.

The Circular Economy Promotion Law came into force in 2009. The CE Promotion Law is the world's first national law to make the circular economy a national strategic focus of economic and social development, thereby differing greatly from the traditional linear economic models. The ultimate objective of the CE approach is to achieve decoupling of economic growth from natural resource depletion and environmental degradation. The CE Promotion Law is very broad and far-reaching. Its enforcement therefore requires supporting regulations.

The Land Administration Law was implemented in 1999 to protect environmentally sensitive and agricultural land, and to coordinate the planning and development of urban land. The law reinforces farmland preservation efforts by stipulating that the total amount of cultivated land within each administrative area needs to remain unreduced (Lichtenberg and Ding, 2008).

The 12th Five Year Plan (FYP) also includes a legally binding energy intensity reduction target of 16 per cent, slightly lower than the target of the previous FYP. The Chinese government is also in the process of developing a range of market mechanisms to complement existing regulations and standards in the energy sector. The key market mechanisms proposed in the 12th FYP include a carbon tax, a natural resources tax, and a carbon emissions trading scheme.[34]

Regulatory Institutions for RE in China

China's RE industry has two basic stages: production or generation, and distribution. In addition, there are three stakeholders involved in its RE development process—the generation companies,[35] the distribution or grid companies,[36] and the regulators. These stakeholders work together to ensure that the RE industry runs smoothly. Although the main regulatory body for China's RE is the National Energy Administration (NEA) which is a unit in the NDRC, China's top economic planning agency, the National Energy Commission (NEC), was officially inaugurated in 2010, to be responsible for coordinating energy planning among different government agencies and formulation of a national energy strategy.[37] Also, the SERC as a government agency is responsible for regulating China's power and electricity industry, including promulgating and ensuring compliance with relevant laws and regulations. Furthermore, for purposes of implementation of its national energy efficiency programmes, China has over seven national institutions and several agencies that work in synergy.

Policy Challenges

Despite its RE accomplishments, China has some challenges. While comprehensive national policies exist, implementation remains difficult. Several policies lack supporting regulations that outline more detailed implementation activities. For example, the CE Development Plan, which will outline the more practical tasks and measures necessary for achieving the implementation of the CE Promotion Law, was only adopted in late 2012. This delay hindered the effective implementation of the CE Promotion Law, which came into force in 2009.

The difficulty of implementing and enforcing resource efficiency policies at the local level of government is one of the major policy challenges to be overcome. In China, the responsibility for environmental compliance and enforcement lies principally at the local level. However, financial and human resources as well as technical equipment at the local scale are often insufficient. Capacities of local government officials need to be strengthened through continuous training and regular information provision on new regulations.

Lessons from China on RE

Overall, the Chinese policy, legal, and institutional framework has been adjudged successful and worthy of emulation for several reasons. It includes a Monitoring, Reviewing, and Verification (MRV) framework, a detailed implementation plan and a set of guidelines and tools to facilitate implementation. The policy encourages private investment and is institutionalized into enterprises and business operations (including dedicated business unit and personnel) to ensure the sustainability of actions. Where a policy is driven by the national government funding and implementation becomes easier and state, local parastatal, and individuals become motivated to share in the national vision. Setting proper goals and regular review of such targets helps to redefine policy, speeds up policy action towards sustainable environment and the assessment of achievements. The top bottom approach tends to make the citizen active stakeholders and able to synergize with the government. That, while a favourable policy environment is crucial for promoting renewable energy technologies, a supporting strategy for indigenous/domestic manufacturers and investment in research and development are crucial for achieving higher long-term targets. Generally, that the regulation of the energy industry can be achieved only with a complex interaction of policies and laws regulating energy and various other sectors such as the environment, agriculture, maritime, welfare, and so on. The mandatory connection and purchase policy coupled with sanction for defaulters is exemplary.

Policy Framework for Renewable Energy in Nigeria

Having discussed some of the RE initiatives in China the monograph examines the Nigerian RE potentials, energy policy, legal, and institutional frameworks available to identify existing gaps and discords if any in the system to provide insight on the status quo of renewable energy policies. This discussion will dovetail into exploring strategies for filling the lacuna or smoothening the discord.

Hydropower in Nigeria

A study carried out in twelve states and four (4) river basins identified over 278 unexploited small hydropower (SHP) sites with total potentials of 734.3 MW (approximately 734 MW).[38] However, in 2012, hydro-power accounted for about 29 per cent of the total electrical power supply.[39] The Nigerian Electricity Supply Company (NESCO) and the federal government have installed eight (8) SHP stations with aggregate capacity of 37.0 MW in Nigeria, most of which are found around Jos at Kwall and Kurra Falls.[40] These rivers, waterfalls, and streams with high potentials for Hydropower, if properly harnessed, will lead to decentralized use and provide the most affordable and accessible option to off-grid electricity services, especially to the rural communities. The hydropower in Nigeria is not without some criticism. It is said to have attracted huge debts and corruption, droughts, and reduced energy generation and harm to wild life and human ecosystems.[41]

Solar Energy in Nigeria

Nigeria lies within a high sunshine belt and as such receives an appreciable quantity of solar radiation.[42] Annually the average total solar radiation varies from about 12.6 MJ/m per day in the coastal latitude to about 25.2 MJ/m^2 per day in the far North.[43] Furthermore, the intensity of the solar radiation ranges from 3.5–7.0 KWhm per day while the sunshine duration ranges from 4.0–9.0 hours/day.[44] In fact study seems to suggest that if solar collectors could cover one per cent of Nigeria's land area, Nigeria would generate 1850×10^3 GWh of solar electricity per year.[45] The acceptance is expected to increase in the future.[46] Some successful research work such as the implementation of solar PV water pump and electrification, solar cooking stoves, crop drying facilities, and so on, have been undertaken out in the Sokoto Energy Research Centre and the national Centre for Energy Research and Development under the supervision of the Energy Commission of Nigeria.

Wind Energy in Nigeria

Nigeria has an annual average wind speeds of about 2.0 m/s at the coastal region and 4.0 m/s at the far northern region of the country,[47] with air density of 1.1 kg/m^3, the wind energy intensity perpendicular to the wind direction ranges between 4.4 W/m^2 at the coastal areas and 35.2 W/m^2 at the far northern region.[48] The country equally has an annual average wind speed of 10m heights which is said to fluctuate from 3 m/s in the coastal areas to 7 m/s in the far North with less vegetation. Notwithstanding the above, as of 2012, there was no commercial wind power plants connected to the national grid.[49] On wind development, the government has completed the installation of the 20 MW wind farm in Katsina state. However, the progress of wind development in Nigeria is still relatively low when compared to other developing countries in Africa. Since the Northern part of Nigeria has high potential for wind energy and the communities have low access to energy, the government must do more regarding the exploitation of wind energy to meet the needs of the populace.

Geothermal Power in Nigeria

Nigeria is yet to develop geothermal energy[50] but there is speculation that there may be potential for geo-pressured systems and hydrothermal systems in some Nigeria Basins such as Benue through the Niger Delta.[51]

Bio-energy in Nigeria

It has been submitted in some quarters that Nigeria is rich in biomass. Bio-energy is derived from biomass like, forage grasses and shrubs, animal wastes crops, and waste arising from forestry, agriculture. This use of food stuff for has negative implications for availability of food and the cost. Biomass has been an important renewable energy source in Nigeria, but its sustainability is still a big issue in Nigeria. The government should continuously use municipal waste, oil palm product, sugar cane, and rice husk for the sustainability of biomass energy production. Also, the government incorporation with the private sector should establish some biogas plants to help the country's energy sector through energy production.[52]

Existing Policy Instruments and Initiatives in Nigeria

Generally government policies precede the laws and form the basis for government sponsored laws and to that extent play regulatory roles in the absence of laws.[53] Policies are therefore important for the development of RE since there aren't many laws regulating this sub-sector presently.[54]

Towards the implementation of the Nigeria's National Policy on Environment,[55] the following relevant specific policies and action plans have been enacted:

1. Draft Rural Electrification Strategy and Implementation Plan (RESIP), 2014: Although the Power Sector Reform team initially prepared the Rural Electrification Strategy and Implementation Plan in 2006 (RESIP, 2006) it was a committee involved in the power sector that reviewed and redrafted the RESIP in 2014.[56] RESIP was expected to establish a clear institutional step-up for the sector and a roadmap projected to create an enabling framework for rural electrification in Nigeria. It also set out to develop access to electricity as rapidly as possible and in a cost-effective manner using various means of electricity supply like the on-grid and off-grid. The draft is awaiting approval of the government.

2. National Renewable Energy and Energy Efficiency Policy (NREEEP), 2014: The policy outlines the global thrust of the policies and measures for the promotion of renewable energy and energy efficiency. The Federal Ministry of Power (FMP) developed the NREEEP in 2014 and is awaiting the approval of the Federal Executive Council.[57]

3. National Energy Policy (NEP), 2003, 2006, 2013: The NEP approved by the Federal Government of Nigeria in 2003 was the first comprehensive energy policy. This Policy which was developed by the Energy Commission of Nigeria (ECN) sets out government policy on the production, supply, and consumption of energy reflecting the perspective of its overall needs and options. The main goal of the policy is to create energy security through a robust energy supply mix by diversifying the energy supply and energy carriers based on the principle of 'an energy economy in which modern renewable energy increases its share of energy consumed and provides affordable access to energy throughout Nigeria, thus contributing to sustainable development and environmental conservation'.

4. Renewable Energy Master Plan (REMP) 2005 and 2012: In partnership with the United Nations Development Programme (UNDP) in 2005, the Energy Commission of Nigeria (ECN) developed the Renewable Energy Master Plan (REMP). The REMP was later

reviewed in 2012. It was anchored on the mounting convergence of values, principles, and targets as embedded in the National Economic Empowerment and Development Strategy (NEEDS), National Energy Policy, National Policy on Integrated Rural Development, the Millennium Development Goals (MDGs), and international conventions to reduce poverty and reverse global environmental change.[58] The REMP stresses the need for the integration of renewables in buildings, electricity grids, and for off-grid electrical systems. Its target is for Nigeria to increase the supply of renewable electricity from 13 per cent of total electricity generation in 2015 to 23 per cent in 2025 and 36 per cent by 2030 so that renewable electricity would account for 10 per cent of Nigeria's total energy consumption by 2025. Laudable as the bill is, the REMP has not been approved by the National Assembly to be passed into law.

5. Multi-Year Tariff Order (MYTO), 2008 and 2012: The Nigerian Electricity Regulatory Commission (NERC), in 2008, developed a 15-year roadmap towards cost reflective tariffs called the Multi-Year Tariff Order (MYTO 1). The first two phases, 2008–11 and 2012–17 were designed to keep consumer prices relatively low, while the final regime is intended to provide the necessary incentives for power producers and investors to operate and maintain electricity infrastructure.[59]

6. Renewable Electricity Policy Guidelines (REPG), 2006: This policy document presents the Nigerian government's plans, policies, strategies, and objectives for the promotion of renewables in the power sector. It was established in 2006 by the Federal Ministry of Power and Steel. The Guideline mandated the Nigerian government to expand electricity generation from renewables to at least 5 per cent of the total electricity generated and a minimum of 5 TWh of electricity generation in the country.[60]

7. Renewable Electricity Action Programme (REAP), 2006: The Federal Ministry of Power and Steel in 2006 developed REAP and set out a roadmap for implementation of the REPG. The document provides a synopsis (gestalt) of the Nigerian electricity sector, reviews government targets and the issues, challenges, and prospects for renewable energy development.[61]

8. Nigerian Biofuel Policy and Incentives (NBPI), 2007: The policy sets out to develop and promote the domestic fuel ethanol industry using agricultural products and in line with the August 2005 government's directive on an Automotive Biomass Programme for Nigeria. The policy thrust is to ensure the gradual reduction of the nation's dependence on imported gasoline, reduction in environmental pollution, while at the same time creating a commercially viable industry that can precipitate sustainable domestic jobs. Overall, the policy is designed to create additional tax revenue, provide jobs to reduce poverty, boost economic development and empower those in the rural areas, while also improving agricultural activities, energy, and environmental benefits through the reduction of fossil fuel related GHGs in the transport sector.[62]

9. The Millennium Development Goals (MDGs): which was a historic and intuitive devise for global mobilization to achieve a set of important social priorities on issues of poverty, hunger, disease, unmet schooling, gender inequality, and environmental degradation. It has been canvassed by some scholars that the low progress in achieving the MDGs was due to the lack of priority for RE as the primary energy source for the country. The view of the International Energy Agency, also echoed in the report of the 6th Annual Meeting of the African Science Academy Development Initiative (ASADI),[63] was that Lack of access to modern energy services is a serious hindrance to economic, social development, and sustainable food security and must be overcome if the UN MDGs were to be achieved.'[64] The MDGs have been extended to another 15 years, renamed Sustainable Development Goals (SDGs) with more comprehensive programme.

10. In addition to the above, the Nigerian government also put in place the National Adaptation Strategy and Plan of Action for Climate Change Nigeria (NASPA-CCN), Strategic Framework for Voluntary Nationally Appropriate Mitigation Action (NAMA) in Nigeria, and the National Climate Change Policy and Response Strategy (NCCPRS),[65] as national documents for implementing climate activities in the country. Some of these activities involve investment in Clean Development Mechanism programmes which is also an aspect of RE.

Legal Framework for RE in Nigeria

Section 20 of the Constitution of the Federal Republic of Nigeria[66] herein after referred to as the Constitution provides that 'the State shall protect and improve the environment and safeguard the water, air and land, forest, and wild life of Nigeria'. Apart from the Constitution the first two laws that have bearing on RE are the Energy Commission of Nigeria (ECN) Act 1979[67] and the Nigerian Electricity Act 1988.[68] While the commission developed centres[69] that have largely functioned as research 'facilities' for the development of RE and its law, the electricity law provides in its Section 2 for the regulation and control of electricity installations, and of the generation, supply, and use of electricity energy.

The Electric Power Sector Reform Act (EPSRA), 2005 was established to ensure the liberalization of the Nigerian power sector. The Act consolidates and gives legal backing to the reforms being undertaken in the sector particularly with regards to the promotion of RE.[70] It emphasizes the role of renewable electricity in the overall energy mix, especially for expanding access to rural and remote areas.[71] EPSRA formed the basis for establishing the Nigerian Electricity Regulatory Commission (NERC) and the Rural Electrification Agency (REA) which have been functioning since May 2005 and March 2006, respectively.[72] It equally replaced the National Electric Power Authority (NEPA) with the Power Holding Company of Nigeria (PHCN). In order to tackle the issue of unstable power generation more effectively, NERC made the Nigerian Electricity Regulatory Commission Regulations for Embedded Generation 2012,[73] pursuant to EPSRA Act.[74] The law not only streamlines and encourages private electricity generation, but also in its chapter V recommends feed-in-tariffs (FITs) approved by the Commission for energy produced by RE embedded generators fixed for a specified period, subject to periodic reviews and approval by the Commission. Although this appears to be a step in the right direction in that it rewards stakeholders for the use of RE, it has been criticized in some quarters as benefiting only large scale stakeholders as against every scale of RE power generation, that is both large and small.

Institutional Framework

In addition to the Federal Ministry of Power (FMP) that is supposed to be the main actor in the sector responsible for policy formulation, planning, coordination, and monitoring, the Presidency also decided to get directly involved in the power sector reform with the creation of the Presidential Action Committee on Power (PACP) and the Presidential Task Force on Power (PTFP). The Federal Ministry of Power (FMP) is one of the institutions that are currently undergoing a restructuring process to adapt to the new structure of the liberalized power sector with the newly created Department of Renewable and Rural Power Access in the Ministry.

Apart from the above institutions there are also RE research and development centres of the ECN which organize conferences at some selected Nigerian Universities. However, most of these existing research and development centres and technology development institutions are not adequately strengthened to support the shift towards an increased use of renewable energy. Capacity development, critical knowledge, and know-how transfer should be the focus for project development, project management, monitoring, and evaluation. The preparation of standards and codes of practices, maintenance manuals, life cycle costing, and cost-benefit analysis tools are yet to be assessed as priority areas for immediate attention.

By way of review of the entire policy, legal, and institutional framework in RE in Nigeria one can say that over the recent years, increasing progress have been made on the policy, legal, and regulatory framework on renewable energies. Seemingly, the framework for medium scale interconnected generation (embedded generation) is also being elaborated at a 'fast' pace due to the importance of this aspect for the Electricity Distribution Companies (DISCOs) and overall energy supply. On the contrary, most of the strategy and policy documents for rural electrification, particularly off-grid, and also energy efficiency remain in draft form and still need to be developed considering inherent realities associated with the sector.

A cursory look at the RESIP 2014, NREEEP, NEP 2013, and REMP 2013 shows that they are all very recent draft legislations that would be expected to be very reflective of the current global issues on RE. However, they have not been allowed to see the light of the day three years after they were drafted because the legislative arm of the government is yet to carry out their assigned responsibilities. In fact, in the case of Renewable Energy Master Plan (REMP) 2012, its first target set for 2015 has elapsed while the draft is still before the National Assembly. This brings to the fore the issue of political support of the Nigerian government for its policies and laws. There must be a correlation between government policies and action. As borne out of the China experience, the government must be on the driving seat to be able to galvanise other stakeholders to key into the programme.

Furthermore, it has been argued that a major reason for the relative inefficiency and low productivity of the RE sector is the lack of clear institutional step-up that should have created and implemented an enabling framework for rural electrification in Nigeria. There is, however, a gap between the policy framework and their operationalization through the formulation and effective implementation on plans with a clear roadmap.

Challenges for RE Development in Nigeria

There are some principal challenges that affect RE development in Nigeria. These include:

Policy Framework

1. The challenge for Nigeria is not one of lack of policy as is evident from the earlier discussions,[75] but one of implementation.[76] Implementation has a lot to do with the focus of the government. Since 2005 there has been several policies on Energy in Nigeria; some inconsistent while others are complementary. Most governments in Nigeria have the unprogressive attitude of not continuing with policies of previous government. The result is that the government's plan for development of RE is not consistently addressed. In addition, the follow-up and active implementation of this policy is lacking from the Nigerian government.

2. Harmonization of Policy: One remarkable challenge born out of the discussions above is the fact that the energy sector is replete with several policies leading at times to confusion among various stakeholders. The challenge is however how the Nigerian energy sector will harmonize rather than fragmentize the policies. This challenge is not peculiar to the energy sector. Nigerian policies and laws in areas of agriculture and forestry, energy efficiency, low carbon energy production, climate change, and finance for emission reduction are not synergized or designed to promote RE development.

RE should be developed through various policies that cut across diverse areas of the economy. This contrasts with the practices in China.

3. Narrow scope of Policy: Another major challenge is the restriction of RE development to the electricity sector. Although the Nigerian Biofuel Policy and Incentives 2007 covers the transportation sector, it does not extend to sectors like agriculture and other industrial sectors.

Legal and Regulatory Framework

1. Framework: Achieving adequate energy supply where renewables play a role necessitates the creation of appropriate policy framework of legal and regulatory instruments that would attract domestic and international investments.[77] The challenge is that clear rules, legislation, roles, and responsibilities of various stakeholders along every stage of the energy flow from supply to end-use are generally key elements of the overall policy framework needed to promote renewable energy technologies.[78] A further challenge is that such policy, legal, and institutional frameworks are at their beginning stage in Nigeria and are being developed under the reform programme.

2. Power Purchase Agreement: Currently there is no Power Purchase Agreements (PPA) plan for RE generation to the national grid. Grid-based renewables[79] usually require legally binding long-term PPA as a formal system of rational expectations between the grid operators and the renewable electricity producers. In the case of Nigeria, this is lacking. The PPA set the terms by which power is marketed and/or exchanged, determines the power characteristics, the delivery encouraging the expansion of renewable electricity development through investments.[80] This does not boost the investors' confidence.

3. Institutional Framework: In Nigeria, coordination between government ministries and agencies responsible for rural development and RE development is weak and rather complex. Unlike oil and gas, no agency has a clear mandate to oversee the development of RE in Nigeria. The implication of such a lack is that the sector is bereft of the needed driving force for its growth and development. The new Electricity Law is expected to facilitate the establishment of a Rural Electrification Agency and a Rural Electrification Fund. The expectation is that these developments will aid RE development.

4. Capacity: Presently in Nigeria, human and institutional capacities needed for the development of RE is under performing. Capacity building in some areas is most lacking, namely, training of manpower to install, operate, and maintain RE technology, development of manufa turing capabilities, development of critical mass of scientists, engineers, and economists, to design an effective and functional institutional framework. Human and institute onal capacity building at all levels would be required to sustain the scientific, engineering, and technical skills relevant for the design, development, fabrication, installation, and maintenance of RE technology in Nigeria.

5. Public Awareness: Public awareness of renewable energy resources and technologies in Nigeria and their benefits, both economically and environmentally are generally low; thus,hampering the development, application, and dissemination of RE resources and technologies in the national energy market. A substantial part of the challenge is how to create awareness of the enormous opportunities offered by RE and their technologies among the public and private sectors.

6. Technology and Funds:[81] Most RE technologies have high up-front capital cost compared to their conventional energy alternatives.[82] Apart from the higher capital costs most RE technologies face the barrier of being perceived as untested technologies. Given these twin barriers to RE technologies, investors face higher risks and uncertainties when making investment decisions. Therefore, in a capital constrained economy like Nigeria, where there are many competing demands for available scarce capital resources, the promoters of RE technology face the problems of high transaction costs and restricted access to capital. On the other hand, the end users of RE technology, especially the poor, face problems of access to credits. Lack of access to micro financing, high interest rates, poor business development skills by RE system vendors, and unsupportive climate for investments are some of the primary barriers to market growth.[83] Small scale hydropower, central and residential solar photovoltaic technologies, and so on, have not penetrated Nigeria's energy supply systems, because of their relatively high investment cost.[84] Presently, technology imports for conventional electricity production carry a much lower tariff than RE electricity technologies. This hinders the growth of RE technologies.

Strategies for Improving RE Development in Nigeria

It has been argued in some quarters that the knowledge of some peculiar circumstances of Nigeria is necessary before suggesting the kinds of regulation needed to promote RE in Nigeria. Ladan[85] identifies some essentials of RE policy and regulation to include that RE regulations must—promote rural applications of efficiency and RE in all its ramifications; maintain fair, just, and reasonable rates for rural electricity consumption; ensure uninterrupted electricity to rural areas; promote rural energy efficiency; promote technological innovations and transfer of RE technologies to rural communities (and to the people who would directly benefit from their use); facilitate and encourage effective competition, education, training, and public participation; improve people's lives and livelihoods; and meet goals of sustainable development (SD), including obligations, and norms per multilateral environmental agreements such as those for greenhouse gas emissions reductions.

There is no doubt that Ladan's points reflect the peculiarities of Nigeria namely, that large ever growing population with increasing need to provide accessible RE energy in the rural areas at competitive price and that RE enhances SD and international best practices. Nigeria needs to consider adopting these essentials in order to achieve a workable RE policy framework of legal, fiscal, and regulatory instruments that would provide for the basic energy needs of the citizens and, at the same time, attract domestic and international investments. The legislation, rules, roles, and responsibilities of various stakeholders at every stage of the energy flow from supply to end-use, are key elements attracting attention. Some of the strategies to mitigate the identified RE development challenges include:

1. Policy: The Nigerian government should develop the political will to ensure continuity and consistency of policies notwithstanding change in political governance. In the same vein, the will to ensure holistic implementation of policies and regulations should be overtly manifest. Furthermore, the government must ensure that policies complement one another and be part of a whole process of achieving SD for the country. Therefore, identified RE policies and other complementary policies should be amended and/or expanded to encourage development in RE production. For example, the

Nigerian Biofuel Policy and Incentives 2007 should be amended to extend to other sectors, and prospective biofuel policies and laws should cover wide sectors. The government must be on the driving seat of the programme.

It is important to harmonize the Nigerian energy sector rather than fragmentize it. Mutual cooperation between government agencies and ministries is the first step in a successful RE development. The government and policymakers need to explore RE options such as small hydropower plants across the country, especially in the rural areas. This will transform the communities into semi-urban centres.

2. Legal and Regulatory Framework: For proper and consistent development of the RE sector there must be appropriate policy, legal, and regulatory instruments. Taking a cue from China, the Federal Government of Nigeria should draw policies that mandate the connection of RE producers to the National grid and the purchase of energy produced from renewable resources. In addition, measures must be taken by the Nigerian government at all levels to ensure the establishment and strengthening of stable and predictable regulatory and investment environments necessary to foster RE climate.

   The Power Purchase Agreement plan for RE generation to the national grid must be clear; there should be penalties for failure to comply with these policies, and the penalty should provide for a fine that covers, at least, the economic loss of the RE producer. Federal, State, and Local government should also be mandated by law to purchases a minimum unit of RE produced in the country and its regions. Such policy and legal framework would also attract and encourage domestic and international investments and equally boost stakeholder confidence.

3. Public Awareness: There is need for dissemination of information on RE resource availability, benefits, and opportunity to the general public in order to raise public awareness and generate activities in the area. Such process is paramount to building public confidence and acceptance of RE technology. Providing information to selected stakeholder groups like the investors can help erase some cultural/religious beliefs and mobilize financial resources needed to promote RE technology projects.[86] Furthermore, effective communication and robust awareness potentially thwarts any market gap that could result in higher risk perception for potential RE projects. The draft

Renewable Energy Master Plan proposes the set-up of a National Renewable Energy Development Agency (NREDA), which can assist in increasing public awareness and providing information and assistance to interested stakeholders. This is to be done together with non-governmental organizations (NGOs).[87]

4. Technology and Funds: The challenge posed by the capital intensive nature of the technology required in this sector can be resolved overtime through investment subsidies. These are financial concessions from the host government targeted at spurring the investor to invest more. The Nigerian government should be encouraged to buy into this programme. Under such programme, the government supports the investment effort of the investor, for example, by granting capital assets in cash or kind, reduce the cost of registration of the company of an investor, wave certain requirements which must otherwise be met, reduce the cost of the property to be used as the principal place of business, assign a pioneer status to the company, tax incentives,[88] and so on.

   The government should be able to overcome this by developing the necessary political will to consistently implement its policy and pursue its target of achieving 20 per cent RE by 2035. Thus, although the transition will be expensive in the short term due to the huge investments in renewable energy investment, and the transfer of technology and knowledge, the long-term benefit of this move will outweigh the short-term disadvantages.

5. Political Support: In order to achieve an expanded RE market, the creation of a favourable legal and regulatory framework is essential. Thus, the development of RE technologies requires a strong political support through an easy bureaucratic procedure. Clear and simple regulations about the required license for the construction and operation of RE power plants is very important. In general, the government should improve Nigeria's rating on 'ease of doing business'. Another means of political support is through the setting of national reviewable target as was carried out in the case of China.

***

Thus this paper throws light on the legal issues/impediments to the development of RE in Nigeria and recommends to draw from the success experiences of China in RE development in order to articulate strategy to enable the government of Nigeria attain RE objectives by including harnessing the RE resources through effective policy implementation and regulation, enhancing energy efficiency, and integration of RE into the energy supply mix to chart a new energy future for Nigeria. This is sequel to Nigeria's endowment with abundant resources of fossil fuels as well as RE resources like solar, wind, biomass, and small hydro. Presently, the demand for fossil fuel is higher than the national output over the years. Nigeria has a very high potential for RE production and distribution which would have been used to fill in the energy gap and avert possible energy crises. Regrettably, the RE sector is bedeviled by some legal challenges which have impeded its development over the years. Many of the RE policies including the master plan are still being processed in National Assembly for over three years and are yet to be enacted into laws. These have not augured well for the RE sector.

The monograph has examined these impediments and proffered strategies aimed at resolving these legal barriers. The strategies proffered are borne out of the lessons from China's RE experiences which are also based on international best practices like the need for national regulation for the RE sector, effective implementation of monitoring, reviewing and verification of legal/policy framework, sharing mechanism, and introduction of various incentives for various stakeholders. In order to enhance the energy security of Nigeria and establish a sustainable energy supply system, the policy makers should make RE development a priority policy statement of government at federal, state, and local government levels. Furthermore, lawmakers should accord priority to the development of appropriate legal and institutional frameworks to enhance RE development, reduce emphasis on over-dependence on fossil fuels and address the challenges posed by the present energy crisis. In addition, this monograph asserts that proper implementation of relevant environmental policies and laws remain a sine qua non for RE development. Therefore, measures must be taken by the Nigerian government at all levels to ensure the establishment and strengthening of stable and predictable regulatory and investment environments necessary to foster RE climate.

Arising from the aforementioned discussions and conclusions, it has become imperative for Nigeria to attain its RE objectives. It is therefore recommended that Nigeria should ensure the passage of the RE master plan bill and the various bills on RE implementation before the National Assembly; Establish Power Purchase Agreement to boost stakeholders confidence; Demonstrate political support for RE implementation by taking the lead as in the case of China; Create awareness on renewable energy and energy efficiency; Provide incentives as part of tax and investment policy, subsidies for rural users tax, profit and ensure proper implementation of same; Establish an institution or agency for implementation of RE policies and laws; Renewable energy policies should come as a 'package' and not a 'stand-alone' kind of policies, the packaged policies works in an interactive mode, where the success or failure of a single policy will depend on the effectiveness of other complementary policies; Develop and imbibe energy efficiency technologies; Government at all levels should make adequate provision for research grants in renewable energy resources to make production cheaper and generation more efficient.

If these suggestions are implemented properly, it will go a long way in establishing and strengthening of stable and predictable regulatory and investment environments necessary to foster RE climate.

Notes and References

1. R.J. Heffron, 'Energy Law: An Introduction, Springer Briefs in Law', DOI 10.1007/978-3-319-14191-6_1.

2. C. Chauhan, 'India a Fast Growing RE Country', *Hindustan Times,* available http://www.hindustantimes.com/News-Feed/India/India-a-fast-growing-renewable-energy-country/Article1-770129.aspx., accessed on 6 August 2016.

3. B. Zhu, 'Exploitation of Renewable Energy Sources and Its Legal Regulation', *Journal of Sustainable Development,* 3(1) (2010), pp. 116–19, 116, available http://www.ccsenet.org/jsd, accessed on 27 August 2016.

4. Zhu, 'Exploitation of Renewable Energy Sources and Its Legal Regulation', p. 118.

5. Department of Petroleum Resources (DPR) Nigeria, (2007), available http://www.DPR.gov.ng, accessed on 5 August 2016.

6. ECN Oka for and J.U. CKA, 'Challenges to Development of Renewable Energy for Electric Power Sector in Nigeria', *International Journal of Academic Research,* 2 (2) (2010), pp. 211–16, available at Google Scholar, accessed on 7 August 2016.

7. Adati Ayuba Kadafa, 'Oil Exploration and Spillage in the Niger Delta of Nigeria' *Civil and Environmental Research* 2(3) (2012), p. 44, available

pakacademicsearch.com/pdf-files/.../38-51%20Vol%202,%20No%203%20 (2012).pdf.

8. Ayuba Kadafa, 'Oil Exploration and Spillage in the Niger Delta of Nigeria'.

9. Energy Commission of Nigeria (ECN): National Energy Policy. Federal Republic of Nigeria, Abuja (2006) available at Google Scholar; Energy Commission of Nigeria (ECN): National Energy Policy. Federal Republic of Nigeria, Abuja (2008), available, accessed on 7 August 2016.

10. Developing a Plan for Policy; ctb.ku.edu.table of content; Policy–Wikipedia, the free Encyclopedia, available https://en.wikipedia.org/wiki/Policy, accessed on 10 September 2015.

11. 'What Is Energy?', available www.scholarschools.net, accessed on 27June 2016.

12. 'What Is Energy?', available www.qrg.northwestern.edu/projects/vss/docs/space.../1-what-is-energy.htm, accessed on 27 June 2016.

13. W. Mingyuan , 'Government Incentives to Promote Renewable Energy in the United States', *Temple Journal of Science, Technology, and Environmental Law*, XXIV(1) (2005), pp. 355–66, esp. 356.

14. O. Awogbemi and C.A. Komolafe, 'Potential for Sustainable Renewable Energy Development in Nigeria', *The Pacific Journal of Science and Technology*, 12(1) (2011), pp. 61–169, 161.

15. 'What Is Development? Sustainable Measures', available www.sustainablemeasures.com/Training/Indicators/Develop.html, accessed on 27 June 2016.

16. 'What Is Development? Sustainable Measures'.

17. 'What Is Development? Comhlamh', available www.comhlamh.org/issues-to-consider/what-is-development/, accessed on 27 June 2016.

18. 'Introduction: What Is Development?', Globalization101, available www.globalization101.org › issues in depth › development accessed on 5 August 2016.

19. E. Rummey, 'China Is the Worlds Largest Investor in Renewable Enregy', available www.public financeinternational.org, accessed on 27 August 2016.

20. Zhu, 'Exploitation of Renewable Energy Sources and Its Legal Regulation', p. 116.

21. J. West, et al., 'Resource Efficiency: Economics and Outlook for China' (Bangkok, Thailand: UNEP, 2013).

22. West, et al., 'Resource Efficiency: Economics and Outlook for China'.

23. West, et al., 'Resource Efficiency: Economics and Outlook for China'.

24. China pledged to reduce its carbon intensity (the amount of carbon dioxide emissions (CO_2) emitted per unit of gross domestic product (GDP)) by

40 to 45 per cent by 2020 based on 2005 levels at the United Nations Climate Change Conference in Copenhagen in December 2009. The country also pledged to raise the share of non-fossil sources of energy (that is, renewables and nuclear) to 15 per cent of its primary energy consumption by 2020, from about 9.9 per cent as of the end of 2009. China acted in this direction by passing the RE Law on 28 May 2005 which became effective on 1 January 2006 (amended in December 2009) and its associated implementing regulations, which have played a major role in the rapid growth of China's renewables resources in the last few eight years. The *RE Law* created four mechanisms to promote the growth of China's RE supply: (1) a national RE target; (2) a mandatory connection and purchase policy; (3) a feed-in tariff system; and, (4) a cost-sharing mechanism, including a special fund for RE development. See, S. Schuman, 'Improving China's Existing RE Legal Framework: Lessons from the International and Domestic Experience'. Natural Resources Defense Council White Paper (2010), pp. 1–3, available https://www.google.com.ng/url?sa=t& rct=j&q=&esrc=s&source=web&cd=1&cad=rja&ved=0CCsQFjAA&url=h ttps%3A%2F%2Fseors.unfccc.int%2Fseors%2Fattachments%2Fget_attachme nt%3Fcode%3DLPSYUD2USW7ZV7AI6PB0IU7SACB6DJNN&ei=tozVUfnf HYiu7AaUw4Eo&usg=AFQjCNHFNzswc0C8uxkjY-2lcmg2oOiwbw&sig2=c XuOB6uHJzr0ApEOgoxfpw, accessed on 4 July 2013.

25. Policies for Reducing Energy Intensity in China. Sources: (UNDP 2010, Cao and Xu 2010, ADB 2010).

26. Hicks and Dietmar, 2007.

27. The Renewable Energy Law of the People's Republic of China, 2005.

28. The State's Council Energy Department is the energy body under the NDRC which was previously called NDRC Energy Bureau. The National Energy Administration was formed in March 2008 and replaced the NDRC Energy Bureau. See, E.S. Downs, China's New Energy Administration. China Business Review, November–December, available www.brookings.edu/%7E/ media/Files/rc/articles/2008/11_china_energy_downs/11_china_energy_ downs.pdf, accessed on 4 July 2013.

29. S. Schuman, 'Improving China's Existing RE Legal Framework: Lessons from the International and Domestic Experience', Nrdc White Paper (2010), available http://seors.unfccc.int/seors/attachments/get_attachment%3Fcode %3DLPSYUD2USW7Z V7AI6PB0IU7SACB6DJNN, accessed on 29 June 2013.

30. Schuman, 'Improving China's Existing RE Legal Framework'.

31. Schuman, 'Improving China's Existing RE Legal Framework'.

32. On 28 May 2005, the Standing Committee of the National People's Congress (NPC) promulgated the RE Law, which became effective on 1 January 2006. See, The RE Law of the People's Republic of China, 2005. Retrieved on 3 January 2013, available http://www.npc.gov.cn/englishnpc/special/

combatingclimatechange/2009 -08/25/content1515301.htm., accessed on 29 June 2016.

33. M. Prakash, 'Promoting Environmentally Sustainable Transport in the People's Republic of China', (Philippines: Asian Development Bank, 2008).

34. A new Climate Change Law is expected in the next two to three years to draw together existing climate-related policies and to lay a legal foundation for future institutions.

35. For purposes of power generation, five large state-owned power generators namely, Huaneng, Datang, Huadian, Guodian, and CPI, often referred to as the big five, are all active in developing RE projects through their subsidiaries, although there are other power generators contributing a smaller portion of production. See Schuman, 'Improving China's Existing RE Legal Framework: Lessons from the International and Domestic Experience', p. 2.

36. The two main transmission and distribution companies that provide coverage for the whole of China are State Grid Corporation of China and China Southern Grid both of which are owned by the government. State Grid's coverage area accounts for about 80 per cent of China's total electricity consumption, and it is one of the largest utilities in the world. State Grid has five wholly-owned subsidiaries with coverage areas in Northern China, Northeastern China, Eastern China, Middle China, and Northwestern China, each of which operate through their subsidiaries in the covered provinces, municipal areas, and autonomous regions.

37. See, Schuman, 'Improving China's Existing RE Legal Framework'.

38. T.T. Onifade, 'Legal and Institutional Framework for Promoting Environmental Sustainability in Nigeria through Renewable Energy: Possible Lessons from Brazil, China, and India'.

39. I. V.-Akpu, 'Renewable Energy Potentials in Nigeria'; *Energy Future: The Role of Impact. Assessment,* 32nd Annual Meeting of the International Association for Impact Assessment, 27 May–1 June 2012, Centro de Congresso da Alfândega, Porto–Portugal, available http://www.iaia.org/conferences/iaia12/Final_Paper_Review.aspx?AspxAutoDetectCookiesSupport=1#, at 2012, hydropower accounted on 12 June 2013.

40. Akpu, 'Renewable Energy potentials in Nigeria'.

41. Onifade, 'Legal and Institutional Framework for Promoting Environmental Sustainability in Nigeria through Renewable Energy: Possible Lessons from Brazil, China, and India'.

42. A. Idris, et al., 'An Assessment of the Power Sector Reform in Nigeria', *International Journal of Advancements in Research & Technology,* 2(2): 1–37, available http://www.ijoart.org/docs/An-Assessment- of-The-Power-Sector-Reform-in-Nigeria.pdf., accessed on 30 December 2013

43. A. Idris, et al., 'An Assessment of the Power Sector Reform in Nigeria'; Akpu, 'Renewable Energy potentials in Nigeria'.

44. Idris, et al., 'An Assessment of the Power Sector Reform in Nigeria'.

45. C.E. Nnaji, et al., 'Renewable Energy Penetration in Nigeria: A Study of The South-East Zone', *Continental Journal of Environmental Science*, 5(1) (2011): 1–5, 1, available http://www.wiloludjournal.com/pdf/environsci/2011/1-5.pdf., accessed on 30 December 2013

46. A.I. Shehu, 'Solar Energy Development in Northern Nigeria', *Journal Eng. Energy Resources*, 2(1) (2012).

47. Akpu, 'Renewable Energy potentials in Nigeria'.

48. A.S. Sambo, 'Strategic Developments in Renewable Energy in Nigeria', *International Association for Energy Economics*, 2009, 3.15–19, at 15.

49. Akpu, 'Renewable Energy potentials in Nigeria', p. 3.

50. Nnaji, et al., 'Renewable Energy Penetration in Nigeria: A Study of the South-East Zone', referencing O. Adegbuyi, A.A. Adipe, and M.A. Rahaman, 'Hot Dry Rock (HDR) Geothermal Energy Resource Potentials in Nigeria', *J. Geothermal Sci.Technology*, 1 (1990), pp. 1–9.

51. Nnaji, et al., 'Renewable Energy Penetration in Nigeria', pp. 14–21.

52. R. Opeh, and U. Okezie, 'The Significance of Biogas Plants in Nigeria's Energy Strategy', *J Phys Sci. innov* (2011), p. 3

53. See, E.G. Orie, 'Clean Development Mechanism as a Tool for Sustainable Development a Case for Regulatory Action', Unpublished PhD Thesis, Nigerian Institute of Advanced Legal Studies, University of Lagos, 2013.

54. Onifade, 'Legal and Institutional Framework for Promoting Environmental Sustainability in Nigeria through Renewable Energy'.

55. See, The National Policy on Environment.

56. Rural Electrification Strategy and Implementation Plan (RESIP) (2014), Power Sector Reform Team, available www.power.gov.ng/National%20Council%20on%20Power/Rural%20Electrification%20Committe%20Recommendation.pdf., accessed on 6 August 2015.

57. National Renewable Energy and Energy Efficiency Policy (NREEEP) (2014). Energy Commission of Nigeria (ECN) and Federal Ministry of Science and Technology (FMST), available www.energy.gov.ng., accessed on 28 August 2016.

58. Renewable Energy Master Plan (REMP) (2005), Energy Commission of Nigeria (ECN) and United Nations Development Programme (UNDP), available www.spidersolutionsnigeria.com/wp-content/uploads/2014/08/Renewable-Energy-Master-Plan-2005.pdf; Renewable Energy master Plan (REMP) (2012), Energy Commission of Nigeria (ECN) and United Nations Development Programme (UNDP), available www.energy.gov.ng, accessed on 28 August 2016.

59. N.V. Emodi and S.D. Yusuf, 'Improving Electricity Access in Nigeria: Obstacles and the Way Forward', *International Journal of Energy Economics and Policy*, vol. 5, no. 1 (2015), pp. 335–51.

60. A. Iwayemi, et al., 'Towards Sustainable Universal Electricity Access in Nigeria. CPEEL' (2014), available www.cpeel.ui.edu.ng/sites/default/files/monograph-2.pdf, accessed on 26 August 2016.

61. Renewable Electricity Action Program (REAP). Federal Ministry of Power and Steel (2006). Federal Republic of Nigeria, Retrieved 6 August 2015, available www.iceednigeria.org/backup/workspace/uploads/dec.-2006-2.pdf, accessed on 30 August 2016.

62. N.V. Emodi and N.E. Ebele, 'Policies Enhancing Renewable Energy Development and Implications for Nigeria', *Sustainable Energy*, 4(1) (2016), pp. 7–16, available http://pubs.sciepub.com/rse/4/1/2, accessed on 28 August 2016; Nigerian National Petroleum Corporation (NNPC) Draft Nigerian Bio-Fuel Policy and Incentives, Nigerian National Petroleum Corporation, Abuja, 2007.

63. N. Tijjani, et al., 2013 referencing Linda Nordling, 2010, Report of a paper presented at the 6th Annual Conference of African Science Academy Development Initiative, Improving energy Access in Africa at Somerset West, Cape Town South Africa, November 2010.

64. N.Tijjani, et al., (2013).

65. The policy framework established in September 2012 by the Federal Executive Council (FEC) is designed to guide economic and social response of Nigerians to the global trend of climate change. In addition, the policy will also explicitly itemize the comprehensive national goals, objective, and strategies towards mitigating the consequences of climate change.

66. The Constitution of the Federal Republic of Nigeria, 1999. CAP. C.23, Vol. 3, LFN 2004.

67. ECN Act 1979. No. 109, Laws of the Federation of Nigeria (LFN) 2004.

68. The Nigerian Electricity Act, Cap.E7 Vol. 5, Laws of the Federation of Nigeria, 2004.

69. The centres include Centre for Energy Research and Training, Ahmadu Bello University, Zaria, Kaduna State; Sokoto Energy Research Centre, Usman Dan-Fodio University, Sokoto, Sokoto State; Centre for Energy Research and Development, Obafemi Awolowo University, Ile-ife, Osun State; National Centre for Energy Research and Development, University of Nigeria, Nsukka, Enugu State.

70. M.T. Ladan, 'Policy, Legislative, and Regulatory Challenges in Promoting Efficient and Renewable Energy for Sustainable Development and Climate Change Mitigation in Nigeria'.

71. In Part IX under Rural Electrification, Section 88 (9) stipulates that information shall be presented to the President by the Minister of Power and Steel on, among others: (a) expansion of the main grid, (b) development of isolated and mini-grid systems, and (c) RE power generation. See, Electric Power Sector Reform Act 2005, at Part IX.

72. The Federal Government of Nigeria (FGN) (2007) Electric Power Sector Reform Act (EPSRA), available www.nercng.org/index.php/nerc-documentfunc-startdown/35/, accessed on 5 July 2015.

73. Nigerian Electricity Regulatory Commission Regulations for Embedded Generation NERC (2012), available http://www.nercng.org/index.php/document-library/func-startdown/4/, accessed on 12 June 2015.

74. 'Overview of the NERC Regulations on Embedded Generation & Independent Electricity Distribution', *Business Day*, Thursday, 31 May 2012, available http://www.businessdayonline.com/NG/index.php/law/legal-culture/38733-overview-of-the-nerc-regulations-on-embedded-generation-a-independent-electricity-distribution-networks, accessed on 12 June 2013.

75. Nigeria has many ongoing policy and strategy initiatives which, if backed by the appropriate relevant laws and properly implemented, can serve as effective measures to boost RE generation.

76. See, Orie, 'Clean Development Mechanism as a Tool'.

77. C. Chineke, et al., 'Much Ado about Little: and Renewable Energy and Policy, *Journal of international Scientific Publication*, Vol. 9 (2015), available http://www.scientific-publication.net/get1000015/1432901545984409.pdf.

78. E.L. Efurufmibe, 'Barriers to the Development of Renewable Energy in Nigeria', *Scholarly Journal of Biotechnology*, vol. 2 (2013), pp. 11–13.

79. J.A. Ajobo and A.O. Abioye, 'An Expository on Energy with Emphasis on the Viability of Renewable Energy Sources and Resources in Nigeria: Prospects, Challenges, and Recommendations'.

80. World Bank, available http://www-wds.worldbank.org/external/default/WDSContentServer/WDSP/IB/2006/04/27/000012009_20060427113507/Rendered/PDF/36986.pdf, accessed on 6 November 2016.

81. I. Oghogho, 'Solar Energy Potential and Its Development Sustainable Energy Generation in Nigeria: A Road Map to Achieving this Feat', *International Journal of Engineering and Management Sciences*, 5(2) (2014), pp. 61–7.

82. See, 'Barriers to Renewable Energy Technologies', http://www.ucsusa.org/clean_energy/smart-energy-solutions/increase-renewables/barriers-to-renewable-energy.html#.VjIbnSsbjMw, accessed on 5 August 2016.

83. E.L. Efurumibe, 'Barriers to the Development of Renewable Energy in Nigeria', *Scholarly Journal of Biotechnology*, 2 (2013), pp. 11–13.

84. Efurumibe, 'Barriers to the Development of Renewable Energy in Nigeria'.

85. Ladan, 'Policy, Legislative and Regulatory Challenges in Promoting Efficient and Renewable Energy'.

86. Efurumibe, 'Barriers to the Development of Renewable Energy in Nigeria'.

87. Renewable Energy (REMP) Energy Commission of Nigeria Masterplan (ECN) and United Nations Development Programme (UNDP), available www.energy.gov.ng, accessed on 6 August 2015.

88. Producers of RE are often exempt from certain taxes such as carbon taxes in industrialised countries, or taxes for imports of RE equipment in developing countries. They justify tax exemptions on the basis of unfair competition with conventional energy sources which occurs when negative external costs are not internalized. In India, an accelerated depreciation policy enables RE investors 100 per cent depreciation in the first five years of operation. The effect of accelerated depreciation is similar to the effects of investment tax credits. See Onifade, 'Legal and Institutional Framework for Promoting Environmental Sustainability in Nigeria through Renewable Energy', p. 102.

9

ABDUL HASEEB ANSARI

Sustainable Energy for Sustainable Development

World View on Nuclear Energy with Special Reference to India

The traditional sources of energy, fossil fuels, became popular because the other sources of energy, renewable and alternative, required technological expertise and financial resources. They became popular also because of their easy availability. The countries with plenty or easily obtainable fossil fuels resorted to them as energy sources without making proactive efforts for the non-traditional sources of energy. It is for this reason that even today the share of energy in the energy mix from fossil fuels is much more than the other sources of energy.[1] Since fossil fuels are exhaustible natural resources and are not clean, the quest for cleaner energy sources were augmented. As a result, biofuels such as, biodiesel and ethanol, hydropower, solar, and wind power, and nuclear energy are already in use. Initially, among all the alternative energy sources, nuclear energy became popular because a reactor has the capability to give continuous supply if it gets sufficient uranium fuel, and because it is a clean energy source. However, this source of energy is mainly limited to developed countries because, (a) it requires sophisticated reactor technology, (b) it should have regular supply of uranium, (c) enrichment of uranium is a high-end on nuclear technology, (d) it needs a comprehensive safety infrastructure and a competent group of scientists and technocrats competent to run it safely, (e) it requires a competent legal and institutional framework, and, (f) it warrants treatment of nuclear wastes. Some developing countries, notably India and China, have already ventured into nuclear energy, but a good number of them have plans to have it in the near future. But chances for most of them to have it are bleak because they have failed to earn the trust of the countries with nuclear power, and there is a persistent inhibition that the technology might go in hands of terrorists, which might yield disastrous results.

India, Pakistan, and China have an ambitious plan to increase the number of reactors, but face resistance from their people, they do not get technology transfer from the countries with nuclear power, they lack enough financial resources, they lack competent infrastructure to run the reactors, and they do not have capability to safely collect nuclear wastes generated form reactor(s). It is notable here that as of 5 February 2015, under the United Nations Framework Convention on Climate Change 1992, 188 countries had given their voluntary carbon emission reduction targets, with or without technological and financial support from developed countries, including India, China, and Malaysia, keeping in view their capabilities.[2] The Paris Agreement was adopted on 12 December 2015 at the 21st Session of the Conference of Parties of United Nations Framework Convention, which was open for signing on 22 April 2016. Currently, there are 180 signatory countries to the agreement.[3] Out of them, 129 countries have already ratified. The agreement has come into force on 4 November 2016.[4] The energy performance of these countries will be reviewed after five years. The chances of achieving the targets set by developing countries are bleak because barring some, they cannot actively involve in renewable energy sources, for example, they need to develop capability to fabricate their own photovoltaic panels and high-capacity windmills. For these reasons, they might not get the chance to resort to nuclear energy. However, the problem with most of the developing countries, especially of Africa and Asia, is that their population will keep on growing until 2050.[5] The growing population and fast urbanization keeps on increasing the energy demand. In order to solve their energy problems, they should be allowed to resort to nuclear power with renewable energy sources, but the nuclear rich countries are sceptical about it; they have even agreed to allow them to have it even in controlled conditions. It seems their ambitions will not be granted unless they themselves become capable of it, like Pakistan, China, and India. For example, Indonesia aspired and tried hard to make its ambitions to have nuclear reactor(s) true, but even today there are no positive indications. It is for these reasons that Malaysia does not have plan to have nuclear power in the near future. So is the case of some other fast developing economies, except Taiwan and Vietnam. It is true that some of them, like Malaysia and Thailand, need to resort to nuclear power also not later than 2030. It is because the path of development in fast growing populous countries goes through nuclear power plants. Nuclear power will remain intact in the middle of the power mix, with fossil fuels in one hand and renewable energy in the other hand. When fossil fuels recede, the renewable sources will increase in that ratio. This is an ideal situation, but the question is, will it really be achieved?

There is no restriction on peaceful use of energy. Rather, there is International Atomic Energy Agency (IAEA) for facilitating this. The IAEA is working, based on agreement between it and the United Nations Organization in 1957 (INFCIR/ 11), as an agency of the UNO.[6] It monitors the promotion of information, methods, and technologies to enhance safety via continuous improvements in communication and knowledge sharing between the network participants and provision for training. It encourages countries to engage in peaceful use of atomic energy but the task has not been so accomplished because of the many safety concerns echoed in its annual reports, the Nuclear Technology Review and Nuclear Safety Review.[7] Safety concerns are true to some extent, but not serious.[8] It is true that accidents might be disastrous, as experienced in cases of the Chernobyl nuclear blast of 1986 and the Fukushima Daiichi nuclear disaster of 2011. It is worth noting here that the Chernobyl accident was the result of a flawed reactor design that was operated with inadequately trained personnel.[9] Likewise, the Fukushima disaster was due to *vis major* (act of God) for which tsunami was responsible.[10] If both are compared, there were a large number of casualties in Chernobyl but there was no loss of life in Fukushima. It is submitted that both the reactors had decades old technology. Now, reactor technology is quite safe. According to the World Nuclear Association (WNA), nuclear power generation now is quite safe. It says,

That the technology now is safe and there is good level of awareness about accidents; that the present industrial design is quite safe; that only 3 accidents have so far taken place among the 16,000 reactors operative in 33 countries[11]; that the risk is on the decline; that the chances of terrorist attacks are minimal.[12] These facts are considered to be true by a large number of scientists. Russ Bell, director of new plant licensing at the industry's Nuclear Energy Institute in Washington, maintains that the new plants will be extraordinarily safe.[13] This has also got support from the IAEA's report 'Nuclear Technology Review, 2015'. The report says, 'Safety improvements have continued to be made at nuclear power plants (NPPs) throughout the world. These have included identifying and applying lessons learned from the accident at the Fukushima Daiichi Nuclear Power Plant, improving the effectiveness of defence in depth, strengthening emergency preparedness and response capabilities, enhancing capacity building, and protecting people and the environment from ionizing radiation.'[14] Moreover, IAEA annual reports are warning in advance for adopting further preventive measures. The tasks that IAEA undertook in Iran have been immensely appreciated. However, looking at the Iraq and Iran episodes, it may be said that the problem is that the IAEA allegedly works under tremendous amount of influence of the countries, which are dominating in the United Nations.[15] This is a puzzling issue. The question here is can, in the near future, IAEA take its decision independently?

India and China are exceptionally ahead among all developing countries as they are technologically advance, fast growing, and populous countries. Their energy needs are much higher than other developing countries. This is compounded by the fast urbanization in them. It is for this reason that they are amongst the top carbon emitters. They are under pressure to resort to alternative and renewable energy sources, including nuclear energy, but they are not sufficient to encourage them to reduce carbon emission by reducing power generation using fossil fuels. They succumbed to the international pressure although their even B grade cities are facing power cut problems impeding their economic growth. Since they already have nuclear enrichment technology, they can easily increase the number of reactors, and uranium rich countries have trust in them. Right now, the ratio of the nuclear energy in the total energy mix in these countries is not substantial, but with commissioning of the proposed reactors, it will be a robust part in that.

Among the two, China is ahead, but the ambition of India to have few reactors of 1,000 and more Megawatt capacity is looming in the garb of uncertainty because it is not yet member of the Nuclear Suppliers Group (NSG). The questions here are, will they be able to have more than fifty per cent of clean energy, and how? Will they be able to treat their nuclear wastes?

In order for nuclear energy programme to go ahead in a country, there has to be a group of competent scientists, a capable legal and institutional framework for safe running of reactors, and ensuring civil remedy to sufferers from nuclear radiation, capability to store and treat or make arrangement to treat nuclear wastes generated from nuclear reactors. Making a legal framework is not difficult but to have competent trained personnel, scientific, and support personnel is really arduous. It will much depend on the political will on the part of the developed counterpart of the developing countries because it will not be like dream come true unless nuclear technology rich counties come forward to support them in finalizing a viable reactor design and later building it, changing the size of their grid system, ensuring supply of uranium for running the reactors, and continue to provide technological support in running the reactor and safe management of nuclear wastes generated from the reac- tor. There are a large number of developing countries aspiring for nuclear power, but only those, which are technologically advanced and get favour from at least one nuclear rich country of the Nuclear Supplies Group (NSG), will be able to have it. Right now, courtiers capable to extend help building nuclear plants in developing countries, notably UK, US, France, Russia, and China, are suffering from lots of inhibitions leading to scepti- cism. They need to grow over these inhibitions and come forward to help developing countries build nuclear reactors for power generation. Their reactors may be put under intense supervision of the IAEA. They may not be allowed to engage with enrichment activities; rather, enriched ura- nium should be provided to them. The question here is, will developed countries come forward to do so? They wish that developing and least developed countries should work only for maximizing their capabilities with respect to renewable energy sources.

The chapter will critically examine the questions posed above. Before that in order to understand the magnitude and possibility of expanding the use of nuclear energy, we need to know the nuclear energy profile at global level.

World Energy Profile

The first nuclear reactor was commissioned in 1950 followed by several others in 1950s. In 2014, there were 438 commercial reactors with total power generation capacity of 376.2 GWe. With one permanent shut down and some additions, now as of June 2016, there are 446 commercial reactors with power generation capacity of 388 GWe. Two reactors have been shut down after long use, and 63 are under construction.[16] There are 31 countries around the world enjoying the benefit of the atomic natural resources. If the plan to have more reactors goes well, in the near future, there will be 63 more reactors, which are under construction. In the world, total energy mix, nuclear power contributes 11 per cent. There are 240 research reactors in 56 countries around the world, including one of Malaysia. There are 140 ships and submarines having some 180 nuclear reactors. There is a possibility that the number of nuclear reactors will further increase because it is a clean energy source and the supplies are uninterrupted. Moreover, the supply of the enriched uranium is always more than the demand. It is notable here that supply of uranium, equipments related to that and technology are controlled by the Nuclear Suppliers Group (NSG) of 48 countries, including China. The NSG countries, which control the uranium supply, can have business only with the countries that are party to the Nuclear Non Proliferation Treaty (NPT). A non-NPT country can be a member of the NSG only when the aspiring country fulfils the entry conditions and all the members of the NSG agree for it.

The top ten nuclear power generating countries are: US, France, Russia, China, South Korea, Canada, Germany, Ukraine, UK, and Spain with the generating capacity of 798, 419, 182.8, 161.2, 152.2, 95.6, 86.8, 82.4, 63.9, and 54.8 Billion KWH respectively.

In 2012, the ratio of the nuclear power was 10.8 per cent of the world electricity production. It rose to 11.5 per cent in 2016. The figure looks low, but the growth of the nuclear power has risen in spite of the fact that renewable energy has risen at a considerably higher rate. The World Nuclear Association still proposes nuclear energy to be 25 per cent by 2050 of the total power generation. The author is of the opinion that this projection may not be true as renewable energy sources need to grow faster and by then it may become 30 to 40 per cent, and the rest will come from fossil fuels and nuclear source. [17]

Nuclear Energy Aspirants in Asia

The nuclear energy aspiring countries are there around the world, but their number in Asia, especially in South Asia and Southeast Asia is relatively higher. The aspiration for resorting to nuclear power, as stated above, are mainly due to increasing energy demand because of fast urbanization and industrial development. It is because of this reason that the number of nuclear energy aspiring countries will keep on growing. However, the aspiring countries have to be technologically advanced and economically sound. Asia is the main region in the world where nuclear power in future will grow faster than other regions of the world. It is the future hot-spot of nuclear power. The momentous growth is going to be in India, China, and South Korea.[18] The emerging nuclear energy countries in Asia are, UAE, Saudi Arabia, Qatar, Kuwait, Yemen, Israel, Syria, Jordan, Azerbaijan, Kazakhstan, Bangladesh, Sri Lanka, Indonesia, Philippines, Vietnam, Thailand, Laos, Cambodia, Malaysia, Singapore, Myanmar, and North Korea. Some of them already have power reactor under construction, some have signed agreement with nuclear rich countries, and some others have plan to be finalized.[19]

The UK, US, France, Russia, and China are prominent among those countries, which are providing help to other countries in building nuclear reactors. Among them, Russia and China have taken the lead. Worth noticing here is that a nuclear power aspiring country must enter into an agreement with a well established country in this area in order to have a close cooperation with respect to, (a) Pre-project phase; (b) Project decision-making phase; (c) Construction phase; and, (d) Operation phase. Among the nuclear power aspiring countries in Asia, Vietnam has successfully developed the plan. It has entered into an agreement with Russia and hoped to have two operational reactors by 2026. Nevertheless, before construction starts, it has to successfully prepare a comprehensive nuclear power plant security plan as well as a management plan for spent fuel. In order to meet the international safety standards, the country is working closely with IAEA. Indonesia might be the second country after Vietnam to have nuclear power followed by Malaysia, Philippines, and Thailand. Malaysia wants to have nuclear energy by 2030 fulfilling about 15 per cent of its total energy demand. Thus, 'as Vietnam comes closer to completing its first nuclear power plant, and with Indonesia and Malaysia both considering the prospects for a nuclear energy future, there is significant interest among the countries to strengthen nuclear governance ... and strictly uphold nuclear security, safety and safeguards.'[20]

Any country aspiring to have a nuclear reactor must work to develop a legal framework on nuclear safety, security and safeguards; to have a workable nuclear literacy programme; to develop a competent resource training programme; to have an effective food testing system, especially with respect to exportable food; to have cooperation on continuous supply of enriched uranium; to establish a nuclear disaster management centre; to hold frequent emergency exercises; to have regional cooperation with respect to safety, security and safeguards; and to explore possibility to have regional management capability with respect to 'Spent Fuel'.

Nuclear Energy Politics

Although nuclear energy is a clean energy source and IAEA is there to facilitate this source of energy, it suffers from unwarranted international politics. The nuclear states, International Atomic Energy Agency (IAEA), Nuclear Suppliers Group (NSG), and the UN Security Council (UNSC) have lot of say in liberalizing or restricting use of nuclear technology for power generation. One inherent and potent fear is that if a country proceeds with the task unfettered, it might engage in nuclear enrichment leading to making nuclear weapons. It is for this reason that nuclear countries and the Security Council have adopted a restrictive and cautious approach that an aspiring country can have it only if it has a workable design approved by the contracting nuclear power state and has attained certain level of technology necessary to safely run reactor(s); also, the country should have sound economic indices. They also need to ensure about a competent legal and institutional framework so that in case of any eventuality the country could take necessary measures. These facts are ensured by unrestricted inspections conducted by the IAEA. The system seems to be easy and smooth, but it is not so. Because of very restrictive approach of some of the nuclear rich countries, the ambitions of several developing countries to resort to nuclear power generation are only at the plan stage. Perhaps, as stated above, because they do not want them to develop faster and compete with them in international markets, and they are apprehensive that nuclear materials might go in hands of terrorist and they can make dirty bomb with that. As stated above, only those, which have been successful in getting a support from a nuclear rich county and have entered into agreements with them, have been able to really venture. Among the top ten nuclear rich countries, the approach of Russia and China is quite liberal; whereas, others, namely US, France, South Korea, Canada, Germany, Ukraine, UK, and Spain, seem to be skeptical. For their restrictive approach, there is no discernable reason. Perhaps, they do not want other courtiers to develop faster and compete with them, or they hold the opinion that a liberal approach might provide opportunity for the terrorists to steal uranium and make dirty bomb. Possibly, they also understand that in absence of enough safeguards, technical and other, terrorists might get a chance to sabotage the nuclear installations causing wide spread and long-term damage to properties and lives. Such fear has been expressed by the Secretary-General Ban Ki-Moon in his address to the East-West Institute entitled 'The United Nations and security in a nuclear–weapon-free world' on 24 October 2008. He said, '...The main worry is that this will lead to the production and use of more nuclear materials that must be protected against proliferation and terrorist threat'.[21] His words actually reiterate such fear expressed in the Security Council Resolutions 1373 and 1540, which highlighted the threat of nuclear terrorism and nuclear proliferation and call for national, regional, and international cooperation to strengthen the global response to these challenges.[22] The point is that if all sufficient measures suggested by the IAEA from time to time are taken, there will be no possibility of such an act. It is to be noted here that if a state is engaged only with power generation and for that the country gets enriched uranium from a contracting state party and other NSG countries, there will be no chance for terrorists to make dirty bomb, except the threat of causing sabotage resulting in nuclear radiations. Moreover, until now, there is no reported atomic terrorism. Moreover, under the Treaty on the Non-Proliferation of Nuclear Weapons (NPT), the IAEA conducts on-site inspections from time to time to ensure that nuclear fuels are being used only for peaceful purposes. In order to address the fear of nuclear terrorism, there are, the Convention on the Physical Protection of Nuclear Materials, 1998, and the Convention for the Suppression of Acts of Nuclear Terrorism 2005. All necessary measures have to be taken under the first Convention by the states engaged in producing, storing, and transporting nuclear materials; whereas, under the second Convention all required measures have to be taken in order to ensure that nuclear materials do not go in terrorists hands. Here also, the IAEA plays a supervisory role. The fear is there when a state pursues its nuclear ambitions secretly, like Iraq, Syria, and Iran. They were not fully open to the IAEA inspectors. So, their pursuits were doubted and they were classified as rogue countries.

The IAEA is the nuclear watchdog with the objective to 'seek to accelerate and enlarge the contribution of atomic energy to peace, health, and prosperity throughout the world. It shall ensure, so far as it is able, that assistance provided by it or at its request or under its supervision or control is not used in such a way as to further any military purpose' (IAEA Charter). The role of the IAEA has been significant in the Iraq and Iran episodes. Iraq's ambitions brought this country under foreign invasion and then to devastation under the garb of the civil war. Iran was pursuing its nuclear enrichment programme secretly and did not show all its activities to the IAEA inspectors. It is for this reason that the country had to face sanctions imposed by the Security Council, European Union (EU), and the United States (US). Irony is that for all sanctions, US was instrumental. Supporters of Iran were of the opinion that US had soft attitude towards Israel. It is because of it only that Israel now has even atom Bomb(s). In view of this, it is concluded, while the IAEA is an important institution with some power, the real international power lies with the individual countries that produce nuclear energy ... These countries play a major role in nuclear energy facilities.[23] Among these countries, the US is the most powerful and most instrumental. If we take into account the Iraq and Iran episodes, the US played a proactive role. Perhaps, it is so because it wanted to make Israel the only powerful nuclear country in the Middle-East so that it could do anything, legal or illegal in the region. It is worth mentioning here that US' dual policy has been conspicuously in favour of Israel. It is evident from the number of the Vetoes excised by it for protecting Israel. This seems to be accentuated by the July 2016 Sir Chicot Report on United Kingdom's position in the Iraq War.[24] Iran faced atrocities due to the sanctions. It could be out of it only when on January 2016, the IAEA reported that Iran had taken a series of initial nuclear-related measures as called for by the Security Council Resolution 2231 (2015). Resultantly, the previous resolutions of the Security Council got terminated. This could also be possible because of a proactive role played by the US. It is both an exporter of nuclear technology and a major factor in restricting the technology. Some countries got favour from it, and some others could not earn its blessings. In fact, the country is the major factor in almost all international politics.[25]

Nuclear power generation politics conspicuously surfaced when India sought its entry into Nuclear Supplies Group (NSG), in order to ensure supply of uranium for its reactors and to venture into bigger nuclear reactors having capacity to generate 10,000 megawatts (MW) and more. There are companies ready to help India to fulfil its aspiration. It is needless to say that to sustain growth and fulfilling electricity needs of its rising population and fast growing urbanization, the only way for India is to maximize its renewable energy sources and to supplement it with sufficient nuclear energy. Looking at the pressing energy needs, India put all its efforts and also did lobbying for it. The US, France, Russia, and UK were agreeable to India's entry. In spite of the fact that India has very good track record with respect to non-proliferation of nuclear technology and peaceful use of atomic energy, its entry into the NSG was blocked by China, Brazil, Australia, New Zealand, Ireland, and Turkey. In order to garner China's support, Indian Prime Minister Narendra Modi met with Chinese President Xi Jinping. China and five other states, which opposed India's entry into NSG, took the plea that since India is not a member of the nuclear non-proliferation Treaty (NPT), its membership cannot be positively considered. Actually, China and its allies found a pretext in order to ban entry into NSG. It may be noted here that to be a member of NPT is not a condition precedent; it is simply a regulation, thus desirable. Actually China in order not to be left alone also did its lobbying. China's bone of contention, though based on surmises are:

India is not eligible to become a member of the NSG as it is not a member of the nuclear non-proliferation treaty (NPT), adherence to which latter is necessary for membership in the former. China has also averred that for non-NPT members some definite criteria should be evolved rather than granting country specific waivers. At other times, it has stated that Pakistan also has similar credentials to join the NSG and that if India is admitted, Pakistan should also be admitted simultaneously. China has also maintained that there are several countries, which have reservations about India's membership of the NSG. Further, if only India were to be admitted, it would disturb the nuclear-arms balance in South Asia as India will engage in a massive nuclear weaponization programme. Finally, China has stated that India's membership will 'jeopardise' China's national interests and touch a 'raw nerve' in Pakistan.[26]

It is notable here that China succeeded in blocking India's entry in spite of the fact that India is a member of the Missile Technology Control Regime since 7 June 2016, and China strived hard its entry into it in 2004 but failed. India could get into it because of its good track record with respect of non-proliferation of nuclear and missile technology.

The above paragraphs reveal that nuclear technology is a subject matter of political determination. States smart in making an effective lobby succeed, others fail. States are also facing opposition from their political rivals; thus, failure or success depends on reaching a political consensus in favour of setting up nuclear reactor. In early development of nuclear energy, the political left in capitalist countries opposed nuclear energy except France. The countries, where anti-nuclear forces are strong, chances of resorting to nuclear energy are bleak. The Chernobyl disaster contributed in making this lobby stronger all over the world. In the US, this lobby has succeeded in stalling building new reactors, except for those, which are under construction. It is hoped that in future, since states and their people have realized that technologically advanced but populous countries, like India, China, Thailand, and Indonesia, must resort to nuclear power in order to fulfil their energy needs. It is surprising that Vietnam will soon have nuclear reactors and in this the US is extending its help, perhaps because it was also responsible for ruining Vietnam's economy and now it needs impetus with the help of enough energy generated by means of nuclear reactors.

It has been stated above that the nuclear rich countries—with whatever motivations they have—are skeptical about liberalizing atomic power generation in developing countries, mainly because of inadequate legal and institutional framework and safety concern. For instance, the attitude of the rest of the nuclear countries about India is not favourable enough to encourage it to go ahead with the proposed two imported nuclear reactors by 2017. In view of this, it is necessary to examine general law governing atomic energy and law pertaining to safety and civil liability in case of damage. For this, India may be the most suitable country.[27]

Nuclear Power and Law in India

India's nuclear power programme is as ambitious as that of China. The 2015 edition of the Bharat Petroleum's Energy Outlook projected India's energy production rising by 117 per cent to 2035, while consumption would grow by 128 per cent. The energy demand in India is increasing rapidly because of fast urbanization, ever growing industrial demands, and ever expanding railway network. In the energy mix of the country, contribution of nuclear power is not so significant. The 12th Five-Year Plan for 2012–17 proposed to an addition of 94 gigawatt (GW) over the period. Of this, nuclear power will be only 3.4 GW, provided the 2 proposed nuclear reactors of 1000 MW capacity each and 2 indigenous nuclear reactors of 700 MW each, become operative. It is projected that by 2032 the total installed capacity of 700 GW will meet 7–9 per cent of the GDP growth, and this will include 63 GW from the nuclear source.[28] This can be possible only when the local production of indigenous uranium is optimized and import of uranium is maximized keeping in view the uranium needs of the reactors, existing and proposed. The blocking of entry of the country into NSG is quite discouraging. In future, if India fails to garner the support of the members of the Group, its ambitions will fall short and country's economy might suffer.

Nuclear power reactors in India are safe and well regulated. All operating units are ISO 14000 and IS 18001 certified. They are working under the Nuclear Power Corporation of India Limited. Under its supervision, all protective and safety measures such as ensuring protection from any radiation by following global standards based on 'As low as reasonably achievable' (ALARA) and frequent review conducted by the Atomic Energy Regulatory Board (AERB), conducting proper and regular surveillance done by the Health Physics Unit (HPU), and Environment Survey Laboratories (ELSs) under the authority of the Bhabha Atomic Research Centre (BARC), following world-class operational and maintenance standards, adopting a high standard waste management methods, periodical rehearsals for emergency preparedness and disaster management by the National Disaster Management Authority (NDMA) established under the Disaster Management Act, 2005 by adhering to the zoning—exclusion zone (1.5 km radius), sterilized zone (5 km radius), and emergency planning zone (16 km radius) and taking all emergency measures such as proper estimation of magnitude and appropriate and advance notification and warning—conducting regular medical examination of persons engaged with the plant, and other necessary procedures. According to R. Deolaikar,

> All nuclear facilities are sited, designed, constructed, commissioned, and operated in accordance with strict quality and safety standards. Principles of defence in depth, redundancy, and diversity are followed in the design of all nuclear facilities and their systems/components. The regulatory framework in the country is robust, with the independent Atomic Energy Regulatory Board (AERB) having powers to frame the policies, laying down safety standards and requirements, and monitoring and enforcing all the safety provisions. The AERB exercises the regulatory control through a stage-wise system of licensing. As a result, India's safety record has been excellent in over 277 reactor years of operation of power reactors.[29]

In the whole process, the crucial aspect is to have an efficient national policy governing regulation of nuclear and radiation safety.

The Atomic Energy Act, 1962,[30] with an objective to provide for the development, control, and use of atomic energy for the welfare of the people of India and for other peaceful purposes, and for matters connected therewith, confers on the Government of India powers and responsibilities to framing rules and issuing notifications for implementing provisions of the Act.[31] This includes, safe production, development, use, and disposal of atomic energy and radioactive substances; control over radioactive substances; control radiation generation plants in order to prevent radiation hazards; secure public safety and safety of persons handling radioactive substances or radiation generating plant; and ensure safe disposal of radioactive wastes. The government is also authorized under the Act to conduct inspections and take appropriate legal action against anyone contravening the provisions of the Act, including provisions pertaining to safety in the activities related to use of atomic energy. The Act has been amended by the Atomic Energy (Amendment) Act, 2015. The 'government company' after the amendment includes such joint venture public companies as may be formed between the Nuclear Power Corporation of India Limited (NPCIL) and other Public Sector Undertakings (PSU). The Act originally provided only that a 'government company' is that which is allowed to produce atomic energy in which at least 51 per cent of the paid-up share capital is held by the central government. It may be noted here that the requirement of 51 per cent shares to be held by the government has been lifted. Thus, the scope of the definition of 'government company' has been widened. Now, it would include also companies where the whole of the paid up share capital is held by one or more government company and whose articles of association empower the central government to constitute its Board of Directors. This will facilitate for forming of joint ventures between Nuclear Power Corporation of India Limited and other government companies.

By an executive order, the central government has created the Atomic Energy Regulatory Board (AERB) to carry out necessary regulatory functions under Sections 16, 17, and 23 of the Atomic Energy Act. Sections 16 and 17 of the Act have provisions about controlling radioactive and safety in producing, handling, use, and disposal of radioactive substances. Section 23 deals with the administration of the Factories Act 1948 in the factories established and owned by the central government. Thus, the AERB is mainly responsible for safety related matters.

> The AERB has also been empowered to perform the functions under the Sections 10(1) (Powers of entry) and 11(1) (Powers to take samples) of Environmental Protection Act, 1986 and Rule 12 (Agency to which information on excess discharge of pollutants to be given) under the Environmental Protection (Amendment) Rules 1987. Further, Rules 2(b) and 3 of the Manufacture, Storage, and Import of Hazardous Chemicals Rules 1989 under the Environmental Protection Act 1986 has notified AERB as the authority to enforce directions and procedures as per the Atomic Energy Act 1962. Under the Civil Liability for Nuclear Damage Act, 2010 and the Civil Liability for Nuclear Damage Rules, 2011, AERB also has the responsibility of notifying the occurrence of any nuclear incident.[32]

It has issued Safety Codes and Standards covering the areas of regulation of nuclear and radiation facilities and related activities, taking into account contemporary scientific and technological capabilities and world-class working standards and practices. It is notable that in the Safety Codes and Standards, principles remain the same but the means and criteria go on changing, which has to be taken into account. The AERB, in 2008, has also issued a Code of Ethics, which reflects the core values to be adhered to in the whole process.

The rules, regulations, codes of conducts and standards, and code of ethics are in line with the policy requirements for the safety pertaining to the nuclear and radiation facilities of the IAEA enshrined in 'Governmental Legal and Regulatory Framework for Safety (IAEA GSR Part-1)'[33]and the IAEA standard 'Fundamental Safety Principles (IAEA-SF-1)'.[34]

In July 2014, the AERB designed policies governing regulation of Nuclear and radiation safety. The salient features of the policies are summarized as:

1. The fundamental objective of the AERB is to ensure safety.
2. It will do safety reviews from time to time and conduct inspections. However, it does not diminish the safety responsibility of the consentee for safety.
3. The regulations shall have the objectives that: (a) permitted practices are only for the social benefit, (b) radiation protection is optimized, (c) radiations do not go above the prescribed limit, (d) accidental possibilities are low.
4. Decisions related to regulatory consent shall be based on review and assessment by the AERB.
5. The regulatory process shall provide for review and assessment.
6. The regulatory control shall follow a general graded approach.
7. All facilities shall be in accordance with requisite quality assurance programmes.
8. All facilities shall implement appropriate radiation protection programmes.
9. The radioactive wastes shall be managed in safe manner.
10. All facilities and activities shall have arrangements for development of adequate plans and preparedness for responding to emergency situation(s).
11. There shall be safe decommissioning of used equipment(s).
12. The regulatory body will be responsible for enforcement of laws.
13. Natural form of radiation shall be out of regulatory control.
14. The objectives of the 'National Policy on Safety, Health and Environment at Work Place, February 2009', issued by the Ministry of Labour and Employment, Government of India and the provisions of the Atomic Energy (Factories) Rules 1996, shall prevail.
15. The Regulatory Body shall take steps to keep the public informed on safety issues. It shall also be responsible to notify on the

'extraordinary nuclear events' as mandated by the Civil Liability for Nuclear Damage Act 2010.

16. In the conduct of regulatory activities, the Regulatory Body shall be governed by the provisions of the Right to Information Act 2005, as acceptable to the 'public authority'.[35]

On 27 March 2015, the Integrated Regulatory Review Service, a body under the IAEA based on the Integrated Regulatory Review Services (IRRS) Guidelines for the Preparations and Conduct did first IRRS Review in India. After conducting review in India, upon the request of the Indian government in wake of the Fukushima accident, on 16–17 March 2016, it submitted a Report to the AERB. This is considered to be the most significant transparency gesture shown by the Indian Government, which may be considered that the Government will now resort to a new public engagement model practiced by the US. The IRRS Report appreciated some of the aspects of the Indian practices. The team leader said,

> India's atomic Energy Regulatory Board is an experienced, knowledge-able, and dedicated regulatory body for the protection of the public and the environment. It continues to enhance its regulatory progarmme to face the current and future challenges in regulating nuclear safety, such as reinforcing the exiting nuclear facilities, monitoring ageing, and decommissioning as well providing oversight to the construction, com-missioning, and operation of new nuclear power plants.

At the same time, it criticized some other practices and offered suggestions for improvement in the Indian regulatory system. Certain corners of people stressed on making thede facto autonomy of the AERB to a de jure one. On this matter the Report said, '… the professionalism and integrity of the Atomic Energy Commission (AEC), National Power Corporation India Ltd (NPCIL), and AERB senior staff towards ensuring the regulatory decision-making processes/arrange-ments were completed independently and did not notice instances in which de facto AERB independence was compromised.' However, the Report urged, '…for making an independent regulatory body. It should promulgate a nuclear policy for making strategy for safe and unharmed radioactive waste management … It should consider increasing the frequency routine on-sight inspections … it should develop its own internal emergency arrangement, including detailed procedure to fulfil its emergency response'.[36] The author is of the opinion that it will be better to make AERB a de jure authority for the sake of confidence on the part of both the Board and the people. Another important aspect pointed out in the Report is the appeal from the AERB. Presently, it lies with the Atomic Energy Commission (AEC) and its decision is final. The IRRS referred to it but did not offer any suggestions. This aspect has to be addressed by the Government because most of the aggrieved people now go to the civil courts for redress. If the procedure at the AERB and the AEC are efficient, smooth, and less time-taking, people will certainly follow the system rather than going to the regular courts. About the Fukushima accident, the Nuclear Accident Intendant Investigation Commission concluded that '... the Fukushima accident was the result of the persistent collusion between the government the regulators and TEPCO (Tokyo Electric Power Company), and the lack of governance by said parties. They effectively betrayed the nation's right to be safe from nuclear accidents.'[37] The Indian authorities should learn from it and should ensure that the relationship between the regulators and the government authorities remains synchronized and smooth.

The AERB has accepted the review report's recommendations. It considered it as an opportunity to enhance the regulatory framework and decided to work on it based on the detailed action plan prepared for it. Among the suggestions given by the IRRS, establishment of an independent statutory (de jure) regulatory body by delinking the AERB from the Atomic Energy Commission is crucial. The Nuclear Safety Regulatory Authority Bill 2011 was introduced in the Parliament on 7 September 2011, but the Bill could not be taken up because of the dissolution of Fifteenth Lok Sabha. With suitable modifications, the Nuclear Safety Authority Bill 2015 was introduced in the Parliament but could not get through. It is said that the Bill is expected to be re-introduced in the Parliament after the completion of necessary pre-legislative formalities. It might be a crucial step towards converting the functional independence of the countries atomic regulator into a de jure body.[38] If the Bill becomes an Act, the AERB will become like the United States' Nuclear Regulatory Commission, the French Nuclear Safety Authority, the Canadian Nuclear Safety Commission, and the Nuclear Directorate within the United Kingdom Health & Safety Executive. These bodies enjoy complete autonomy with respect to designing and implementing safety measures, keeping the public fully informed on the nuclear safety matters, sharing of joint responsibility to ensure physical security of installations, punishing offenders, taking quick appropriate measures in case of any accident resulting in radiation, and publishing annual reports. The author is of the opinion that most of the functions these bodies exercise are also excised by the AERB, but making it a de jure fully autonomous body will generate confidence in it. It will also make the AERB completely independent as desired by the IAEA. Presently, there are 14 reactors out of 22 reactors in operation under the commissioning under the IAEA safeguard requirements. The AERB as an independent body will be able to bring all the present 22 reactors and the proposed reactors under the IAEA safeguards. It is important for this, that India should invite the IAEA inspectors from time to time. It will help maintaining international safety standards and change them as and when it is required. It will, apart from other things, develop confidence among the NSG counties which will, in turn, make easier for the country to ensure regular and sufficient supply of uranium, to obtain foreign direct investment in building high capacity nuclear reactors and high-end technology.

Liability for nuclear damage is yet another issue, which provides redresses to all kinds of damage to be compensated by civil legal means. It is based on the strict liability principle that nuclear power plants will be liable for any damage caused by them, even if there is no fault on their part. In order to cover it, many countries' laws provide them to get insured to meet the future civil liabilities. In case of trans-boundary damage, national law should have specific provisions to supplement the international legal regime comprising international conventions, IAEA's Vienna Convention on Civil Liability Damage 1963 (came into force in 1977 and is operative in outside Western Europe); the OECD Paris Convention on Third Party Liability in the Field of Nuclear Energy 1960 (came into force in 1963 and applicable to the Western European countries barring a few. The Brussels Supplementary Convention1963 bolstered it). The Joint Protocol of 1988 linked these Conventions in order to bring together the geographical scope of the two conventions. They purport to cover damage to property, health, and loss of life excluding environmental damage and economic loss. It also does not cover money spent on taking preventive measures. In order to follow the principle of recompense, in 1997, the IAEA Parties adopted a

Convention on Supplementary Compensation for Nuclear Damage (CSC), which came into force on 15 April 2015.[39] Likewise, the Brussels Supplementary Convention 1963 was adopted, which came into force without any delay; it was revised in 2004, which has not yet come into force. Both together with the Supplementary Conventions provide a 3-tier system of compensation, first, to be paid under the original convention; second, to be supplemented by the state; and third, a fund having contributions from the member states. The IAEA Supplementary Convention is independent. Any state without being member of any of the two main conventions can become a member of the Convention on Supplementary Compensation.

The liabilities under the CSC are limited and are normally covered by insurance. If insurance companies are not agreeable to cover nuclear incidents for insurance of nuclear installations, because it is different than ordinary industrial structures, states should provide the coverage. In the whole legal regime, the difficult aspect is to define the damage to be covered by the civil liability. And the law must be based on the following cardinal legal principles, strict liability, exclusive liability, compensation without all kinds of discrimination, mandatory financial coverage via insurance, exclusive jurisdiction of the courts of the state where accident took place, and limited liability.

India joined the Convention on Supplementary Compensation for Nuclear Damage. An official document to this effect was handed over to the IAEA on 4 February 2016. It will certainly provide impetus to the country's ambitions to become a member of the NSG. It will also be instrumental to bring the Civil Liability for Nuclear Damage Act 2010 (CLND Act) within the scope of the CSC. India's ratification to the CSC was appreciated because the CLND Act was considered to be in compliance with the provisions of the international legal regime for civil liabilities (CSC and its Annex). The Act is in line with Price Anderson Nuclear Industries Act 1957 (revised last in 2005), the first nuclear liability law on nuclear damage. It will be better to look into the Indian Act and to discern salient features from it in order to judge its international credibility. The Ministry of External Affairs has already expressed its opinion that the ratification of the CSC is 'a conclusive step in addressing the issues related to civil nuclear liability'. The CLND Act imposes liable for any nuclear accident related damage but without protecting third party suppliers [Section 17 (b)]. The Act prescribes overall pecuniary limits to liability. Operators' liability is to be covered by insurance companies. It is generally understood that the CLND Act is in compliance of the CSC. The notable features of the CLND Act are: 1. The operator will be liable for the nuclear damage caused by incidents [s 4(1)]; 2. There will be strict liability based on the principle of no fault liability [s. 4(4)]; 3. Before the operation begins, there must be an insurance coverage or any financial security covering the liability [s. 8 (1)]; 4. The central government may establish a nuclear liability fund by charging a suitable amount of levy on the operators [s. 7 (2)]; 5.There will be right to recourse against the suppliers [s. 17 b)]; 6. A case for compensation can be brought under statutes other than the CLND Act [s. 46]; 7. Section 46 does not extend to suppliers in violation of the CSC because every statute has to be interpreted in accordance of the intention of the legislature or maker of the statute.[40] The move to include them was rejected by the Parliament.[41] Moreover, a case can only be filed in Indian courts.

A controversy had loomed after the ratification of the CSC whether the CLND Act is in line with the CSC. As noted above, the Ministry of External Affairs claims that it enforces the CSC norms. However, there are certain grey areas that need to be hammered in order to make them fully compatible. Section 17 (b) of the CLND Act is an addition to Article 10 of the Annex of the CSC. Article 10 reads: 'National law may provide that the operator shall have a right of recourse only: 1. if this is expressly provided for by a contract in writing; or, 2. if the nuclear incident results from an act or omission done with intent to cause damage, against the individual who has acted or omitted to act with such intent.' It excludes the supplier or his employee from the recourse action. Contrary to this, the CLND Act holds them liable for supply of equipment or material with patent or latent defects or sub-standard services. Section 17 of the CLND Act reads:

> The operator of the nuclear installation, after paying the compensa-
> tion for nuclear damage in accordance with section 6, shall have a right
> of recourse where: (a) such right is expressly provided for in a contract
> in writing; (b) the nuclear incident has resulted as a consequence of an
> act of supplier or his employee, which includes supply of equipment or
> material with patent or latent defects or sub-standard services; (c) the
> nuclear incident has resulted from the act of commission or omission of
> an individual done with the intent to cause nuclear damage.

By virtue of Section 17, there will be liability on the suppliers of equipments and materials if later they are found defective or sub-standard. It has to be noted here that this provision of the CLND Act was brought into it after a long discussion in the Parliament on negligence-based liability or fault-based liability. According to Arghya Sengupta, 'Parliament thus consciously departed from the CSC provisions in legislating Section 17(b) to adopt a fault-based liability regime for suppliers. Scarcely can dexterous diplomacy bypass such an imperative mandate of the law and parliamentary will.'[42] It is notable here that suppliers' liability is not mandatory, it depends on the will of the operator; he may or may not choose to bring action against the supplier. The author is of the opinion that IAEA should not have any objection against section 17(b), as it carries forward the liability regime in a better way as it like a product liability which is common supplies of equipments and materials and ordinarily part of supply contracts. And Article 10 of the Annex to the CSC has not specifically prohibited it.

Section 46 of the CLND Act provides that the revisions of the Act are in addition to and not in derogation of any other law for the time being enforce and it will not bar actions under any other law. Domestic as well as foreign suppliers have expressed concern over the broad scope of the section. It may be noted that the language of the section is a commonly used in several other legislations. Such a language it provided routinely to underline that other laws will remain operative in their respective domains. It simply does not affect the applicability of other laws. It, therefore, does not exempt the operator from application of other laws relating to matters other than civil liability for nuclear damage. As stated above, it does not authorize any case be instituted in any foreign court. Thus, the section does not extend to suppliers in violation of the CSC. Moreover, this section applies only to operators and not suppliers. It is confirmed by the Parliamentary debates on the section.

***

The momentous process of urbanization and industrialization, especially in developing and least developed countries, with the expected growth of their population will keep on increasing the energy demand. Because of the promises made to the United Nations by around 195 countries under the United Nations Framework Convention on Climate Change 1992, there might be tremendous growth of renewable sources of energy. But they might not replace the fossil fuels and nuclear sources of energy. In the world energy mix, nuclear power will have a considerable ratio. Its present ratio will rather inflate because many technologically advanced, economically sound, and politically stable developing countries will fulfil their energy demands from nuclear power reactors.

Developed countries should not be skeptical of expansion of nuclear energy for political or economic reasons. Rather, they should come forward and extend their helping hands for developing counties to have nuclear reactors, because for them the path of development goes through nuclear power generation. It can well be done under the supervision of the IAEA. On the contrary, they and their people will suffer; it is needless to say that in India, even B-grade cities are facing power cut 8–10 hours daily throughout years, and the economic setback because of this is serious.

India's track record with respect to nuclear power generation has been clean and encouraging. It has the capability to have a very strong legal and institution framework satisfying the international standards supported by a competent team of scientists and technocrats capable of installing and running nuclear reactors safely. This is evident from its ratification of the CSC. The Indian law is compatible with the legal framework prescribed by the CSC and its Annex. The provisions pertaining to suppliers' liability under Section 17 (b) of the Atomic Energy Act is a welcome feature in the Act. It is within the legal framework endorsed by the IAEA. The only thing that is needed to be done is making the AERB an independent regulatory body. It is, therefore, warranted that the 2015 Bill should get through the law-making system soonest possible. India also has capability to fulfil the requirement of civil liability in case of nuclear incidents resulting in damage.

China's proactive role in blocking India's entry into the NSG was politically motivated, as membership to the NPT is not a mandatory requirement; it is simply desired. In order to secure its entry into the NSG, India should consider its entry to ratify the NPT. It should also make its fullest efforts to get support of all the members of the Group.

Notes and References

1. See, 'Green Energy Dropping Out of Mix in Developing Countries', *Sci Dev Net*, available http://www.scidev.net/global/energy/news/green-energy-developing-world-renewable-energy.html, accessed on 10 July 2016.

2. At the last climate conference (COP 20) in Lima in late 2014, it was decided that each of the 195 States would have to set out their roadmap to limit the effects of global warming to less than 2°C by 2100. In view of that, a deadline of 1 October 2015 was set by the UN (the Framework Convention on Climate Change) so that there would be time to analyze these contributions. See at,http://www.cop21.gouv.fr/en/185-countries-have-committed-to-reducing-their-greenhouse-gas-emissions/, accessed on 12 July 2016.

3. See UNFCCC website, available http://unfccc.int/paris_agreement/items/9444.php, accessed on 15 June 2016.

4. See UNFCCC website, available http://unfccc.int/paris_agreement/items/9444.php, accessed on 7 February 2017.

5. United Nations Population Division, *World Population Prospects: The 2010 Revision*, Medium variant (2011).

6. The agreement stipulated that 'The Agency undertakes to conduct its activities in accordance with the Purposes and Principles of the United Nations Charter to promote peace and international co-operation, and in conformity with policies of the United Nations furthering the establishment of safeguarded worldwide disarmament and in conformity with any international agreements entered into pursuant to such policies.' See at, https://www.iaea.org/sites/default/files/publications/documents/infcircs/1959/infcirc11.pdf, accessed on 20 June 2016.

7. Both the 2015 reports can be seen at the official website of the IAEA.

8. See the reports at the official website of the IAEA.

9. The accident happened because of a combination of basic engineering deficiencies in the reactor and faulty actions of the operators: the safety systems had been switched off, and the reactor was being operated under improper, unstable conditions, a situation which allowed an uncontrollable power surge to occur. See at http://www.world-nuclear.org/information-library/safety-and-security/safety-of-plants/chernobyl-accident.aspx, accessed on 20 June 2016; and at http://www.greenfacts.org/en/chernobyl]/, accessed on 20 June 2016.

10. See at http://www.world-nuclear.org/information-library/safety-and-security/safety-of-plants/fukushima-accident.aspx, accessed on 20 June 2016.

11. This figure includes all kinds of reactors. See at http://www.world-nuclear.org/information-library/safety-and-security/safety-of-plants/safety-of-nuclear-power-reactors.aspx, accessed on 10 August 2016. Of them, as of 04 February 2016, nuclear power reactors were, 442 under operation and 66 under construction. See at https://www.euronuclear.org/info/encyclopedia/n/nuclear-power-plant-world-wide.htm, accessed on 12 August 2016.

12. See at http://www.world-nuclear.org/information-library/safety-and-security/safety-of-plants/safety-of-nuclear-power-reactors.aspx, accessed on 20 June 2016.

13. See at http://e360.yale.edu/feature/the_nuclear_power_resurgence_how_safe_are_the_new_reactors/2287/, accessed on 20 June 2016.

14. IAEA, *Nuclear Technology Review*, 2015, p. 1. See at https://www.iaea.org/sites/default/files/ntr2015.pdf, accessed on 24 June 2016.

15. Iran is of this opinion, see http://edition.presstv.ir/detail.fa/169080.html, accessed on 12 May 2016.

16. IAEA-PRIS, available https://www.iaea.org/pris/, accessed on 10 May 2016.

17. http://www.world-nuclear.org/information-library/current-and-future-generation/nuclear-power-in-the-world-today.aspx, accessed on 2 January 2018.

18. B. Linton, 'Nuclear Power is Booming in Asia', *Power Engineering*, at http://www.power-eng.com/articles/npi/print/volume-7/issue-5/nucleus/nuclear-power-is-booming-in-asia.html, accessed on 12 May 2016.

19. See, http://web he.googleusercontent.com/search?q=cache:http://www.world-nuclear.org/information-library/country-profiles/others/asias-nuclear-energy-growth.aspx&gws_rd=cr&ei=wpMqWMT_ForgvAS1tI6oBQ, accessed on 15 November 2016.

20. For a detailed account, see Mely Cabellero-Anthony, Alistair D.B. Cook, Julius Cesar Trajano, and Megareth Sembiring 'The Sustainability of Nuclear Energy in Southeast Asia: Opportunities and a Challenges', NTS Report No. 1, October 2014. The Report is available at https://www.rsis.edu.sg/wp-content/uploads/2014/10/NTS-Report-October-2014.pdf, accessed on 20 April 2016. For ASEAN countries, see, M.C.Anthony, et al., 'The State of Nuclear Energy in ASEAN: Regional Norms and Challenges', *ASEAN Perspective*, 39 (2015): 695–723.

21. See at http://www.un.org/en/globalissues/atomicenergy/, accessed on 12 May 2016.

22. See the Resolution at http://www-ns.iaea.org/security/sc_resolutions.asp, accessed on 12 May 2016.

23. International Politics, 'Nuclear Energy', available https://sites.google.com/a/ncsu.edu/nuclear-energy/politics/foreign, accessed on 10 June 2016.

24. The Report concludes, among others, that Saddam Hussain was on an eminent danger, that the US' actions were not supported by intelligence, that the aftermath of the war was not anticipated.

25. 'PRIS Home Page', IAEA: Nuclear Fusion Web (27 September 2010), available http://www.iaea.org/programmes/a2/index.html, accessed on 14 May 2016.

26. A. Sajjanhar, 'Why is NSG Membership Important for India?', available http://www.idsa.in/idsacomments/why-is-nsg-membership-important-for-india_asajjanhar_210616, accessed on 25 June 2016.

27. For legal and institutional framework, see http://www.nucleartourist.com/operation/nrc1.htm, accessed on 13 May 2016.

28. World Nuclear Association, 'Nuclear Power in India' [http://www.world-nuclear.org/information-library/country-profiles/countries-g-n/india.aspx], accessed on 12 April 2016.

29. R. Deolaikar, 'Safety in Nuclear Power Plants in India', *Indian Journal of Occupational and Environmental Medicine*, 12(3) (2008), pp. 122–7; H Pandve, P.A. Bhuyar, and A. Banerjee, 'Indo-US Nuclear Deal: A Challenge for Occupational Health', *Indian Journal of Occupational and Environmental Medicine*, 11 (2007), pp. 47–9

30. Act No. 33 of 1962.

31. The following rules and regulations have been made: the Atomic Energy (Radiation Protection) Rules 2004, the Atomic Energy (Working of Mines Minerals and handling of Prescribed Substances) Rules 1984, the Atomic Energy (Safe Disposal of Radioactive Wastes) Rules1987, the Atomic Energy (Factories) Rules 1996, the Atomic Energy (Radiation Processing of Food and Allied Products) Rules 2012.

32. Government of India, Atomic Energy Regulatory Board, 'Policies Governing Regulation of Nuclear and Radiation Safety' (July 2014), available http://www.aerb.gov.in/AERBPortal/pages/English/prsrel/policies.pdf, accessed on 20 May 2016.

33. Available at http://www-pub.iaea.org/books/IAEABooks/10883/Governmental-Legal-and-Regulatory-Framework-for-Safety, accessed on 10 May 2016.

34. Available at [http://www-pub.iaea.org/MTCD/publications/PDF/Pub1273_web.pdf], accessed on 12 April 2016.

35. *Supra* note 22.

36. See A. Mohan, 'Nuclear Safety and Regulation in India: The Way Forward', ORS Issue Briefs and Special Reports (26 June 2015).

37. See at https://nuclear-news.net/2016/01/04/report-of-indias-integrated-regulatory-review-services-irrs-on-atomic-energy-board/, accessed on 12 May 2016.

38. A. Sasi, 'New Atomic Regulator: Nuking the Autonomy Red Flag', *Indian Express* (29 July 2015).

39. 'It aims at increasing the amount of compensation available in the event of a nuclear incident through public funds to be made available by the Contracting Parties on the basis of their installed nuclear capacity and UN rate of assessment. It also aims at establishing treaty relations among States that belong to the Vienna Convention on Civil Liability for Nuclear Damage, the Paris Convention on Third Party Liability in the Field of Nuclear Energy or neither of them, while leaving intact the 1988 Joint Protocol that establishes treaty relations among States that belong to the Vienna Convention or the Paris Convention.' Quoted from the IAEA website at https://www.iaea.org/

newscenter/news/india-joins-convention-supplementary-compensation-nuclear-damage, accessed on 12 July 2016.

40. Supreme Court in *M/s. Turtuf Safety Glass Industries v. Commissioner of Sales Tax U.P.*, 2007 (9) SCALE 610; and *State of Kerala & Anr v. P.V. Neelakandan Nair & Ors*, 2005 (5) SCALE 424.

41. Ministry of External Affairs, 'Frequently Asked Questions and Answers on Civil Liability for Nuclear Damage Act 2010 and Related Issues', available http://www.mea.gov.in/press-releases.htm?dtl/24766/Frequently_Asked_Questions_and_Answers_on_Civil_Liability_for_Nuclear_Damage_Act_2010_and_related_issues, accessed on 10 May 2016.

42. See at http://thewire.in/21587/indias-nuclear-liability-regime-is-still-up-in-the-air/, accessed on 12 June 2016.

10

C.M. JARIWALA

The Indian Energy Conservation Law

A Critical Overview

Energy is the life line of human beings and that of any nation. The total dependence on non-renewable energy has put India to the fourth position in the energy consuming countries of the world. It is even predicted that by 2020 the consumption will increase by 3–4 times than the present needs. One of the solutions is the conservation of energy. If a green energy approach is adopted it will result in the surplus of energy and in the long term help in mitigating the problem of environmental pollution and that of climate change. In this direction the present legislation tries to control the inefficient and unsustainable use of energy and aims at the conservation of energy. The inefficient and unsustainable use of energy finally got the legal control in 2001.

In India, the energy resources attracted the attention of the legislature since 1887 but the concern for the conservation of energy drew attention of the government only in 1974. It was only after three decades that a specific legislation on conservation of energy was enacted in 2001. The present study mainly focuses its attention to the select provisions of the Energy Conservation Act, 2001.

The conservation of energy law is an area which has yet to attract a serious attention by the law academics. Even the *International*

Encyclopaedia of Laws, Energy Law, running into four volumes deals with the energy laws of only twenty-two countries of the world including India, and also the *Energycopia* published in four volumes, have hardly dealt with the energy conservation law. In this scarcity of literature, even the judiciary or the Tribunal has yet to make any important contribution in this micro fine specialized area. However, a humble attempt is made in this paper to explore the least explored field in this branch of law.

In order to study the position of Indian law, it becomes necessary to examine the contributions of the comparative energy conservation law. Such a study will help a research scholar to find out as to where India stands, and what efforts may be made to equip Indian law to success-fully march forward in future. The main focus of the present study is, first, the constitutional scenario, and second, the appreciative and criti-cal perspective of the Indian energy conservation law.

Coming to the Constitution of India, it has yet to come out of the traditional contour and change with the changing time. It has yet to open its window to see what some of the world constitutions have done in this regard. The questions of a balanced energy federalism; the fundamental right to consume energy for domestic or commercial purposes; the fundamental right to comfortable life against the acute scarcity of energy, the right to property over the energy resources, still remain unanswered here. The role of government's fundamental obligations and the citizens' fundamental duties relating to the conser-vation of energy, have not been explored. These are the questions and missing links which the present work tries to answer in the light of the comparative constitutional law perspective, and also in relation to the existing norms in the Constitution of India.

The next part is devoted to a detailed critical study of some of the *select provisions* of the Act. It will examine the retrospective background and efforts made in planning at the governmental level, the shape given by the Joint Select Committee of both the Houses of Parliament, delib-erations of the Members in Parliament, with a note on what has finally emerged. The critical study aims to find out whether the legislation, its amendments and rules framed thereunder, have succeeded or failed to attend its mission. In this regard the present study will critically examine the important components of the legislation. Will such provisions be able to do justice with the conservation of energy? These are some of the questions which need appropriate answers. Unfortunately, the pres-ent work is unable to include the examination of the judicial approach as there is hardly any leading case law directly dealing with the Act of 2001.

Finally the work closes with the Epilogue dealing with conclusion and reforms needed in the Supreme Law of the Land and also the legislative exercises undertaken in this regard so as to attain in future the efficient and sustainable conservation of energy in India.

Constitutional Perspective

Before dealing with the Constitution of India it is important to know what has been done by the world constitutions in the field of energy.[1] The general direction is that they have not given due constitutional recognition to such an important subject. A bird's eye view of the constitutional scenario is that most of the constitutions have not given any specific place to energy in their constitutions.[2] However, there are only few constitutions wherein the expression 'energy' and/or 'natural resources' find some place.[3] The provisions of these constitutions may be divided into four categories. First, generally, the legislative power in this regard is given to federal or central legislature.[4] However, the Constitution of Germany puts the nuclear energy under the concurrent jurisdiction.[5] It may be pointed out that nuclear energy being a subject of national importance cannot be left to the dual control except the local matters. The Constitution of Brazil rightly puts this subject specifically only in the federal domain.[6]

The Constitution of Switzerland deserves a special mention because it may be said to be the first constitution of the world which provides detailed constitutional provisions relating to energy. Chapter 2, Article 6 devotes exclusive treatment to 'energy and communication'. The Constitution puts energy under the legislative powers of both, the Federation and Cantons. However, Article 89 of the Swiss Constitution carves out some legislative subjects relating to energy for the federation which include, use of energy for installation, vehicles and appliances; development of technology in the field of energy saving and renewable energy;[7] the principles of economical and efficient use of energy in domestic and renewable energy.[8] The nuclear energy[9] and the transportation of energy are exclusively given to the federation.[10] A part of energy which goes in the exclusive power of the Cantons is its use in buildings. The Swiss Constitution also gives constitutional recognition to the energy policy under Article 89. It requires that the federation, while following a cooperative approach in the energy policy, shall take into account, first its effect on the Cantons, the municipalities, and the economic circles; and second, the conditions in various regions and the limitation of what is economically feasible.[11]

The second direction of the energy conscious constitutions is that the energy and the natural resources—surface or underground—are treated as national properties.[12] It is for the State to decide about its use and regulate activities relating to energy. The State may delegate its power in this regard to the corporates or private agencies as it may consider necessary. The third aspect is that all the activities dealing with energy shall be regulated by Law[13] and not by the executive orders.[14] Further it is the duty of the State to see that those who exploit the natural resource shall restore the degraded environment. The State shall lay down precise plan of management of energy. The Constitution further imposes a detailed obligation on the State to strive to ensure sufficient, economical, and efficient energy supply compatible with the environment.[15]

Coming to the Constitution of India, it nowhere specifically uses either the term 'energy' or the expression 'energy conservation' though it deals with some of the energy resources. Thus, the Supreme Law of India at the initial stage, unfortunately, remained blind in this specific aspect. However, the 'Non-conventional energy' has been now placed in the Eleventh Schedule to the Constitution.[16] Further, the Seventh Schedule to the Constitution deals with some of the energy resources, for example, Parliament has been given exclusive legislative power with respect to atomic energy,[17] regulation, and development of oil fields and mineral oil,[18] and mines and minerals.[19] The State Legislature is also given legislative subjects, for example, regulation of mines and minerals[20] subject to the concerned Entry in List I and 'gas'.[21] The concurrent legislative power has no reference to energy or its resources except the forest.[22] The Report on Energy Conservation Legislation, 1988 prepared by the Indian Law Institute, and submitted to the Advisory Board of Energy unfortunately suggests that Parliament was competent to enact such law under List I and List III,[23] whereas the true position is that there is no specific entry dealing with 'energy' or 'conservation of energy'.

It was only in 1992 that the Constituent Power came out of hibernation and provided 'non-conventional energy sources' as a subject in

Item 15 of the Eleventh Schedule. Item 15 read with Article 243G confers the legislative power on Panchayats in this regard as may be endowed by law by the State Legislature. But unfortunately, List II, the State List, nowhere specifically provides for such a subject, and therefore, the State Legislature, having no legislative power cannot endow Panchayats to enact such a law and thus Item 15 creates an absurd situation. Further, will the members of Panchayats be able to handle successfully this complex subject is a doubtful proposition.

In the present case, Parliament, in order to enact the Act of 2001, has taken recourse to the residuary legislative power under Article 248 read with List I, Entry 97. However, such a source of power has attracted litigations in the other fields and the courts have adopted a conflicting approach.[24] In order to avoid any controversy and to fill the existing vacuum, it is suggested that List I needs amendment to the effect that a new entry be added to List I dealing with 'Nuclear energy and other energy as may be declared by Parliament by law to be of national importance'. This, in turn, will need a corresponding amendment in List II to include a subject with respect to 'Energy subject to the provisions of the corresponding Entry of List I'. This is will provide a balanced approach in dealing with energy.[25]

The Constitution of India, unlike the Constitution of Switzerland, nowhere provides in Part IV any fundamental obligation of the State relating specifically to energy. It is suggested that a specific Article may be introduced in Part IV dealing with the following provision, 'the State shall, within the limit of environment and sustainable development, make effective provisions for the efficient and sustainable consumption, production, sale transportation, storage, and conservation of energy and also management of energy wastes, and, in particular, safeguard the non-renewable energy'. Such provisions will help the State in taking any action in the above matters and the courts may not adopt a rigid but a harmonious approach in interpreting the other provisions of the Constitution including the fundamental rights in this regard.

An efficient and sustainable use of energy and its conservation requires active participation of the people. If there is no coordination in their activities whatever laws or executive orders are passed, they will remain only a paper work. This takes us to the discussion of the fundamental duties enshrined in Part IVA of the Constitution. Unfortunately none of the existing duties take care of energy. One approach in this regard may be that the expression 'energy' may be given a place in Article 51A(g) itself dealing with the duty to protect and improve the environment. But such an addition will give energy a secondary place allowing citizens to perform or not to perform their duty in this regard. In view of the problems of growing concern for the consumption and conservation of energy in the industrialized India, it is submitted that the citizens' duty cannot remain at the back seat. It is suggested that one more fundamental duty may be added to Part IVA to the effect, 'to safeguard efficient and sustainable consumption of energy and its conservation in all its sectors'.

Coming to the right to property though the energy and the natural resources are the properties of the nation but once a private entity comes in, the claim of property right starts. Fortunately the fundamental right to property is repealed otherwise it would have created inroads in dealing with energy. Now the property right has been shifted to Article 300A making it only a constitutional right to be enforced through the high courts under Article 226.[26] Thus any energy law affecting the property right may attract Article 300A provided it does not follow the due process of law.

The other Article which may attract the constitutional validity of the energy law is the fundamental right to carry on the business or trade guaranteed under Article 19(1) (g).[27] Any business or trade in energy has to fulfil the norms prescribed in Article 19(6) of reasonable restriction in the interest of the general public. For example, in order to control any crisis in supply of fossil fuel, if the State rations its supply, it shall be protected under Article 19(6). Further in order to avoid cut throat competition in the energy market, if the State takes over in its own hands or authorizes the government corporates to control production, supply, transportations, storage, etc., of any energy resource then such an energy law will be protected under Article 19(6)(ii).

There is one more problem that needs to be considered, it is the settlement of litigations relating energy and energy resource. The energy or energy resources litigations need technical handling. The question is can the judges alone decide such disputes? The Supreme Court of India, while handling the issues involved in the environmental cases, have time and again pointed out that they are of complex, technical, and scientific nature which needs a separate court having scientific and technical inputs in this regard.[28] In 1976, the Constitution was amended, and a new Part as Part XIVA, was added to it to provide for the establishment of tribunals. Article 323A provides for the administrative tribunal and Article 323B provides for the tribunals for other matters which are divided into ten areas. But the technical inputs in the administration of justice in energy litigations are left out. It is suggested that a new sub-article to Article 323B in place of Article 323B (j) may be brought in giving 'a residuary power to Parliament to constitute tribunal on any other matter not enumerated in Article 323B'. Such an added provision will then allow Parliament to constitute tribunal even on any other non-specified matters including energy.

The Energy Conservation Law

The energy as such did not attract any attention of the Legislature either during the British raj or at the commencement of the Constitution of India. However, there existed large number of the energy resources laws since 1887. As time passed, the consumers of the energy started making excessive use of energy, uncontrolled consumption, mismanagement of the energy resources, and further heavy dependence on the fossil fuels for commercial use. All these attracted the attention and concern of the government in 1974 when the Fuel Energy Policy set the ball rolling. The working Group on Energy Policy, 1979 pushed the State's concern for energy further. The Inter-Ministerial Group on Energy, in 1983, considered for the first time, the problems of energy conservation in the different sectors of energy. A demand for legislation in this regard was raised by the Advisory Board of Energy. In 1987 the Board referred the matter to the Indian Law Institute, New Delhi to prepare a draft bill which was provided to the Board in 1988. A similar need was highlighted in the Ninth Five Year Plan. However after a long delay over which even the Member of the House of Parliament showed concern,[29] it was only in the year 2000 that a Bill was introduced in Parliament, and finally the Energy Conservation Act was enacted in 2001 to deal with the efficient use of energy and its conservation. Indian law deserves appreciation, as there are not many countries where they have such a specific law.[30] In the following pages a humble attempt is made to critically evaluate the select provisions of the Act. The study will further explore their directions and also try to answer the question—What is left out and what is needed now?

Objective of the Legislation

The Statement of Objects and Reasons has an usual pattern which normally is not followed in case of other legislations. It is not only the lengthiest Statement but also Clause 3(a), (b), and (c) of the Statement literally copy down the provisions of Sections 13, 14, and 15 of the Act, which is a sign of bad drafting. The two main reasons for introducing a comprehensive legislation were, first, the increasing demand of energy which needs conservation and its efficient use in various sectors; and second, creation of a statutory body to take measures in this regard. The Statement of Objects and Reasons in clause (4) says that the Legislation is 'enabling in nature'. Even in the Rajya Sabha, the Minister of Power, Sri Suresh Prabhu explained that the Bill introduced enabling provisions. It is submitted that looking to the provisions of the present legislation, it cannot be said to be an enabling statute.[31]

Extent and Commencement of the Legislation

Section 1(2) extends the application of the Act to whole of India except the State of Jammu and Kashmir. The question is why the State of Jammu and Kashmir should be left out? The legislation dealing with the environment, for example, the Environment (Protection) Act, 1986 and the Air (Prevention and Control of Pollution) Act, 1981 are applicable to the whole of India, and no exception is made in favour of the State of Jammu and Kashmir. The environment and energy are two sides of a coin. Moreover, some of the resources of energy, for example, nuclear energy cannot be left to the State Legislature to handle as per its own norms. The conservation of energy is also an important international obligation in combating climate change, and India is a party to the Conventions and Protocol on Climate Change, can the State of Jammu and Kashmir be left out to handle such matter at its own level? The answer cannot be positive.

Coming to the commencement of the Act, Section 1 sub-section 3 authorizes the central government to decide the date of commencement of the provisions of the Act. It is true that the government will require some time to provide infrastructure and other facilities for implementing the legislation. However, there are cases wherein though such power was given to the central government, it remained inactive in this regard.[32] This raises the question—should it be left to the absolute discretion of the government or some time schedule be prescribed, or some monitoring authority be established to see that the efforts are not flushed out in the drains by the non-issue of a notification?

Section 1(3) authorizes the central government by notification to appoint different dates for different provisions.[33] This will give further laxity to the government to take its own time to commence different provisions at different dates. On the contrary, in both the Houses of Parliament, during the debates on the 2001 Bill, the Members pointed out that such legislation should have been enacted long ago; aiming at the urgency of such measure[34] still the Houses surprisingly approved such provisions. It may be pointed out that the Israel Law[35] allowed the government to bring into force its provisions at the discretion of the government. The unfortunate part was that within four months of the enactment of the Act only one subsection (b) of Section 93 was brought into force, and it took more than three years for the commencement of other provisions. There are sixty-two sections in the 2001 Act which are inter-related, and if a piecemeal approach is adopted, it will be like providing oxygen to some parts and leaving others yet to be brought to life, a delayed process or a half-hearted approach.[36] On the contrary, the Standing Committee on Energy in its Report[37] suggested a time frame for implementing each of the twenty three activities envisaged under the Bill ranging from minimum three months to maximum thirty-six months. But such time schedules were hardly followed. However, the positive part on the side of the central government was that the government made operative the provisions on time.

The Definition Clause

Section 2 of the Definition Clause defines twenty words or expressions. Clause (4) of Section 2, also takes the help of the Electricity Act, 1910, the Electricity (Supply) Act, 1948, and the Electricity Regulatory Commission Act, 1998, to give the same meaning to the words and expression used in the Act of 2001. An approach depending on the definitions in other legislations indicates short cut approach which cannot be said to be a case of good legislative drafting.

Coming to the word 'energy', defined in Section 2(j), to mean any form of energy derived from fossil fuel, nuclear substances or materials, hydro-electricity, and includes electrical energy or electricity generated from renewable sources of energy or bio-mass connected to the grid.[38] It further puts more stress on electricity. In this regard the definition given in the Chinese Energy Law of 1997 is more detailed. In India, the definition simply uses 'materials', whereas the Chinese definition specifies it in detail. The expression 'energy', defined in Article 2 of the Chinese Law, refers to 'coal, crude oil, natural gas, power, coke, coal gas, heat, processed oil, liquefied petroleum gas, biomass energy, and other resources from which useful energy can be derived directly or through processing or conversion'.

Further, though the Act of 2001 deals with conservation of energy, unfortunately it nowhere defines the basic expression 'energy conservation'. In this regard further help may be taken from the Chinese legislation. Article 3 defines 'energy conservation' to mean 'enhancing management in the use of energy adopting measure which are technologically feasible, economically rational, and, by reducing loss and waste at every stage from production through consumption of energy, environmentally and socially acceptable in order to use energy more efficiently and rationally'. It may be pointed out that production and consumption are not the only areas which require conservation. There are other areas which also need attention, and they may include, for example, establishment, transmission, conversion, transportation, storage, and waste disposal. This will, it is submitted, provide a comprehensive approach to energy conservation. The concept of energy efficiency and its conservation cannot be complete unless its sustainability also is maintained. The Definition Clause also misses the term 'sustainable energy'. The expression is defined in the German Energy Law[39] to mean 'energy that is capable of meeting the needs of present generation without compromising the ability of future generation to meet their own energy needs'. Such provisions, it is submitted, may also find a place in Section 2 of the Act.

The Bureau of Energy Efficiency

Section 3 of the Act provides for the establishment and incorporation of a Bureau—the Bureau of Energy Efficiency. Even the ILI *Draft Bill* gave the title as 'Nodal Energy Conservation Organization'.[40] The term 'Bureau' simply means an office for the transaction of business which is generally confined to subordinate and inferior executive department or a subsidiary of an executive department, or usually a commercial agency.[41] On the contrary the word 'authority' has an authoritarian force which means a body having power to enforce obedience created by Statute to carry out governmental or quasi-governmental functions or a body having ability to affect legal relation by virtue of its powers vested by law.[42] Looking to both the terms and also the position and functions of the Bureau, it is suggested that the title should have been the 'Energy Conservation Authority'.

The Bureau, it is appreciated, has been established within the year 2002. The Bureau takes over all the functions, assets, and liabilities of the then existing Energy Management Centre which was set up under the erstwhile Ministry of Energy and registered under the Societies Registration Act, 1860.[43] The management is vested in the Governing Council. It shall consist of the Chairperson, who shall be the Minister-in-charge of the Ministry of Power, Government of India, and twenty five other members. It is suggested that the Council should have a reasonable number of not more than ten Members as provided in the Greece Law.[44] The six members shall be secretaries to the Ministries/Departments related to different energy sectors in the Government of India but the Ministry of Environment/its Department has not been included. A person from such a section could have better contributed in functioning of the Bureau. Seven Members shall be from the administrative heads of related research institutes or concerned companies; not more than four members shall be representing industry, equipment and appliance manufacturers, architects, and consumers, and not more than two members nominated by the Governing Council. The Director General of the Bureau shall be the member-secretary.

Out of these members, thirteen are ex officio members, a bureaucratization of the Bureau or what Sri Natchiappan, an Hon'ble Member of Lok Sabha calls 'bureaucratic asylum'.[45] This should have been avoided and a real professionalism in the constitution and working of the Bureau could have been brought in. It is interesting to note that the Chinese Energy Law in Article 21 imposes a negative obligation on the government saying that no person who has no education and training in conservation of energy shall be engaged. Such a negative mandatory obligation will bring expertise operations in the matter. The Greece Energy Law[46] requires that the member should be a distinguished scientist having scientific training, professional ability, and specialist experience. Further, in case of the above non-exofficio members, no minimum qualification has been prescribed, giving a free hand to the government to appoint any one whom it wants to give favour, giving a scope for discrimination and corruption.[47]

Section 7 gives power to the central government to remove any non-ex officio member on the grounds specified therein. However, there is no specific ground for the removal of such a member if it is adjudged that he had any vested interest in the energy or any energy resource sector. The Greece Energy Law, 1999[48] requires that the members shall make a declaration that they have no interest or stake in any energy sector. To avoid any scope for corruption, the law even enjoins upon their spouse to also furnish such a declaration. It is submitted such a ground be added in Section 7 so as not to allow any member to take benefit of his position. The Rules of 2004 adds a disqualification for the Secretary who had married a person while having a spouse living or having a spouse marry another person. But this is already an offence under the Indian Penal Code and it has no direct relation with the conservation of energy.

The Director-General of Bureau[49] shall be appointed by the central government for a term of three years or until he attains the age of sixty years. As the longevity of life is on increase in India, it is submitted, the age limit may be extended to at least sixty-five years. Section 9(1) lays down, fortunately, qualifications in case of the Director-General who shall have ability and standing with adequate knowledge and experience in dealing with the energy related matters. The term 'adequate' is very vague and even a mediocre may get in as the Chief Executive of the Bureau. Fortunately, now in the Bureau of Energy Efficiency Appointment and Terms and Condition of the Service of Director General (Amendment) Rules 2003 have come out with detailed qualifications for the appointment of the Director General. It includes a graduate in Engineering or different disciplines in Sciences, Energy Studies, Energy Management, or Energy Economics with twenty-five year experience and adequate knowledge and experience in energy sectors. The Rules further lay down desirable qualification and experience which includes post-graduate in Engineering or doctorate gegree in related field of energy. These qualifications will in turn allow a deserving candidate to hold the highest position in the Bureau provided no manoeuvring is done. The above

Rules deserve appreciation as none of the countries' Energy Laws have provided such detailed qualifications.[50]

Coming to the powers and functions of the Bureau, Section 13 enumerates twenty functions and powers. However, they have been made unfortunately optional by the words 'may perform' and not mandatory. They may be broadly divided into two, first, are of recommendatory nature in which comes for example, standard norms for processes and energy consumption display of label on equipment and appliances; and identify the designated consumers.[51] The remaining are action based directives which include, for example, education, awareness, and training in efficient use of energy and its conservation; finance matters related to energy and energy efficient processes and personals. Clause (u) of Section 13(2) provides for any residuary power or function as may be prescribed. The term 'prescribed' is not qualified and, therefore, the question is—prescribed by whom and in which way—by Order, Rules, or Regulations framed under the Act which needs to be clearly specified in the above clause.

The mission of the Bureau is to reduce energy intensity with the participation of all the stakeholders. For the functional distribution, the Bureau has divided its functions into two, the main functions which include, for example, laying down the energy performance standards, developing label designs, specifying building code, and the promotional functions which include for example, creating awareness, training the personnel, supervising pilot project, and encouraging use of energy friendly appliances and equipment.[52]

Coming to the resume of the contributions of the Bureau, taking cognizance at least of its last Report,[53] it has operationalized many schemes to enhance sustainable development, notified fuel consumption standards, ensuring mandatory labelling, particularly for passenger cards, continued enhanced energy efficiency in the energy intensive industry sectors, establishment of two more funds so that the government may deepen the financial market for energy efficiency; and last but not least, develop new energy efficiency models.[54] In spite of all its efforts, the Bureau has yet to move further and the government data do not show any positive sign.

Furthermore, in order to keep the Bureau accountable to its duties and functions, Section 47, an important provision, authorizes the central government to supersede the existing bureau if it is unable to discharge its functions; made defaults in complying with the directions issued by the central government; or, such action is necessary in the public interest. The government may reconstitute the Bureau. Such a provision will be a deterrent to the acts of Bureau, such as neglecting responsibilities and committing fraud or corruption for personal gains.

Powers and Functions of Governments

Section 14 divides the powers into twenty-two subjects which broadly include, energy-friendly equipment and appliances, production and consumption norms and standards, energy audit, inspection and information, energy conservation, building code, labels on equipment and appliances, and so on, which provides the areas over which the central government shall have power. Such powers have to be performed by notification so that the authorities concerned and persons know about it. The central government, in performing its functions, is required to consult with the Bureau and it is only consultation and not concurrence. There are some matters which include matter of local interest where the central government shall also consult the state government concerned. It is submitted that the details in Section 14 could have been left to the rule making power under Section 56(2) because there are some powers mentioned under Section 14 which are not reflected under Section 56(2) which means that the central government shall not have rule making power thereon.[55]

Section 55 further confers powers on the central government to issue directions to both the state government and the Bureau for the execution of the provisions of the Act. This section is silent about the case of non-compliance of the Central directives, and thus it adopts a soft approach in such a matter which needs, it is submitted, a special attention of Parliament. Furthermore, the energy sector, for its conservation, cannot be only confined to consumption, the areas like transportation, conversion, storage, and energy wastes also have to be taken care of which do not find any place in the powers of the central government. Further Section 14 does not include the power to give incentives or awards at the central level to recognize the energy friendly services of designated consumers, energy intensive industries, and other related persons and agencies and also to encourage energy conservation activities.

Coming to the government's actions, particularly that of the central government, a resume will show that the government is making efforts to promote energy efficiency and its conservation. Time to time new schemes, models, and programmes have been introduced, for example, National Programme for LED based home and street lighting, National Mission for Enhanced Energy Efficiency, Energy Standards and Labelling Programme, and Energy Conservation Building Code. However, not much concern is shown to the village and poverty sectors. The central government has also framed detailed Rules to implement the provisions of the Act from time to time. Apart from this, the Planning Commission[56] has been suggesting in its plans large number of sectors to be brought under the Act, allowing private players to play their role and the establishment of funds to finance the functioning of the authorities. Further, it has also suggested a number of programmes to promote energy efficiency and its conservation. Recently, new schemes have also been introduced for example, the Perform, Achieve and Trade Scheme, Super-efficient Equipment Programme, *Bachat Lamp Yojana*, and so on, and in view of their success story, a total out lay of Rs 775 crore has been approved.[57] However, the above actions, schemes and programmes may have moved the process of energy efficiency and its conservation a little ahead but the statistics given above shows that the government has yet to go miles to achieve its mission. It is interesting to note that in the Lok Sabha questions were raised time and again on the issue of contributions, made by the central government in the energy efficiency and conservation sector.[58] It is a matter of further research to find out which of the powers and functions of the central government saw the light of the day and how far the government has succeeded in achieving its mission and which one still remains in hibernation, how can they be fruitfully put into operation, and what does the present time need are some of the questions for further study to provide answers.

The state government likewise is also given powers to facilitate and enforce efficient use of energy and its conservation. Section 15 requires that these powers may be exercised in consultation with the Bureau. The powers may be exercised on nine areas prescribed in Section 15(a) to (j) which broadly covers matters for example, the building code, energy audit, inspection energy awareness, and dissemination information and Section 57 confers on the state government the rule-making powers. The Act by giving such powers to the state government ensures balanced energy federalism.[59]

Appellate Tribunal

Initially in the Draft Bill of 1988 prepared by the ILI and also the Bill of 2000, no such authority was envisaged. The Act of 2001, in Chapter IX, now provides for the Appellate Tribunal for Energy Conservation. Section 30 authorizes the central government to establish the Tribunal by a notification therefor.

It shall hear appeal against the order of the adjudicating officer, central government, state government, or any other authority under the Act. A further example of bad legislative drafting can be seen in Section 27 which says the State Commission shall appoint any of its members to be an adjudicating officer. The 2001 Act nowhere talks about the State Commission; a vacuum is created. Further, the proviso to this section provides that in case no such commission is constituted in any state, then a secretary dealing with legal affairs of an equivalent rank shall be appointed as adjudicatory officer. The question arises—can a member of a Regulatory Body act as a court and administer justice and impose penalty? This dilutes the scope of judicial function and brings bureaucratization in this regard which should have been avoided.

The right to appeal is given to any aggrieved person provided that he shall first deposit the amount of penalty so imposed. However, the Tribunal may relax such requirement subject to such condition as it may prescribe to safeguard the realization of penalty. The provisions of Section 44(1) deserve appreciation because it does not rigidly follow the locus standi doctrine. It allows even a legal practitioner or any accredited energy auditor[60] to present his case before the Tribunal. Further in Section 44(2), the concerned government is authorized in case of need, to appoint one or more legal practitioners or officers as a presenting officer to help the appellant. These provisions, it may be appreciated, ensure the fundamental right to legal aid.[61]

The appeal provision is time framed of forty five days from the date of order passed against the aggrieved person subject to relaxation by the Tribunal. Further, time schedule is prescribed in Section 31(5) for disposition of the matter by the Tribunal which is within six months from the date of receipt of appeal. In case the Tribunal is not able to follow the maximum period of six month, it shall record in writing the reasons therefor. This provision is in the line with the right to speedy justice.[62] But the question remains—will the Tribunal, like other Tribunals, ensure speedy justice? The aggrieved person is given a right to appeal against any decision or order of the Tribunal to the Supreme Court of India within sixty days from the date of communication of the decision or order of the Tribunal, however, this time limit may be relaxed by the Supreme Court.

Sections 32 to 43 of the 2001 Act provide provisions relating to Tribunal for example, composition, qualifications of members, term and service condition, resignation or removal of members, procedure to be followed, and last but not the least, the mode of decision. However, these sections have been repealed by Section 11 of the Energy Conservation (Amendment) Act, 2010 and in their place the provisions of Sections 120 to 123 of the Electricity Act, 2003 shall apply mutatis mutandis. It may be mentioned that conservation of energy is a general expression in which electricity is one of the resources of energy. And, therefore, to regulate the provisions of a general legislation through a specific legislation is not only inappropriate but also not desirable. Further, in Section 30 the Appellate Tribunal established under the Electricity Act, 2003 shall be the Appellate Tribunal for the purpose of the 2001 Act. This means that the Tribunal of the Electricity Act shall hear appeals as if it is a part of this Act. Furthermore, Section 123 of the 2010 Amendment Act replaces Section 43 of the Act of 2001, though the language is literally same. Thus, all these have hardly changed the existing position. This has brought in an unusual situation.

The 2010 Amendment Act, now provides for—first, the Tribunal shall consist of a Chairperson and three other Members[63] in place of formerly four members. Over the years, when energy law will become a full-fledged independent branch of the legal discipline, the question will arise, will such a small number of members cope with the litigations? The provision to Section 112 (b) provides that each Bench shall include at least one judicial Member and one technical Member will such a Division Bench be able to handle the future increasing litigations? However, the provisions deserve appreciation, as the Tribunal will provide expertise justice which is not possible in case of courts exclusively manned by judges.

Second, there are changes in the qualifications for the appointment of the Chairperson and other Members.[64] In case of Chairperson the same qualification as prescribed in the 2001 Act continues. However in case of other members, only three qualifications remain out of the previous six qualifications—First, he is or has been or is qualified to be, a judge of a high court. A legal practitioner can also be a member of the Tribunal. A legal practitioner is a more socialized person than a judge. Moreover, he may have handled cases of designated consumers or energy intensive industries. And, therefore, naturally he will be subject to outside pressure. Further, he has hardly any experience in administering energy justice. It is better to delete such qualifications. Second, he is or has been a Secretary for at least one year in the Ministry/Department of the central government dealing with the energy resources. Induction of such a member will bring bureaucratization of justice. And the last qualification is for the technical member which provides that he is or has been a person of ability and standing having adequate knowledge or experience in dealing with the matters relating to energy production and supply, energy management, standardization, and efficient use of energy and its conservation, and dealing with problems relating to engineering, finances, commerce, economics, law, or management.

It may be mentioned that the 2001 Act had provided in Section 33, three more qualifications. First, a member of Indian Legal Service of Grade I rank with at least three years' experience; second, the Chairperson of the Central Electricity Authority for at least one year; and the third, the Director-General of Bureau, Central Power Research Institute or Bureau of Indian standards for at least three years. It is good that a large bureaucratic indulgence in the Appellate Tribunal is kept at rest.

In order to avoid any corruption, Section 113 (4) provides that before appointing the Chairperson or any Member, the central government shall satisfy itself that such a person does not have any financial or other interest which is likely to affect his functions prejudicially. The Malaysian Energy Law[65] has similar provisions but it requires a declaration by the members in that regard. However, in the Greece Energy Law even the spouses have to make such declaration. The Indian law, in view of the prevalent corruption and scams, must also incorporate such provision. This will control them from gaining any undue benefit of their position. However the question still remains—will this really control the deeply infected disease of corruption prevalent in all sectors of energy?

The Act provides that the decision of the Tribunal shall be according to the opinion of the majority of the members. The problem comes when the bench consists of two members and the members differ with each other. Such process should normally be avoided as it causes undue delay in the decision. In such situation, the Act provides that the Chairperson shall either hear the points himself or refer the matter to one or more members of the Tribunal, and the matter, shall be, then, decided by the majority. However, the question remains whether the Tribunal will be in a position to deliver court like judgment or simply a governmental order. Taking a clue from the judgment of the Appellate Tribunal for Electricity, it can be said that it normally cannot be like a judgment of a court.[66]

Section 121 has brought in an absurd position which deals repeatedly with the 'Appropriate Commission' but no such Commission exists under the Act of 2001. It is suggested that the Act of 2001 should have been allowed to operate independently rather than walking on the crutches of the other misfit legislation. The Standing Committee on Energy had strongly objected to such an approach.[67] Lastly, the Act of 2001 takes care against the multiple window clearance of any suit. Section 29 bars the jurisdiction of the civil court in respect of any matter which the Tribunal is empowered of and the court shall not issue injunction in such matter. Thus the Tribunal has been made only one forum to hear appeal. However, a party can approach the Supreme Court in appeal.

Penal Sanction

Chapter VIII of the Act provides for penalty. Section 26 prescribes the penalty which shall not extend to more than ten thousand rupees, and in case of continuing failures thereof, the additional penalty may extend to one thousand rupees for every day during which such failure continues. The energy and energy resources are the very life line of development and basic necessity for the enjoyment of life. A non-compliance of any action or order even of the select respective clauses of Sections 14 and 15 will create inroads in the efficient use of energy and its conservation, and ultimately affect the environment and abate climate change, endangering the life of human beings. In view of such far reaching consequences, the penalty is too soft which will result in pay and continue not to abide by orders and further exploit illegally energy and energy resources.

It is interesting to note that the environmental legislations over the years have increased the penalties from three months to seven years imprisonment and/or fine of which may extend to twenty-five crore.[68] Even the Draft Bill prepared by the ILI in Sections 29, 30, and 31 enhances the penalty to Rs 1,000,000 and in case the contravention continues, then Section 32 enhances the penalty by twice of the aforesaid amount which was reduced by the Energy Conservation Bill, 2000 to one lakh rupees. Even in the Lok Sabha, the Members had argued for stricter penalty.[69]

Those who are dealing with energy have mostly their business and industrial establishments for whom rupees ten thousand is a very meagre amount to pay, and its imposition cannot be a lesson for the similarly situated criminals. The Malaysian Energy Law[70] provides stricter punishment; of five years imprisonment and fine not exceeding of Ringgit 5,000,000. Even the Japanese Energy Law, in Article 27, imposes punishment of imprisonment not exceeding one year with hard labour and fine not exceeding one million yen. This is one of the objectives of imposing criminal sanctions. The liberal treatment of the Act of 2001 will bring more problems in conservation of energy.[71]

The appreciative part of the Amendment Act, 2010 is that now nearly after a decade, it was realised that providing a soft law in the name of so called enabling statute was wrong. Now the penalty is increased to ten lakh rupees with an additional penalty extending to ten thousand rupees for every day during which such contravention continues. However, the Amendment Act of 2010 has surprisingly not provided for the punishment by way of imprisonment.

Section 53 of the Act provides an exemption from criminal sanction in favour of the notified designated consumers by the central or the state government, as the case may be for a maximum period of five years. The Original Bill of 2000 did not provide for this but the Members of the Standing Committee on Energy suggested that in view of less awareness in the people of India about the efficient use of energy and its conservation and further to provide the infrastructure required under the Act will take its own time, and therefore, no liability should attract to their action.[72] But the basic principle is that, ignorance of law is no excuse. Further is it possible that, within a period of five years will all the infrastructure be provided and the concerned people be adequately educated and trained in this branch of science and technology?

The Act takes care of any contravention by a company. The title to Section 48 uses the expression 'default by companies', whereas, the environmental legislations uses the expression 'offences by companies'.[73] The use of the term 'default' indicates the intention of the Legislature to adopt a soft approach. In the present Act any contravention results in punishment, and therefore, the second expression is, more suitable. Section 48 brings the vicarious liability of those, first, who were in charge of and responsible to the company for the conduct of its business; and second, who committed contravention with consent or connivance shall also be responsible and such categories of persons include the director, manager, secretary, or other officer of the company. However, a liberal approach is adopted wherein a company can prove that the contravention was committed without knowledge or due diligence was exercised to prevent the contravention[74] then in such a case it will not be subject to any criminal sanction. On the contrary, if an individual or group of individuals contravene the provision, no exception is provided, a case of discriminatory approach. Further, the question is how far such a position stand in view of the *M.C. Mehta* case[75] wherein the doctrine of absolute liability was evolved. Moreover, the above exceptions will prolong the litigations and give a chance to the culprit to escape through such unwanted window and thereby continue to create problem for the conservation of energy.

There is another point which has been left out and needs attention. The environmental legislations have not spared even the government departments and its head from the penal sanctions under the concerned laws.[76] Even the Draft Bill prepared by ILI included in Section 35 provisions for the defaulting government department. But the present Act does not provide anything in this regard. There may be inaction or corruption[77] in handling the matter relating to the efficient use of energy and its conservation by the government. In such cases they will go scot free. The Chinese Energy Law in Section 49 provides that the criminal sanction shall be attracted even against government actions. Such a vacuum, it is submitted, needs to be filled in. Section 61, on the contrary, surprisingly authorizes the central government by notification to exclude the operation of the Act in case of the Ministry or Department of the central government dealing with defence, atomic energy, or such other Ministries or Departments as the central government may consider necessary. Defence or no defence atomic energy or other energies, the law must speak the same language, if not, such Ministries and their Departments can play a casual, negative, or inactive role in conservation of energy.

It is interesting to note that the American Energy Law[78] provides tax concession and the German Energy Law provides financial assistance for the energy caring actions. In India also tax exemptions are given in income tax, customs duty, and excise duty.[79] Further, the Government of India presents National Energy Conservation Awards every year and the success of the Awards is supported by the fact that in 1999 there were only 123 participants and by 2014 the number has increased to 1010.[80] The above incentives will encourage others to adopt energy friendly approach and also provide attractive marketability for their products. The other energy friendly exercises include certification pro-gramme, strict standards requirement, labelling of the products, energy audit which has been made optional, building code, and identification of more energy intensive industries. These efforts however, cannot reach to the ground level till the time the problems of energy illiteracy, inaccessibility of friendly technology, lack of trained manpower, and inappropriate budgeting are solved. In all these incentives and recogni-tion political manoeuvring should not enter.

$$\star\star\star$$

India has been placed at the fourth position amongst the greatest energy consuming countries. The main problem revolves around the fact that a large number of people do not know about energy in the way they should know, and also the energy conservation law. The government and the educators, the law academics, and in particular, the law schools, in their constitutional responsibility, must see that all those dealing or handling and consuming energy must be educated about the energy and energy law, so that a green energy discipline and conservation cul-ture are internalized in their behaviour and conduct.

The Constitution of India is totally blind on the conservation of energy. It is time to adjust with the changing techno-science needs. In the energy conservation sector, a balanced energy federalism must find a place in the centre-state legislative relations with no exclusive power given to the local authority as was done in 1999. This will allow a coop-erative approach in handling this subject. The right to energy security and the right to commercialization of energy and many more rights in this regard may emerge as fundamental rights in future. And, therefore, the Legislature and the Judiciary must evolve a harmonious balance between the maintenance of sustainable energy and the enjoyment of fundamental rights. The rights will also need support of a specific fundamental duty and fundamental obligation of the state, a *triveni sangam*, which will not only help in conserving energy for the present and future generations but also enable the Constitution of India to stand tall among the constitutions of the world in this regard.

India deserves appreciation for enacting a timely legal control on inefficient use of energy which has yet to be provided by many countries of the world. It may be appreciated that this law has seen many exercises before it was enacted but the bad legislative drafting and at places energy illiteracy need reform in the legislation. It is time that the Ministry concerned evolved a pre-enactment mechanism consisting of expert legislative draftsmen and experts in the concerned field before finalizing a Bill.

It has defined large number of terms and expressions used in the Act. However, definition of 'energy' is incomplete, and it is more electricity based, and, therefore, a comprehensive definition is the need of the time. Further, the definition clause must define the most important left out expressions, 'conservation of energy' and 'sustainable use of energy'. The Act in fact, targets the conservation of non-renewable energy but no reference has been given in the Act. The definition clause must give such an expression a place in it. Energy is of national and international importance, and therefore, any legislation on this subject must apply to entire India including the State of Jammu and Kashmir.

The regulatory body, the Bureau for Energy Efficiency, must effectively coordinate with the activities of different role players for the efficient use of energy. In order to look after the management of the Bureau, the General Council is envisaged but it is basically a bureaucrats' Council. A select body of qualified persons with eminence in the field of energy and law along with the Minister of Power Sector in central government to be the Chairperson and a joint secretary of non-conventional energy resource of the central government as its member-secretary could have been its appropriate constitution. Further, all its members along with the members of the Appellate Tribunal should have been required to submit a written undertaking to the effect that they have no interest or stake in any process or sector of energy.

The powers and functions of the central and the state government and also their rule making powers take care of energy federalism with central dominance. The central government, the state government, and the Bureau have been given a long list of powers and functions out of which some have unfortunately yet to see the light of the day. However, those which have become functional have only moved a little further. On the contrary, the government data shows still a long way to go in achieving the objectives of the legislation. It is time a monitoring authority is constituted to make regular audit of their output. The provisions dealing with the Appellate Tribunal deserve special appreciation for, firstly, to bring mechanisms for expertise energy justice; secondly, to provide speedy justice within the prescribed time schedule; thirdly, to introduce public interest litigation scheme; and lastly, to open avenues for legal aid. The Appellate Tribunal needs more judicious care to be handled by persons having judicial experience and long expertise in handling the energy issues instead of the bureaucratic input. The greatest flaw in the Amendment Act, 2010 is that it has brought in the dominance of the Electricity Act, 2003 over the self-contained code of 2001 which will now walk on the crutches of the 2003 Act. No legislations have followed such an approach and, therefore, a status quo position should be retained. In the penalty sanction a soft approach of 2001 has fortunately been made little harder in 2010. However, the penal sanction needs further reforms on three counts—Firstly, the penalty of imprisonment should have been provided for; secondly, the absolute liability against offending companies cannot be avoided; and thirdly, in case of the defaulting government department, the penal sanction must also apply to it.

So what is finally needed? There should be sincere and honest efforts to conserve energy for the present and future generations. If this is not done, then whatever laws, policies, or programmes are made, they will hardly attain the desired goal.

Notes and References

1. All the information is based on M.V. Pylee, *Constitutions of the World*, Vols 1 and 2, Third Edition (Delhi: Universal Publishing Co., 2006).

2. See, for example, the Constitutions of Australia, Canada, Finland, France, Italy, Japan, New Zealand, Norway, South Africa, US, and so on.

3. See, for example, Afghanistan, Algeria, Bahrain, Bangladesh, Brazil, Cambodia, Germany, Greece, Ireland, and Switzerland.

4. See, for example, Constitutions of Algeria, Brazil, Cambodia, and Greece.

5. Art 11a.

6. Art 22, XXVI.

7. Art 89(3).

8. Art 89(2).

9. Art 90.

10. Art 91(1) and (2).

11. Art 89 (5).

12. See, for example, Afghanistan—Art 9 (1); Bahrain—Art 11; Bangladesh—Art 143; Cambodia—Art 58; Ireland—Art 10.

13. See, for example, Afghanistan—Art 9 (2); Brazil—Art 225 (6); Greece—Art 18(1).

14. See also, the Constitutions of Afghanistan—Art 15; Bahrain—Art 11; Cambodia—Art 59.

15. Art 89(1).

16. The Constitution (Seventy third Amendment) Act, 1992, Item 15.

17. List I, Entry 6, Constitution of India.

18. List I, Entry 53, Constitution of India.

19. List I, Entry 54, Constitution of India.

20. List II, Entry 23, Constitution of India.

21. List II, Entry 25, Constitution of India.

22. List III, Entry 17A—taken away from the exclusive subject of the State Legislature by the Constitution (Forty-second Amendment) Act, 1976.

23. Hereafter referred to as the Draft Bill by ILI.

24. See, for example, *Dhillon v. Union of India*, AIR 1972 SC 1061 and *International Tourist Corporation v. State of Haryana*, 1981 SC 774.

25. See, H.M. Osofsky and H.J. Wiseman, 'Dynamic Energy Federalism', *Maryland Law Review*, 72 (2013), pp. 773–843.

26. *Bishamber v. State of U.P.*, AIR 1982 SC 33.

27. See for example, *Sushila Saw Mills v. State of Orissa*, AIR 1995 Ors. 256; *Abhilash Textiles v. Rajkot Muncipal Corpn.*, AIR 1988 Gujarat. 57; *D.S. Rana v. Ahmedabad Municipal Corporation*, AIR 2000 Gujarat 45.

28. See for example, *M.C. Mehta v. Union of India*, AIR 1987 SC 965; *A.P. Poll. Control Board v. M.V. Nayadu*, AIR 1999 SC 812.

29. V.P. Singh Bannore, Member of Lok Sabha pointed out such legislation was 'long overdue' *Lok Sabha Debates*, 16 August 2001, 7/17; Basu Deb Acharia, *Lok Sabha Debates*, 17 August 2001, 1/29.

30. All the references are based on R. Blanpain (ed.), *International Encyclopaedia of Laws Energy Law,* Vols 1–4 (Netherland: Kluwer Law International, 2010).

31. A.S. Oppe, *Wharton's Law Lexicon,* 14th Edition (Delhi: Universal Law Publishing Co., 2006), p. 368; see also, D. Greenberg, *Stroud's Judicial Dictionary of Words and Phrases,* Seventh edition (London: Sweet and Maxwell, 2006), p. 835.

32. See for example, the National Environmental Tribunal Act 1995 and the amended provisions in Article 22, of the Constitution of India.

33. Section 60, The Energy Conservation Act, 2001.

34. V.P. Singh Badnore and B.D. Acharia, *Lok Sabha Debates,* 16 August 2001, 7/17 and 1/29 respectively.

35. *Encyclopaedia of Energy Law,* Israel, Vol. 1, 391.

36. Such an approach was criticized by P. Rashtrapal, a Member of the Lok Sabha and suggested one go approach for implementation of all the provisions, *Lok Sabha Debates,* 16 August 2001, 10/7.

37. The *Ninth Report of the Standing Committee on Energy* (2000): 110–12.

38. The Draft Bill by Indian Law Institute (ILI), in Section 2(I) and (g) provided detailed definitions of these expressions which hardly find any special mention in the present Act.

39. A.J. Bradbrook, et al. (eds), *The Law of Energy for Sustainable Development* (Cambridge: Cambridge University Press, 2005), p. 323. See also, *the Fifth Report of the House of Lords Environment Audit Committee,* A Sustainable Energy Strategy (2001–2), p. 582 which puts a special emphasis on this subject.

40. See Section 7.

41. *Webster's New Explorer Encyclopaedia Dictionary,* p. 238; *The Oxford English Dictionary,* Second Edition, Vol. II (Oxford: Clarendon Press, 1991), p. 665; Y.V. Chandrachud (ed.), *Aiyar's Advanced Law Lexicon,* Book 1 (Nagpur: Wadhwa & Co., 2005), p. 627.

42. The *Oxford English Dictionary* Second Edition, Vol. 1 (Oxford: Clarendon Press, 1991), p. 798. See also *Som Prakash v. Union of India* (1981) 1 SCC 449; *Raj. State Elec. Board v. Mohan Lal,* AIR 1967 SC 1887.

43. The Resolution No. 7 (2) 87-EP (Vol. IV), 5 July 1989. See also Section 12 of the Act.

44. The Greece Energy Law provides for only seven members and its output is reported to be satisfactory, *Encyclopaedia of Energy Law,* Greece, Vol. 1, pp. 28–30. Even the Draft Bill by ILI provides for only three bureaucrats—Section 7(3).

45. *Lok Sabha Debates,* 16 August 2001, 1/17.

46. *Encyclopaedia of Energy Law,* Greece, Vol. 1, pp. 28–9.

47. See for such an apprehension, the *Ninth Report of the Standing Committee on Energy,* 4 (2000). See also, A.R. Kidwai, *Rajya Sabha Debates,* 28 August 2001, SS/2.40/2P-1, 2/10. D.P. Yadav, *Lok Sabha Debates,* 10/17.

48. *Encyclopaedia of Energy Law, Greece*, Vol. 1, pp. 28–30.

49. The Government of India vide their Notification No. GSR 761/E dated 24 September 2003 provided detailed rules for the appointment and related matter with respect to the Director General.

50. See the *Encyclopaedia of Energy Law*, Vols 1,2, 3, and 4 (2010).

51. As defined under Section 2(g).

52. The Report of the Director General, *The Annual Report* 2014–15, The Bureau of Energy Efficiency, pp. 8–9.

53. The *Annual Report*, 2014–15, The Bureau of Energy Efficiency, pp. 6–7.

54. The *Annual Report*, 2014–15, The Bureau of Energy Efficiency, pp. 8–9.

55. A similar position exists in the powers of the state government and its rule making powers: Sections 15 and 57 respectively.

56. See, the *Five Year Plans,* particularly the Sixth Plan (1982–7), the Seventh Plan (1987–92), the Ninth Plan (1997–2002), and the Eleventh Plan (2007–12).

57. The *Annual Report*, 2014–15, The Bureau of Energy Efficiency, p. 45.

58. *Lok Sabha Unstarred questions*, Nos 2873, 1366, 1405, 2598, and 5442 answered on 24 September 2009; 5 March 2010; 12 March 2010; and 27 April 2015, respectively.

59. Wiseman and Ostofsky, 'Dynamic Energy Federalism'.

60. Section 13(2) (p).

61. *Hussainara v. Home Secy.* AIR 1979 SC 1360; *M.H. Hoskot v. State of Mah.*, AIR 1978 SC 1548.

62. *Hussainara v. Home Secy.*, AIR 1979 SC 1360; *Raghubir Singh v. State of Bihar* (1986) 4 SCC 481; *Common Cause v. Union of India* (1996) 6 SCC 775.

63. Sec 112 of the Electricity Act, 2003.

64. Section 113 of the Electricity Act, 2003.

65. *The Encyclopaedia of Energy Law*, Malaysia Vol. 2, 35 and Greece, Vol. 1, pp. 38–9.

66. See for example, *NTPC Ltd. v. Centre. Elec. Regul. Comm.*, 2015 Indlaw APTEL 73. See also, *NTPC Ltd. v. Power Purchase Centre, Haryana*, IP USER 1.

67. *The Ninth Report of Standing Committee on Energy*, 2000, p. 12.

68. See, Sections 277 and 278, the Indian Penal Code, 1860 and Section 26 of The National Green Tribunal Act, 2010.

69. Natchiappan and P. Rasthrapal, *Lok Sabha Debates*, 16 August 2001, 6/17 and 12/17, respectively.

70. *Encyclopaedia of Energy Law*, Malaysia, Vol. 2, p. 39.

71. P. Rashtrapal, a Member of Lok Sabha also suggested stricter penal provisions, *Lok Sabha Debates*, 16 August 2001, 12/17.

72. *The Ninth Report*, p. 46.

73. The Environment (Protection) Act, 1986-Sec 16; the Air (Prevention and Control of Pollution) Act, 1981-Sec 40; The Public Liability Insurance Act, 1991 Section 16, and so on.

74. Section 48(1) Proviso.

75. *M.C. Mehta v. Union of India*, AIR 1987 SC 965.

76. The Environment (Protection) Act, 1986 Section 17; the Air (Prevention and Control of Pollution) Act, 1981–Section 41. The Water (Prevention and Control of Pollution) Act, 1974 section 48, and so on.

77. The Chinese Energy Law in Section 49 provides criminal sanction shall be attracted against such actions.

78. *Encyclopaedia of Energy Law*, US, Vol. 4, pp. 13–20.

79. *The Annual Report* 2014–15, The Bureau of Energy Efficiency, pp. 49–50.

80. *The Annual Report* 2014–15, The Bureau of Energy Efficiency, p. 34.

11

SHIVANANDA SHETTY
SURENDER KUMAR

Information Disclosure

A Policy Tool for Managing Environmental and Energy Challenges

One of the key challenges faced by policy makers across the world and especially in a developing economy is ensuring economic growth with minimum impact on the environment. The key drivers for economic growth are the production of goods and services and the expansion of infrastructure which not only produce goods for consumption but also provide employment to the people. However, economic development comes with an environmental cost which has adverse impact on the economic growth itself not only in the long run but in the short term as well. India's economy is impacted by more than five per cent due to environmental degradation.[1]

Despite having extensive environmental regulations in India, the compliance is low due to lack of enforcement.[2] The inevitable increase in population and economic development have serious implications for the environment, because energy generation processes (for example, generation of electricity, heating, cooling, or motive force for transportation vehicles and other uses) are polluting and harmful to the ecosystem.[3] Energy consumption is also a leading factor of carbon emission in a developing country as bulk of the energy is produced by the coal fired power plants. The high growth rate also increases demand for energy for industries, domestic consumption, and infrastructure requirements. Hence, in order to ensure a sustainable growth it is imperative that the regulators ensure that the industries and consumers are sensitive towards the environment, energy, and climate change impact of their processes and lifestyle.

Environmental compliance behaviour depends on various factors which are external and internal to firms. Key external factors which have direct influence on the firms are regulators, customers, NGO, and internal factors are the firms' management competency and vision, organizational capability to implement the processes and allocation of the requisite resources, and so on. Role of regulators is very critical as they have direct as well as indirect impact on the firms. Enforcement of regulations and the threat of penalties have direct influence on the firm's compliance behaviour. Enforcement of the regulations and regular monitoring of the firm are constrained by the availability of resources with the regulators even in developed economies like the US and EU.

Note that financial market considers the firms' environmental risk in the market valuation which also drives the firms to comply with the regulatory requirements. This has led to the regulators seeking alternative policies to help in enforcing compliance behaviour by the firm. These alternative policies are considered complementary to the existing regulations and use market forces to influence the firms to not only meet the regulatory standards but also over comply, wherever possible. Such policies are considered to be cost effective since these do not require extensive resources like the traditional command and control (CAC) policies. One such policy option is the 'Information disclosure' which has not only been used in developed countries like the US but also in developing countries like Indonesia, China, and Philippines where it has been effective in reducing the firm level environmental pollution.[4] The information disclosure involves releasing information on the firms' environmental performance which includes the emissions to air and water and compliance to regulations. The underlying assumption behind the public disclosure is that the firms may be sensitive about their reputation at the market place or the communities where they are located and they try to enhance their reputation by improving their environmental performance. Thus, information disclosure can be used as an effective policy tool to motivate the firm to improve their environmental performance.[5]

We look at the current compliance situation in India and review the status of the information maintenance and disclosure across major industrial sectors in India in the area of environment and energy. This will help in understanding the gaps in the existing processes and identify suitable measures to make it more effective.

The rest of the chapter is structured as follows: section II provides the conceptual framework which explains the factors that influence the firms to improve their environment performance under the condition of disclosure. The second section covers the disclosure programmes across the world for improving the energy and environmental efficiency. The third section details the energy and environment regulations in India. The fourth section gives the status on the availability of energy and environment information in India while the last section identifies the gap in the existing system. We further analyse the implication of the information disclosure policy mechanism wherein we suggest the way forward in implementing the information disclosure policy.

Conceptual Framework

The implementation of abatement technology to manage the pollution during the production process requires resources like capital, manpower, and so on, which will increase fixed as well as variable cost of the production. When the regulations are enforced stringently, a profit maximizing firm will abate to a level that is required as per the regulation and hence entire firm incurs same marginal cost of abatement. However, in the absence of enforcement, firms have no incentive to abate the pollution since the decision to comply with the regulation depends on the probability of being caught and the severity of penalty for non-compliance. Similarly, in the case of pollution causing inputs such as fossil fuels (coal, oil, and gas), a profit maximizing firm, equalize private marginal cost of energy to its private marginal benefits. Note that marginal social cost of fossil fuels use is higher than private marginal cost, and overuse of such resources happens. The optimal use of energy resources requires government intervention for enhancing efficiency in the use of these resources. Thus, the firm's environmental/

energy compliance strategy is based on the expected cost and benefit of compliance that includes the likelihood of detection.[6]

Following the earlier studies, six channels have been identified as the drivers which influence the firms to improve its environmental performance under public disclosure.

- Output market pressure—disclosure of the firm's environmental performance impacts the demand of the goods by the 'conscious' customers who prefer a green firm.[7]
- Input market pressure—financial market preference for green firms due to business continuity risk and increasing demand for green investment. Also, the employees prefer to work with the firms which are green and less polluting hence a firm's green credential impacts its employee attraction and retention strategy.[8]
- Judicial pressure through the non-governmental organizations (NGOs) and general public who undertake legal process against the polluting firms.[9]
- Regulatory pressure wherein the availability of the information through disclosure provide better insight to the regulators and help them frame and enforce efficient policy measures.[10]
- Community pressure includes the local community groups and the NGOs. NGOs regularly publish reports and attract attention of the media and the regulators. Growing awareness and economic well-being have attracted public attention on the impact of polluting firms' activities on the environment.[11]
- Managerial Information is enhanced as it provides them the valuable information on energy use and pollution level and practices about the industry peers which helps them benchmark and set up improvement targets.[12]

The above drivers impact the cost and benefit to the firm under the information disclosure programme depending on their level of emission and energy use when the programmes are announced. Under the information disclosure condition, in case of cleaner firm the demand for the output increases while the cost for the input declines thereby enhancing the profitability for the cleaner firm and thereby shifting the marginal abatement benefit (MAB) function outward. The high polluting firms will follow suit and implement abatement measure to reduce pollution which will increase the marginal abatement cost for the dirty firm. However, the public disclosure provides the dirty firm's information on the practices of the cleaner firm and helps them to improve their own performance and the Marginal abatement cost (MAC) comes down. Also, regardless of the fact that the firms are clean or dirty, the overall MAB curve moves up and under the equilibrium condition the firm chooses higher level of abatement and cleaner inputs. Thus, the lowering of the marginal abatement cost and improvement of the marginal abatement benefit will ultimately lead to a higher level of abatement and use of clean fuels.

Information Disclosure Programmes

The disclosure of energy and environment are used to create a market pressure to improve the compliance behaviour of the firm. If the communities and the consumer groups are concerned about the environment then the provision of environmental information will cause consumer to adjust their purchase decision. Similarly, the community group will pressurise the firm to reduce pollution below the regulatory requirement and the investors will change their portfolio thereby putting pressure on the firm and the firm will go beyond the legally mandated regulatory standard if it is in their interest to do so.[13]

There are two types of national public disclosure programmes,[14] the first one reports the information without rating or characterising environmental performance. This is a registry of the toxic pollutants which are not normally covered under the conventional regulation. The US toxic release inventory (TRI) is an example of these programmes. Several papers have studied the impact of the TRI on the overall release of the toxic pollutants since and they have reported a fall by at least 45 per cent since the inception of the programme in 1986.[15]

The second type of programme provides the information on the emission data and also rates the individual firm's environmental performance. These programmes have been implemented in countries like China, Indonesia, and Philippines and have been effective in improving the environmental compliance of the firm.

The information disclosure of green rating project on the paper and pulp industry in India has resulted in a significant reduction in pollution level amongst the highly polluting firms. The study also reported that the firms located in in wealthier communities were more responsive to the information disclosure due to community pressure. The information disclosure has also resulted in negative abnormal returns up to 30 per cent for the stock prices of large auto, chlor-alkali firms.[16]

Information disclosure policy has been used in a limited manner in energy efficiency programmes in India.[17] Note that the information disclosure has been used effectively in energy efficiency labelling of appliances wherein the consumer get information on the energy efficiency status of the appliances and makes purchasing decision based on the information. The consumer has incentive of saving the cost of energy while the appliance manufacturers are benefitted through premium pricing and buyers preference for their products. The labelling scheme is designed to inform the customer of the relative energy-efficiency of appliances as compared to their peers and associated potential cost savings through the provision of observable, uniform, and credible standards. These initiative has been used extensively in developed and developing countries like US, EU, Japan, and Australia. The programmes are more focused on high energy consuming consumer goods like air conditioners, heaters, refrigerators, and so on.[18]

Considering the global success in the eco-labelling schemes, the Indian government had launched a voluntary labelling scheme known as Eco-Mark in 1991. The objective was to increase consumers' awareness about environment friendly products in the market.[19] However, due to the lack of consumer awareness and high cost of participation for the manufacturers resulted in failure of the scheme. The labelling scheme was re-launched with a clear focus on energy efficiency of electrical appliances which was known as the Standards and Labelling Program (S&L) and was monitored under the newly created Bureau of Energy Efficiency.[20] The S&L covers agricultural pump sets, distribution transformers, motors, lighting products, refrigerators, air conditioners, and televisions. The labelling requirement is still voluntary, except for frost-free refrigerators (beginning in 2007).[21] Unlike the previous scheme, S&L uses a star rating system to rate appliances. There is so far very limited analysis of the effectiveness and diffusion of energy efficiency technologies through the S&L programme in India.[22]

Similar initiatives have also been undertaken for buildings in several countries (Brazil, India, China, Russia) wherein the large commercial building owners are encouraged to undertake rating of their buildings through the regulatory agency. In India such programme has been managed by the Bureau of Energy Efficiency (BEE) wherein the programme is voluntary and the firms had to pay a registration fees and provide information on the energy consumption to BEE. BEE then validates the same and provides a rating to the buildings. The BEE star labelling is applicable to buildings with the connected load of 100 kW or greater or contract demand of 120 kVA or greater.[23] One to five stars are awarded to the buildings based on their specific energy use with five star label recognized as the most efficient building. A standardized format of data collection of actual energy consumption of the building was developed to collect information pertaining to building built-up area, conditioned and non-conditioned areas, type of building, hours of operation of building in a day, climatic zone, and other information related to facility.[24]

The objective was to provide the occupier or the customer the information on the energy efficiency status of the building as it will help them save the energy cost and the building owners on the other side get a premium rent for their commercial property. At present this rating is applicable to office buildings, business process outsourcing (BPO) buildings, and shopping malls.[25] In future, the BEE may extend the star labelling to hotels and hospitals. These programmes are voluntary in nature and have been criticized for not being monitored post rating as the result of which the buildings do not deliver the potential value. The participation in this initiative has been low due to the high registration fees for getting the star rating of the building and also low demand from the occupiers due to lack of awareness.

Not many disclosure initiatives have been undertaken in the industrial energy sector focused on the energy efficiency of the industries, however recent focus on climate change has brought energy on the centre stage as it contributes significantly to carbon emission. BEE had launched a market based mechanism for energy efficiency popularly known as perform achieve and trade (PAT) wherein the high energy intensive industries have to share the energy consumption per unit of the product with BEE and each of these industries are given an efficiency improvement target which they have to achieve through efficiency programmes or buying the same from markets.[26] The PAT scheme has been provided a legal mandate through amendment of the Energy Conservation Act in 2010.[27] The information provided by the industries on the energy efficiency and the measures undertaken are privy to BEE and is not available with the general consumer. However, as the data is published, the industry can benchmark themselves with their peers and embark on energy efficiency initiative.

One of the disclosure programmes which is evolving in India is the Carbon disclosure project which is a global voluntary initiative that provides a platform for the companies across the world to report their carbon emission. The energy efficiency has been used as a key strategy as a part of the climate change strategy by large Indian companies.[28] Bombay stock exchange (BSE) has signed a MOU with CDP (Carbon Disclosure Project) India for creating awareness about filing sustainability data, environment and social governance data with the CDP. Although voluntary, the CDP has huge leverage as it represents more than 534 investors across the world with over USD 64 trillion worth of assets under its management.[29]

Status of Environment Regulation and Energy Conservation Laws in India

Environmental Regulation

India is one of the few countries in the world wherein the environmental protection, right, and duties are included in the Constitution.[30] In addition to this India has extensive regulatory standards covering policies, standards, and procedure. These provisions include the Water (Prevention and Control of Pollution) Act of 1974[31] which provides regulatory authority to State Pollution Control Boards to establish and enforce effluent standards for facilities discharging pollutants into water bodies. The Water (Prevention and Control of Pollution) Cess Act of 1977[32] is an Act to provide for the levy and collection of a cess on water consumed by persons carrying on certain industries and by local authorities, with a view to augment the resources of the Central Board and the State Boards for the prevention and control of water pollution constituted under the Water (Prevention and Control of Pollution) Act, 1974. Similarly, the Air (Prevention and Control of Pollution) Act, 1981[33] provides for the prevention, control, and abatement of air pollution. With a framework similar to the Water Act, the Air Act gave the central and state board's authority to issue consents to industries operating within designated air pollution control areas. States also prescribe emission standards for stationary and mobile sources.

The Environment (Protection) Act, 1986[34] (EPA) which is an umbrella legislation defines a policy for environmental protection covering air, water, and land and provides a framework for central government coordination of central and state authorities established under previous laws, including the Water Act and Air Act. Under this law, the central government must set national ambient and emissions standards, establish process and practices for managing hazardous substances, regulate industrial siting, investigate and research pollution issues, and establish laboratories, and collect and disseminate information. The Public Liability Insurance Act, 1991[35] makes compulsory for the firm owners operating with hazardous substances to take out insurance policies covering potential liability from an accident and establish Environmental Relief Funds to deal with accidents involving hazardous substances. The Right to Information Act, 2005[36] has been designed to promote greater transparency and accountability of the government and public participation in decision-making.

The medium-specific legislation (the Air Act and the Water Act) empower the Central and State pollution control boards to enforce emission and effluent standards for industries discharging pollutants into air and water. The Water Cess Act, among other things, stipulates the use of fees for water abstraction. The industries are classified into two categories, namely, red, orange, and green depending on their intensity of pollution.[37] The red category industries include[38] Distillery including Fermentation industry; Sugar (excluding Khandsari); Fertilizer; Pulp and Paper (Paper manufacturing with or without pulping); Chlor-alkali; Pharmaceuticals (Basic) (excluding formulation); Dyes and Dye-intermediates; Pesticides (Technical) (excluding formulation); Oil refinery (Mineral oil or Petro refineries); Tanneries; Petrochemicals (Manufacture of and not merely use of as raw material); Cement; Thermal power plants; Iron and Steel (Involving processing from ore/scrap/Integrated steel plants); Zinc smelter; Copper smelter; Aluminium smelter.

Each of these polluting facilities has to get permits of consent to establish (CTE) and consent to operate (CTO). The CTE is required to be obtained before the start of the initial operation of the facility and CTO has to be obtained after submission of plant level environmental information. Once the industry or process plant is established along the required pollution control systems, the entrepreneur is required to obtain consent to operate the unit. This consent is given for a particular period, which needs to be renewed regularly. The CTE is issued to the project after the environmental clearance based on its environmental pollution potential wherein the pollution control measures are clearly prescribed to the industry. The CTO is issued to the industry before the commencement of the operation and it prescribes the emission and effluent limits based on the industry specific standard. The CTO also sets the requirement for self-monitoring and reporting schedules. Additionally, regular inspection is carried out based on the category of the industry.[39]

There are no standard inspection and sampling procedures prescribed either in the Water Act, Air Act, or EPA, or their regulations, and the Central Pollution Control Board (CPCB) and State Pollution Control Board (SPCB) have not issued uniform guidelines. As a result, boards develop and apply their own approaches and methods, which is an inefficient way to use limited agency resources. Generally inspections are conducted based on the size of the firm and the category it belongs to. The industries have been divided in the category of large, medium, and small scale. In the large scale and medium scale industries, there are three sub categories red, orange, and green. They are inspected in three months, annually and once in two years respectively. Similarly, the small scale industries are also divided in red, orange, and green. They are inspected annually, once in two years and once in five years respectively.[40] For example, the lack of the sampling procedure is often quoted as one of the main reasons why courts often rule against the government.[41]

As per the Environment (Protection) Rules of 1986, each facility should submit an environment statement for the financial year (April to March). The frequency of monitoring depends on the category of the industry, also the frequency of the inspection by the regulatory authorities are high for red category industry. Industries identified by Ministry of Environment, Forests and Climate Change (MOEFCC), Government of India as heavily polluting and covered under Central Action Plan. The environment statement provides information on the inputs like quantity of water, raw material, pollution (air, water, and solid-waste) generated and the abatement undertaken along with the mitigation measures. The facility has to submit the environmental statement to the respective SPCB who then reviews and files the same for further actions.[42]

United States Environmental Protection Agency USEPA studied the environmental compliance and enforcement and suggested several measures to improve the process and one of the key suggestions was to implement a standardized and computerised data of the compliance and enforcement across the country.[43] In 2006, OECD conducted a study in close coordination with the Indian regulatory authorities and they have reported several shortcomings in the compliance and enforcement of the environmental regulation in India.[44] They (both OECD[45] and USEPA[46]) have suggested several measures to improve the compliance and some of the key measures were the upgradation of the information management system and establishment of the public information disclosure programme. One of the major challenges faced by the environmental regulators is the enforcement of the regulation, despite having a detailed regulation, compliance by the industries are low due to the lack of enforcement.

Energy Conservation

While reviewing the industrial energy efficiency policies in India during the period 1980 to 2005, Yang has identified six major policies which were implemented during the period,[47] which included the disclosure of company-level energy efficiency information; introduction accelerated depreciation for energy efficiency and pollution control equipment; the establishment of the Energy Management Centre under the Ministry of Energy; removal of price and output controls to promote industrial competitiveness; energy price reforms to guide energy efficiency initiatives and encourage international competitiveness; and, passage of the Energy Conservation Act of 2001 and the Electricity Act of 2003.

The Energy Conservation Act resulted in the creation of the BEE, a statutory body under the Ministry of Power. BEE actively promotes, manages, finances, and monitors energy conservation efforts in the economy through energy audits. It also has the statutory authority to implement mandatory energy efficiency standards but has not yet done so.

The Energy Conservation Act was introduced with the aim of conservation of energy in India. The objective was to bridge the gap between demand and supply side by optimizing the use of energy, reducing environmental emissions through energy saving, and to effectively overcome the barrier. The Act also provided the legal framework and institutional arrangement for driving the energy efficiency policy.[48] Under the provisions of the ECA 2001, BEE has been established in 2002 which would be responsible for implementation of policy programmes and coordination of implementation of energy conservation activities.[49] The key features of ECA amended in 2013 are standards and labelling, enlisting of designated consumers for energy activities, certification of energy manager (EM) and energy auditor (EA), and regulation of compliance from consumers. Under the Act, it is mandatory for the high tension consumers to conduct energy audit by certified energy auditor on a regular basis.[50]

Environmental and Energy Information in India

The importance of the environmental information to regulate the firm is highlighted in the National Environment Policy 2006, which states that,

> Access to environmental information is the principal means by which environmentally conscious stakeholders may evaluate compliance by the concerned parties with environmental standards, legal requirements, and covenants. They would thereby be enabled to stimulate necessary enforcement actions, and through censure, motivate compliance. Access to information is also necessary to ensure effective, informed participation by potentially impacted publics in various consultation processes, such as for preparation of environmental impact assessments and environment management plans of development projects.[51]

Past studies by USEPA[52] and OECD[53] had suggested standardization and implementing information technology in collecting and archiving the compliance and enforcement of the environmental regulation. OECD[54] had also suggested introduction of the information disclosure as a policy measure. There have been no efforts by the regulators to introduce the information disclosure policy and there is no update on the status of standardization of the environmental information across the country. We look at the findings of a case study[55] to understand the status of the environmental information and also review the findings of another study[56] on the quality of the information provided by the firms. Note that the availability and the quality of the information is important to ensure right information to the stakeholders. We review the availability of information in the next section which is based on the outcome of the study by Shivananda Shetty and Surender Kumar as in the section titled 'Status of Energy Information' we look at the quality of the information based on the study conducted in Gujarat by MIT in close coordination with the Gujarat Pollution Control Board (GPCB).[57]

The National Environment (Protection) Rules, 1986 mandates the manufacturing facility to submit an environment statement for the financial year (April to March) to the respective SPCB using the format provided in form V.[58] According to the Rules, each facility has to provide information on the use of the natural resource such as water, raw material, and the quantity of the environment pollutants generated per unit of the product along with the abatement action it undertakes to reduce the impact of pollution.[59] The statement provides the regulator a clear insight into the environmental status and the actions undertaken by the firm to manage the environmental aspects of their production process. The above reporting process is based on the self-declaration by the individual firms who can use their in-house team or subcontract the measurement of the data to an independent third party agency who are approved by the respective state pollution boards.[60] Complete information in the environmental statement is very important for the regulator to monitor and enforce the regulation effectively. The statement can also act as a tool to implement the risk-based inspection, which will reduce the pressure on the existing resources and thus making the CAC effective.

Status of the Availability of the Environment and Energy Information in India

The information on the environmental performance of the firm is available with respective SPCBs as the individual firms have to submit an annual statement at the end of the financial year detailing the quantity of pollutants released and the resources consumed. The only study conducted so far with respect to the availability of the information with the SPCBs has been done by above author and hence for the detail analysis of the environmental information, we refer the reader to Shivananda Shetty and Surender Kumar.[61]

In order to assess the availability of the information with respective SPCBs, Shivananda Shetty had contacted the pollution control board in 2013 seeking annual statement of the facilities submitted for the year 2011–12. However, due to the lack of response and non-cooperation by the SPCBs, the authors had to seek the information under of Right to Information (RTI) Act.[62] The information on the pollution is public information and as per the RTI Act, the SPCBs should make information available to the public within a stipulated period of time regarding the effects of pollution, the need to prevent and control pollution, and to protect the environment.[63]

The study focused on the large polluting sectors and hence authors had prepared a list of facilities which were identified as major industries under the Perform Achieve and Trade programme as these were identified as major consumers of energy and are also identified as red category industries. The study focused on 477 facilities belonging to Power, Cement, Paper, Aluminium, Textile, Chemicals and Fertilizers sectors spread across 23 states which come under the red category of industries[64] and have to mandatorily submit the environmental statement. The authors had sought the copy of the environmental statement using the provision of RTI which were filed with each of the SPCBs with the list of the facilities coming under their jurisdiction. The request was forwarded by the individual SPCB head offices to their respective zonal office under which the facility is located. After the receipt of the payment for photocopying the SPCB provided the documents which were available with them. It is very clear from the study that facility level information is not easily available despite the past studies suggesting standardization of the process and use of information technology to collate and archive the information conducted in coordination with the regulatory authorities like CPCB, MOEF.[65] The response was further analysed and the following shortcoming were reported.

Lack of Availability of Information with the Pollution Boards

The pollution control boards could provide only statements of 199 facilities out of 477 facilities which is 42 per cent of the requested information for the high polluting facilities across 23 states. This also reflects the weak enforcement of the existing regulation despite the fact that the information provided in the statement is a self-declaration and PCB cannot prosecute a facility based on the same. If the information is not available with the pollution control board authorities then it would be challenging for them to monitor the facility considering the resource constraint.

Information Withheld by the PCB at Behest of the Firm

Few SPCBs withheld the information at the behest of the firms despite the fact that the National Environmental Policy has highlighted the need for sharing the information. This clearly reflects the nexus between the PCB and the facilities and the lack of compliance to the principles of the NEP by the regulators themselves.

The Response Rate at State Level

The leading states which provided the maximum information were Madhya Pradesh, Orissa, and Chhattisgarh who provided more than 89 per cent of the requested statements. The next was Gujarat with 78 per cent followed by Maharashtra, Tamil Nadu, Uttar Pradesh, Karnataka, and Andhra Pradesh with 20–30 per cent response rate. Orissa and Chhattisgarh have sponge iron and steel industries while Gujarat and MP have a mix of industries comprising of chemical, steel, power, cement. Tamil Nadu, Maharashtra, and Karnataka have several large MNC firms having their manufacturing base. The Rajasthan and West Bengal Pollution Control Boards did not provide any information.

The Response Rate at the Sector Level

The study covers seven major sectors which are listed under the red category and the response rate based on individual sector is as follows aluminium 100 per cent, steel 63 per cent, Paper 60 per cent, Cement 45 per cent, Chemical and Fertilizer 44 per cent, Power 38 per cent, and Textile 19 per cent.

The aluminium sector is highly concentrated with fully integrated large plants which are regularly targeted by the NGO and inspected by the PCB authorities. Similarly, the textile industry is relatively less polluting which might be the reason of lax attitude by the regulators and the facility managers themselves.

Incomplete Information in the Statement

The environmental statement has three major categories which each facility has to fill in while providing the environmental information. The first category is the Input which covers the raw material like minerals, chemicals, energy, water (process and domestic have to be reported separately). The second category is the quantity and quality of the pollution generated under Air (Sulphur dioxide, Nitrous Oxide, Suspended particulate matter), Water (BOD, COD, TSS, pH), and the solid waste (regular and hazardous waste are reported separately). The third category is the mitigatory actions undertaken by the firm including the investment made in the abatement technology. The section B of the environment statement provides the information on water and raw material that we consider under the input category. The sections C to E detail the information on air pollution, water pollution, and solid waste which we count under pollution category. The remaining sections of F, G, H, and I were considered under the mitigation category.

The information provided under each section was checked for its completeness. If all the information was provided for each of the sections it was considered as complete while in the cases wherein the information was not provided or provided in an ambiguous way were considered as incomplete. The complete section was given a score of 1 and incomplete section was given a score of 0. The score was summed up for each of these three categories and the percentage of the complete response received was calculated. The study further classified the information on the basis of the sector, state, public/private ownership, and the vintage of the firm. The vintage of the firm was grouped under two category, the firms started their operations before the liberalization policy in 1991 was put under one group and the firm which started their operation post liberalization was put under another category.

All the sectors have provided more than 90 per cent of the responses for the input wherein the Chemical sector which represents a combined group of chemicals, fertilizers, and chlor-alkali has provided 100 per cent information on inputs followed by paper, textile, steel, aluminium, power who have reported more than 90 per cent while cement was the last with 86 per cent. However, when it comes to disclosing the pollution related information, aluminium was the last with 52 per cent complete statements, paper was the highest with the 81 per cent complete responses followed by steel 78 per cent, chemical 76 per cent, power 69 per cent and cement 62 per cent. The paper sector provided maximum information on the mitigation actions wherein 76 per cent of the information was available followed by the chemical 67 per cent, steel 66 per cent, cement 64 per cent, aluminium 62 per cent, power 60 per cent. The textile was the worst with only 40 per cent firms disclosing their information on the mitigation actions undertaken by them.

The reporting of the input parameters are quite high amongst the industrialized states excepting Karnataka. In Karnataka, about 67 per cent firms were reporting the complete input data as compared to the national average of 92 per cent. The average reporting of the pollution information is about 65 per cent only. However, we find that the industrialized states like Maharashtra, Gujarat, Madhya Pradesh, and Andhra Pradesh were reporting higher than the national average, while the polluters in Chhattisgarh only reported 44 per cent of their pollution information completely. The reporting of the pollution information is critical for the regulators as it provides them the necessary information on the compliance level of the industry.

There was no significant variation on the environmental reporting based on the ownership type of being a public or private sector industry. In the input category both the groups were reporting 93 per cent complete response. We find the private sector provided 62 per cent complete response on the pollution information while public sector provided 59 per cent complete response in this category. The public sector firms provide 77 per cent while 72 per cent of the private sector firms provide complete information on the mitigation undertaken by the firm.

When we consider the vintage[66] of the industry we find the older industries which have been established before 1991 provide more information on all the three parameters wherein it provides 96 per cent, 73 per cent, and 72 per cent information for input, pollution, and mitigation respectively as compared to the firms established post liberalization who could provide 90 per cent, 50 per cent, and 62 per cent information only.

Status of the Energy Information

In 1988, the Company Act, 1956 (1 of 1956) was amended wherein the Section 217(1) mandated the information on the conservation of energy which introduced the Companies (Disclosure of particulars in the report of board of directors) rules.[67] It was first such initiative by the Government of India to mandate the corporate sector to provide information on its energy conservation initiative[68] including the energy conservation measures undertaken by the company; additional investments and proposals which are implemented to reduce the energy consumption; the impact of the measures at energy consumption and its subsequent impact on the cost of production of the goods; total energy consumption; the efforts made in technology absorption for conservation of energy.

The above Act was mandatory for 21 highly energy intensive industries and there was a provision for criminal sanction in case of non-compliance.[69] A survey of 30 large Indian firms based on their market capitalization for the period of 2006–9 has reported that all the companies provided the information as required by the disclosure requirement.[70] As there is no clear guideline on the energy efficiency the survey was limited to confirming if the firm has filled in the section or not. The Act required the firms to provide company level information and hence the plant or product level was not available for further analysis and it was not possible to relate an initiative at plant or product level. Due to the lack of clear guidelines in reporting, individual firms used their own approach which has led to heterogeneity of the information and it was difficult to compare the firms. The above paper cited an example from the sugar industry wherein some firms reported the bagasse cogeneration under the energy conservation measures and hence it was difficult to compare the firm's energy efficiency initiatives with the other firms in the same sector.

The annual report which is filed by a firm in accordance with The Companies Act, 1956 are prepared by financial professionals like chartered accountants (CA) who take input from the respective energy and environment department of the firm. The CA lack knowledge and expertise in the energy and the environment domain and also due to the lack of clear reporting guideline, each firm uses their own approach to fill the section on energy conservation. Hence, not much importance is given to this section by the management team during their board meetings. Migali Delmas and Neil Lessem have reported that the combination of private and public information is effective in improving the energy compliance behaviour of the consumer.[71]

Quality of the Information Provided by the Facilities

Environment

The process of submitting the annual statement as per the National Environmental (Protection) Rules of 1986 is based on the principle of self-monitoring and reporting wherein the facility can choose a SPCB approved third party agency to conduct the environmental measurement and reporting. This process is questionable as no firm will provide an adverse information of their facility to the regulators although no legal action can be initiated by the regulator on the basis of self-reported information. A study was conducted in close coordination with the CPCB and GPCB along with experts from MIT and Harvard University to assess the effectiveness of the information provision and regulation.[72] They found the information provided by the firms who chose their own auditors and a set of firms wherein the auditors were chosen and paid by the research team. The study reported that the firms who chose their own auditor tend to influence the auditor behaviour who falsely reports the firms to be meeting the regulatory requirement. The study had suggested that the regulators should assign and pay the auditors to avoid the conflict of interest and to ensure correct reporting of the data.[73]

Energy

The information provided by the company in the annual report is reported by the firms to comply with the requirement of the Company Act. Due to lack of standardization and clear guidelines on the way the information is reported, each firm provides this information as per their own convenience and there is no laid down procedure by the Ministry of Corporate Affairs to validate the same. In 2005 the Centre for Monitoring Indian Economy (CMIE) which is a leading provider of business information had carried out a study to interpret the information provided by the companies in the report. The study was focused on analysing the energy conservation efforts across sectors and industries.

The study was inconclusive as the information was at a company level and it was not possible for CMIE to correlate the improvement of energy efficiency with the initiatives undertaken by the company.[74] Due to the lack of reporting guidelines for each sector, the firm level comparison was not possible. However, it was reported that some firms had proactively used the information provided by the peers to bench mark and set up objectives to help them improve their energy efficiency.

The PAT scheme has clearly set the measurement guidelines for each sector and has standardized the reporting process for each sector which helps to overcome the gaps in the system. However, the information is not available to public at large but firm level energy efficiency information is shared by the BEE which helps a firm to benchmark and compare itself to its competitors.

Similarly the Carbon Disclosure Project (CDP) has setup clear guidelines for measuring and reporting the carbon emission however the firms have an option of not disclosing the actual programmes undertaken by them to reduce their carbon emission. Both PAT and CDP do a rigorous analysis of the data provided by the firm.

Gaps in the Existing System

We find that despite existing regulation of informing the environmental and energy data to regulators there is a major gap due to lack of clear reporting guidelines and enforcement of the regulation. This not only impacts the availability of the data but also the usefulness of the relevant data across sectors. There has not been much focus on improving the energy conservation reporting by the Ministry of Corporate Affairs although the BEE which comes under the Ministry of Power has assigned a study to CMIE to analyse and interpret the information provided by the firms. Due to lack of clear reporting guidelines it is not possible for the regulator to penalize the firm for incomplete or incorrect reporting. However, BEE has undertaken a process to streamline and standardize energy efficiency reporting for major sectors as a part of the PAT scheme. Ministry of Corporate Affair can leverage the knowledge and expertise of BEE to roll out a standard process for all the sectors including banks.

In case of environmental data despite clear regulation the pollution boards do not have information of the firms under their jurisdiction.

The quality of the information is also a concern as the firms have to submit self-reported data and there is a clear incentive for the firm to under-report the pollution level data.[75] There have been several studies conducted in the past to study the regulatory enforcement in India and they have suggested use of information technology and standardization of the information across sectors.[76] The OECD study has suggested the introduction of the information disclosure policy however, we have not seen much effort in that direction and in fact, there is no process in place to harmonize the reporting.[77]

Using the conceptual framework we identify five key drivers which influence a firm's compliance behaviour under the information disclosure. These drivers are combination of both internal and external factors that motivate the firm to undertake proactive measures to improve their environmental and economic performance. These drivers which influence the firms are the output pressure which represents the demand of the consumer; Input pressure wherein the financial market influences the decision of the firm to implement environment and energy efficiency; the judicial pressure which proxies the enforcement by the regulatory agencies; community pressure which is the pressure created by the NGOs and the public at large, and last one is the availability of the peer information which puts the pressure on the firm to benchmark and improve their own performance. The firms' environmental performance is not impacted by the output market pressure due to the lack of demand by the consumer; also the input pressure is insignificant due to the lack of availability of information on the firms' environmental performance. The appliance rating programme has been moderately successful due to the active participation by the consumer due the consumer's awareness on the cost benefit in using energy efficient product. However, the building rating programme is yet to take off due to the lack of consumer awareness of the benefit of such programme. The weak enforcement of the regulation has resulted in insignificant impact of the judicial pressure on both the environmental and energy efficiency of the firm. The pressure from the NGO has put pressure on the firms to improve their environmental performance. Similarly, the consumers' demand for energy efficient products has driven the firms to offer energy efficient product in the market.

***

The information disclosure policy approach can be effective in complementing the existing mandatory regulations. The intervention by the regulators in providing an institutional support to manage the programme is a key to ensure the success of the disclosure programme. The appliance rating programme has been successful in India mainly due to demand side pressure created by awareness of the consumer and institutional support provided by the BEE. The recent initiative by BEE to standardize the sector wise reporting process for the PAT scheme is a good initiative which can be useful for introducing the information disclosure mechanism. Similarly, the recent initiative by BSE and CDP will be useful in creating a pressure on the firms through the financial institutions. Above analysis reveals that information disclosure is an effective policy instrument to increase the compliance level in the energy and environment sectors and following steps will help in strengthening the information disclosure, developing a transparent and standardized process for capturing the energy and environment information. This should be done for each sector in close coordination with the respective regulatory authorities like BEE and CPCB; Include the pollutant and energy intensity of the product in the annual financial report wherein product wise energy and environmental information like pollutant, water consumed, and so on should be reported; Make the information available to the stakeholder such that they are clearly understood and can be compared across the firms; create an awareness of the key stakeholders like consumers, financial institutions, and general public will help enhance their understanding and create pressure on the firm.

Notes and References

1. World Bank, 'India: Green Growth Is Necessary and Affordable for India, Says New World Bank Report' (2003), available http://www.worldbank.org/en/news/press-release/2013/07/17/india-green-growth-necessary-and-affordable-for-india-says-new-world-bank-report, accessed on 20 April 2016.

2. Organization for Economic Cooperation and Development, 'Environmental Compliance and Enforcement in India: Rapid Assessment', *OECD* (2006), pp. 1–31; S. Kumar and S. Managi, *The Economics of Sustainable Development: The Case of India* (New York: Springer, 2009), pp. 37–63.

3. I. Dincer and C. Zamfirescu, 'Sustainable Energy Systems and Applications', DOI 10.1007/978-0-387-95861-3_2, (New York: Springer Science Business Media LLC, 2011), pp. 51–91.

4. S. Dasgupta, D. Wheeler, and H. Wang, 'Disclosure Strategies for Pollution Control', in T. Teitenberg and H. Folmer (eds), *International Yearbook of Environmental and Resource Economics 2006/2007: A Survey of Current Issues* (Northampton: Edward Elgar, 2007), pp. 93–119.

5. Dasgupta, Wheeler, and Wang, 'Disclosure Strategies for Pollution Control'.

6. S. Winter and P. May, 'Motivation for Compliance with Environmental Regulations', *Journal of Policy Analysis and Management*, 20(4) (2001), pp. 675–98.

7. T. Tietenberg, 'Disclosure Strategies for Pollution Control', *Environmental and Resource Economics*, 11(3) (1998), pp. 587–602.

8. Tietenberg, 'Disclosure Strategies for Pollution Control'.

9. Tietenberg, 'Disclosure Strategies for Pollution Control'.

10. Tietenberg, 'Disclosure Strategies for Pollution Control'.

11. Tietenberg, 'Disclosure Strategies for Pollution Control'.

12. Nicholas Power, Allen Blackman, Thomas P. Lyon, and Urvashi Narain, 'Does Disclosure Reduce Pollution? Evidence from India's Green Rating Project', *Environmental and Resource Economics*, 50(1) (2011), pp. 131–55.

13. S. Konar and M.Cohen, 'Information as Regulation: the Effect of Community Right to Know Laws on Toxic Emissions', *Journal of Environmental Economics and Management*, 32 (1996), pp. 109–24.

14. Dasgupta, Wheeler, and Wang, 'Disclosure Strategies for Pollution Control'.

15. Konar and Cohen, 'Information as Regulation; M. Khanna and WRQ Anton, 'Corporate Environmental Management: Regulatory and Market-Based Pressures', *Land Economics*, 78(4) (2002), pp. 539–58; L.T. Bui, 'Public Disclosure of Private Information as a Tool for Regulating Environmental Emissions: Firm-Level Responses by Petroleum Refineries to the Toxics Release Inventory' (Working Paper Washington, DC: Center for Economic Studies, US Census Bureau) (May 2005), pp. 1–23; D. Koehler and J. Spengler, 'The Toxic Release Inventory: Fact or Fiction? A Case Study of the Primary Aluminum Industry', *Journal of Environmental Management*, 85(2) (2007), pp. 296–307.

16. B. Goldarand S. Gupta, 'Do Stock Market Penalise Environment Unfriendly Behaviour? Evidence from India', *Ecological Economics*, 52 (2005), pp. 81–95.

17. S. Shetty and S. Kumar, 'Opaqueness of Environmental information in India', *Economic and Political Weekly* (23 July 2016), pp. 15–19.

18. B. Truffer, J. Markardand, R. Wüstenhagen, 'Eco-Labeling of Electricity—Strategies and Trade-Offs in the Definition of Environmental Standards', *Energy Policy*, 29(2001), pp. 885–97.

19. P.S. Mehta, 'Why was India's Ecomark Scheme Unsuccessful?', *International Consumer Unity & Trust Society, India* (2007), pp. 1–64.

20. Mehta, 'Why Was India's Ecomark Scheme Unsuccessful?'.

21. Bhattacharya and Cropper, 'Option for Energy Efficiency in India and Barriers to their Adoption', *Resources for the Future* (2010), pp. 1–38.

22. Bhattacharya and Cropper, 'Option for Energy Efficiency in India and Barriers to their Adoption'.

23. Indo-Swiss Building Energy Efficiency Project, 'Energy Efficiency in Commercial Buildings', available http://beepindia.org/beep-commercial, accessed on 20 April 2016.

24. Indo-Swiss Building Energy Efficiency Project, 'Energy Efficiency in Commercial Buildings'.

25. A. Somvanshi, 'Certified, Not Certain', *Down to Earth*, available http://www.downtoearth.org.in/coverage/certified-not-certain-46900, accessed on 20 April 2016.

26. Ministry of Power, 'The Energy Conservation (Energy Consumption Norms and Standards for Designated Consumers, Form, Time within which, and Manner of Preparation and Implementation of Scheme, Procedure for Issue of Energy Savings Certificate and Value of Per Metric Ton of Oil Equivalent of Energy Consumed) Amendment Rules' (2016), available https://beeindia.gov.in/sites/default/files/Gazette%20Notification%20373%28E%29.pdf, accessed on 02 September 2016.

27. Ministry of Power, 'The Energy Conservation Amendment Rules'.

28. Carbon Disclosure Project (CDP), 'Energy Efficiency: Driving the Climate Change Response in Indian High Performing Companies', available http://www.indiaenvironmentportal.org.in/files/file/CDP-India-200-Climate-Change-Report-2013.pdf, accessed on 20 April 2016).

29. Carbon Disclosure Project (CDP), 'Energy Efficiency: Driving the Climate Change Response in Indian High Performing Companies'.

30. Kumar and Managi, 'The Economics of Sustainable Development: The Case of India'.

31. No. 06 of 1974.

32. No. 36 of 1977.

33. No. 14 of 1981.

34. No. 29 of 1986.

35. No. 06 of 1991.

36. No. 22 of 2005.

37. Ministry of Environment, Forest and Climate Change, 'Classification of Industries for Consent Management', available http://envfor.nic.in/legis/ucp/ucpsch8.html, accessed on 30 August 2016.

38. Ministry of Environment, Forest and Climate Change, 'Classification of Industries for Consent Management'.

39. Shetty and Kumar, 'Opaqueness of Environmental Information in India'.

40. Organization for Economic Cooperation and Development, 'Environmental Compliance and Enforcement in India: Rapid Assessment' (OECD, 2006).

41. Organization for Economic Cooperation and Development, 'Environmental Compliance and Enforcement in India'.

42. Shetty and Kumar, 'Opaqueness of Environmental Information in India'.

43. US EPA, 'Report on Environmental Compliance and Enforcement in India' (December 2005), pp. 1–92.

44. Organization for Economic Cooperation and Development, 'Environmental Compliance and Enforcement in India'.

45. Organization for Economic Cooperation and Development, 'Environmental Compliance and Enforcement in India'.

46. USEPA, 'Report on Environmental Compliance and Enforcement in India'.

47. M. Yang, 'Energy Efficiency Policy Impact in India: Case Study of Investment in Industrial Energy Efficiency', *Energy Policy*, 34(17) (2006), pp. 3104–14.

48. Ministry of Law, Justice and Company Affairs, 'The Energy Conservation Act' (2001), available http://powermin.nic.in/sites/default/files/uploads/ecact2001.pdf, accessed on 02 September 2016.

49. Please see, BEE Website, www.bee-india.com.

50. Ministry of Law and Justice, 'The Energy Conservation (Amendment) Act' (2010), available http://www.prsindia.org/uploads/media/Acts/The%20Energy%20Conservation%20(Amendment)%20Act,%202010.pdf, accessed on 2 September 2016.

51. Ministry of Environment and Forest, 'National Environment Policy' (2006), available http://www.moef.gov.in/sites/default/files/introduction-nep2006e.pdf, accessed on 2 September 2016.

52. US EPA, 'Report on Environmental Compliance and Enforcement in India'.

53. Organization for Economic Cooperation and Development, 'Environmental Compliance and Enforcement in India'.

54. Organization for Economic Cooperation and Development, 'Environmental Compliance and Enforcement in India'.

55. S. Shetty and Kumar, 'Opaqueness of Environmental Information in India'.

56. E. Duflo, et al., 'Truth Telling by Third Party Auditors and Response of Polluting Firms: Experimental Evidence from India', *The Quarterly Journal of Economics*, 128(4) (2013), pp. 1499–545.

57. Duflo, et al, 'Truth Telling by Third Party Auditors and Response of Polluting Firms'.

58. Ministry of Environment and Forest, 'Environmental Policy Rules, 1986', available http://envfor.nic.in/legis/env/env4.html, accessed on 20 April 2016.

59. Ministry of Environment and Forest, 'Environmental Policy Rules' (1986).

60. Ministry of Environment and Forest, 'Environmental Policy Rules' (1986).

61. S. Shetty and Kumar, 'Opaqueness of Environmental Information in India'.

62. Right to Information Act 2005, available http://righttoinformation.gov.in/, accessed on 20 April 2016.

63. Organization for Economic Cooperation and Development, 'Environmental Compliance and Enforcement in India'.

64. Ministry of Environment and Forest, 'List and Classification of Polluting Sector', available http://envfor.nic.in/legis/ucp/ucpsch8.html, accessed on 20 April 2016.

65. US EPA, 'Report on Environmental Compliance and Enforcement in India' (December 2005), pp. 1–92; Organization for Economic Cooperation and Development, 'Environmental Compliance and Enforcement in India'.

66. The vintage was classified on the basis of the firms established before the market liberalization in 1992 and post-liberalization.

67. The rule is called as the Companies (Disclosure of Particulars in the Report of Board of Directors) Rules, 1988.

68. B. Jairaj, 'A Disclosure Based Approach to Climate Change in India? Early Lesson from Business Regulation', Centre for Policy Research and Climate Initiative Working paper (November 2010), pp. 1–35.

69. Section 217(5) of the Companies Act 1956 provides that if any person, being a director of a company, fails to take all reasonable steps to comply with the provisions of subsections (1) to (3), or being the chairman, signs the Board's report otherwise than in conformity with the provisions of subsection (4), he shall, in respect of each offence, be punishable with imprisonment for a term which may extend to six months, or with fine which may extend to Rs 2,000, or with both.

70. Jairaj, 'A Disclosure Based Approach to Climate Change in India? Early Lesson from Business Regulation'.

71. M. Delmas and N. Lessem, 'Saving Power to Conserve Your Reputation? The Effectiveness of Private Versus Public Information', *Journal of Environmental Economics and Management*, 67 (2014), pp. 353–70.

72. Duflo, et al., 'Truth Telling by Third Party Auditors and Response of Polluting Firms'.

73. Duflo, et al., 'Truth Telling by Third Party Auditors and Response of Polluting Firms'.

74. Jairaj, 'A Disclosure Based Approach to Climate Change in India? Early Lesson from Business Regulation'.

75. Duflo, et al., 'Truth Telling by Third Party Auditors and Response of Polluting Firms'.

76. Organization for Economic Cooperation and Development, 'Environmental Compliance and Enforcement in India'; US EPA 'Report on Environmental Compliance and Enforcement in India' (December 2005), pp. 1–92.

77. Organization for Economic Cooperation and Development, 'Environmental Compliance and Enforcement in India'.

12

AMAR ROOPANAND MAHADEW

Assessing the Legal Framework of Mauritius on Sustainable Energy

Is It Robust Enough to Achieve the Dream of 'Mauritius: A Sustainable Island'?

While sustainable development has been conferred more than a hundred of definitions, no question can be raised on the fact that the one that was pioneered in 1987 from the World Commission on Environment and Development's (the Brundtland Commission) report stayed behind as the most recurrently quoted one and has still maintained its position. Hence, hailed as a landmark definition, it goes as follows: 'Development that meets the needs of the present without compromising the ability of future generations to meet their own needs.'[1]

Coming up to the context of Mauritius, the very concept of sustainability and sustainable energy was administered in the Mauritian legal and policy framework on 4 September 2009 following a Steering Committee that was held in the presence of representatives from diverse Ministries and Departments, the Academia and international development partners within the Prime Minister's Office.[2] Basically, the raison d'être of this committee was to synchronize the 'Maurice Ile Durable', translated in English as Mauritius—Sustainable Island (MSI), venture from a more comprehensive standpoint and to bring into line all the efforts devoted to this undertaking while embracing all the miscellaneous features of sustainability. In order to better understand the essence of the MSI venture, the words of the then prime minister, Dr Navin Ramgoolam at the opening ceremony of the workshop on MSI which was held on 1 December 2010 can be instrumental whereby he contended that,

> Maurice Ile Durable is essentially a vision that seeks to transform the environmental, economic, and social landscape of our country. It belongs to each and every one of us. MSI seeks to build up capital, not only for our generation but for generations to come. The MSI vision is embedded in a strategic framework embracing five development pillars (5E's), namely, Education, Environment, Energy, Employment, and Equity.

Subsequently, the Steering Committee was transformed into the Commission on MSI. This Commission is today defunct as a result of a change in regime and re-structuring. However, the concept of Sustainable Island is deeply anchored in policies and law related to energy.

At this stage, it might be pertinent to bring to light that various projects have been materialized by the defunct Commission on the MSI with a firm vision of implementing and promoting sustainable energy in Mauritius.[3] However, it is also fundamental to accentuate on the fact that though the implementation of sustainable energy and the concept of MSI is contingent on policies and relevant Acts of Parliaments, so far, there has been no holistic and encompassing legal document on sustainable development itself. Indubitably, the need of having a Sustainable Development Act (SDA) being sensed at the threshold of the implementation of the MSI and the Economic and Social Transformation Plan (ESTP) lingers as a reality that cannot be veiled.

Against this backdrop, this chapter aims at assessing energy laws and policies in Mauritius whose main aim is to turn the country into a Sustainable Island. Simply put, an attempt will be made to review the existing laws and policies and determine whether they are in line with the concept of sustainability. An energy sector overview in Mauritius is provided to understand the energy demands and requirements of the country. The diverse laws and regulations governing energy in

Mauritius have been thoroughly gauged so as to conclude whether our legal framework either makes provisions for or encompasses in some form or the other the very concept of sustainability and sustainable energy. Some recommendations will eventually follow based on the improvements and amendments that can be brought to achieve the dream of 'Mauritius: A Sustainable Island' through sustainable energy.

Having elucidated on the objective of sustainability which the government wants to achieve under the MSI Commission, the attainment of it will not be possible without various tools such as apposite and efficient linkages among implementation, compliance, shared stewardship, and monitoring (effectiveness and ambient). Here, it will be relevant to state that as the MSI Commission maintained an action-oriented approach to sustainability, it is progressively imperative in signalling areas that may call for compliance and its management. In relation to sustainability governance, reliance is on current regulatory requirements initiated by other institutional actors such as Ministries and Authorities, the introduction of new legislations and regulations, the revision of present laws and regulations, and a shift to action-oriented management. By doing so the MSI Commission will expose and other government agencies will expose themselves to an array of regulatory management regimes under which they will operate.

At this stage, it might be pertinent to highlight that in the quest to pull off more promising results, the preferred mode of operation for MSI Commission calls for both horizontal and vertical integration approaches. To set the scene, there are various elements which will be instrumental in achieving compliance through the sharing of information with relevant agencies. These will be comprehensive consultation (for example, industry participation in developing codes of practice), shared decision-making (for example, use of qualified sustainability/environmental professionals), and shared management (for example, recreational associations assisting in compliance promotion and verification activities).

It is undeniably positive to note that since its inception, the Commission on MSI has been quite prolific in terms of publication of reports on sustainable development associated to energy, environment, employment, education, and equity. It might also be interesting to note here that a green paper has been published with components such as National Programme on Sustainable Consumption and Production

(SCP), National Environment Policy, and Mauritius Environment Outlook Report.[4]

Coming up first to its Final Report of the Working Group on Energy, while mentioning that one of the key objectives of the Maurice Ile Durable project is to make Mauritius, Rodrigues, and the outer Islands less reliant on fossil fuels, the report proposed the way to achieve such a result. It explained that energy consumption for high energy consumers such as transport, land use, buildings, and industrial processes[5] can be highly reduced by increasing energy efficiency. Additionally, the report also advocated for the development of sustainability indicators other than GDP, recommended the promotion of substitutes for modes of transport and that strategies, policies, and regulations be assessed and strictly enforced so as to promote MSI concepts. Finally, it also recommended that wide ranging and integrated approach to transport and land use be promoted in accordance with MSI concepts, that capacity building be given priority and that fiscal policies be present to boost up the use of renewable energy in the transport sector.

The Commission on MSI also published a report known as The Final Report of the Working Group on Environment—Preservation of Biodiversity and Natural Resources whereby the various gaps and inconsistencies in the legislation and fragmentation of jurisdiction were underlined. It also highlighted the dispersed institutional responsibilities, poor implementation, and enforcement and the undermined legal and institutional authority leading to vain attempts of trying to attain the high standards anticipated by government and society. Finally, the report also recommended that the single Biodiversity Act be developed and promulgated.

Amongst the various reports, the Commission also published a report named The Final Report on Employment which accentuated on the need for green jobs that is jobs which trim down the consumption of energy and raw materials, restrain greenhouse gas emissions, lessen waste and pollution, and shield and help ecosystems recover.[6] Furthermore, the existence of The Working Group on Education cannot be overlooked.[7] This report contended that education for Sustainable Development is an upcoming but dynamic concept that looks at education from an absolutely new standpoint as it wishes to empower people of all ages to shoulder the responsibility for creating a sustainable future. On an equally important note, it also underlined that our current lifestyle is not sustainable and that educators, both in the formal and informal sectors, have an instrumental role to play. It is now relevant to have an overview of the energy sector in Mauritius before embarking on the assessment of the legal framework.

Energy Sector Overview

In the absence of oil, natural gas, or coal reserves, Mauritius has to rely on imported petroleum product in view of meeting most of its energy requirements.[8] Biomass, hydro, solar, and wind energy constitutes the local and renewable sources of energy.[9] A by-product of the sugar industry bagasse makes up for biomass energy, contributing to about 22 per cent of the primary supply of energy. There is a minimal use of wood and charcoal.[10] Hydropower plants have a combined capacity of 59 MW.[11] The solar regime of Mauritius is descent with a potential annual radiation value of 6 kWh/m^2/day.[12] An annual mean speed of 8.1 m/s at 30 m above ground level makes the wind regime appealing in some areas over the country.

The Ministry of Public Utilities has the primary responsibility of designing and implementing energy policies related to power sector, water and waste water sectors, and power utility through the Central Electricity Board (CEB). The latter is the electricity regulator with the task of generating and supplying electricity. The CEB Act entrusts statutory responsibility on CEB for the control and development of electricity supply in Mauritius. The CEB is therefore empowered to carry out development schemes with the objective of promoting, co-ordinating, and improving the generation, transmission, distribution, and sale of electricity throughout Mauritius as required.[13] The electricity sector and CEB's operations in Mauritius are regulated by the Electricity Act of 1939 (amended 1991), Electricity Regulations of 1939, and Central Electricity Board Act (1964). The import of petroleum products is guaranteed by the Ministry of Commerce and Cooperatives through the State Trading Corporation. The Ministry also enforces and implements petroleum prices which are regulated by the Petroleum Act of 1970 (amended 1991) and the Consumer Protection Act (Price and Supplies) of 1998.

The granting of prospecting and mining leases for petroleum are regulated by the Petroleum Act despite the fact that till date Mauritius has not discovered any indigenous petroleum sources. Government policy has encouraged the use of local and renewable energy sources to reduce dependency on oil for the production of electricity. Mauritius imports some 900,000 tons of petroleum products, comprising 380 Cubic square ton and 180 Cubic square ton heavy fuel oil, gasoline, diesel, kerosene and LPG, and 225,000 tons of coal annually.[14] The imported petroleum products and coal account for 75.4 per cent of the primary energy requirements.[15]

Solar heaters for domestic water heating have also been encouraged since the country has a decent solar regime. Despite the economic attraction that solar water heating provides, its large-scale adoption by the general public has been hampered by easy access to electricity grid power supply which is reliable. The high prices of the solar water systems have also been an impediment despite loan facilities provided by the Mauritius Development Bank. As for the wind power regime which is considered to be good over certain areas of the Island, installation of wind power system has proved to be tricky especially in the 1980s because of frequent cyclones. However, special designs of turbines by the use of modern technology have reduced this risk today. This has encouraged the setting up of wind farms under the Build-Own-Operate scheme.[16] Bagasse, fuel wood, and charcoal make up the bulk of biomass energy with bagasse being the most available primary energy resource.[17] Sugar industry makes use of bagasse for their electricity and heat generation and any surplus is sold to the national grid.

The Legal Review

The current part of the chapter is based on a review carried out by the author as a consultancy with the UNDP and the University of Mauritius. The reviews are reproduced here for the purpose of discussions and as a foundation for ensuing discussion.[18] The UNDP contends that energy is at the heart of sustainable development and poverty reduction. At the same time, it is irrefutably imperative too as it has a bearing on every aspect of sustainable development ranging from social to economic and to environmental development. As a matter of fact, it is important to underline here that the UNDP does not only focus on physical infrastructure but also on energy affordability, reliability, and commercial viability.

At this stage, it is relevant to note that the Maurice Ile Durable Policy, Strategy, and Action Plan (MSIPSAP) demarcate energy production and transportation as being the core areas which affect sustainable development. It is noted that in terms of energy production, 45 per cent of energy is produced by the state's four thermal power stations and nine hydroelectric plants.[19] The remaining proportion is obtained from Independent Power Productions which use bagasse and imported coal for production.[20] The issues flagged in relation to energy production have been high investment cost for renewable energy, over reliance on external sources of fossil fuels, and limited amount of energy efficiency and conservation measures. Based on the above concerns, the following acts, regulations, and policies have been evaluated as they have a close bearing on the sustainable energy—The Energy Efficiency Act 2011, The Electricity Act 2005, The Utility Regulatory Authority Act 2008, The Central Electricity Board Act 1964, The Finance Act 1973, The State Trading Corporation Act 1982, The Building Control Act, The Finance and Audit Act, and The Central Electricity Board (Mauritius Broadcasting Corporation-Collection of Licence Fees) Regulations 1989. The review is then mapped with the strategies, policies, and indicators provided for in the MSI PSAP. These are as follows in relation to energy: Long Term Energy Strategy, Energy Efficiency Targets, Sustainable Public Transport, Renewable Energy Targets, Energy Security, and Power Sector Reform.

The Energy Efficiency Act, 2011

The Energy Efficiency Act (EEA) is to be read in conjunction with the National Long Term Energy Strategy of the Ministry of Energy and Public Utilities. As a matter of fact, the strength of the EEA is the setting up of the Energy Efficiency Management Office (EEMO) which shoulders the responsibility to translate the objective of the EEA, which is to promote efficient use of energy and to promote national awareness, into concrete actions. As such, the EEA comes in accordance with the MSIPSAP Long Term Energy Strategy in an institutional sense by catering for the EEMO.

The Electricity Act, 2005

As far as the Electricity Act, 2005 (EA) is concerned, it legislates on the licensing of electricity services, obligations of licenses, electricity tariffs, and safety. It is fundamental to note here that the part accentuating on obligations of licensees is pertinent in the sense that it underlines the fact that one of the responsibilities of the EA is in relation to safe, adequate, and efficient energy services. Sustainable development calls for safe, adequate, and efficient production and supply of electric energy. Conversely, a close reading of the EA confirms that it fails to elaborate on what safe, adequate, and efficient would imply. Being the main Act of Parliament for the licensing of electricity providers, it is vital that it brings to light what efficiency would entail. The MSIPSAP has adopted the Energy Efficiency Targets in view of providing a tangible basis for action. It has very clear goals.[21] The National Long Term Energy Strategy of the Ministry of Energy and Public Utilities also makes provisions for renewable energy development strategies. However, it is imperative to at least have wide-ranging and encompassing references to energy efficiency in the EA itself as it will have a more considerable binding force in law compared to such references in guidelines or policy documents.

The Utility Regulatory Authority Act, 2008

Section 5 of the Utility Regulatory Authority Act, 2008 (URAA) can be read as follows, 'Subject to the relevant Utility legislation, the objects of the Authority shall be to ensure sustainability and viability of utility services'. As far as electricity is concerned, it is observed that the EA does not make provisions for criteria for the sustainability of electric energy as discussed above. It implies that the Utility legislations must be brought in line to give genuine effect to sustainability referred to in the URAA. The MSIPSAP identifies renewable energy targets and embraces those targets as one of its policies. Nevertheless, the URAA, as it is now, fails in giving legal effect to the MSIPSAP.

The Central Electricity Board Act, 1964

The CEB Act imparts the responsibility in relation to the general control and development of the electricity supplies in Mauritius. Under the

CEB Act, one of the duties of the board is to cater for new generating stations.[22] Under the Renewable Energy Targets, the MSIPSAP aims at helping the Republic meet its energy needs in sustainable, locally compatible ways, and to reduce dependence on fossil fuel sources of energy without imposing exorbitant economic burdens on the nation. Our goal is to amplify the share of sustainable renewable sources in electricity production of 35 per cent by 2025. The CEB Act should be harmonized with the principles of sustainable development and the MSI principles by way of amendment of Section 10 to further clarify the nature and characteristics that new generating stations should present to be in line with sustainable development.

The Finance and Audit Act, 1973

The Finance and Audit Act (FAA) makes available special funds in its schedule which is divided from the consolidated fund provided by Section 3. Maurice Ile Durable is one of the special funds that are catered for. As noted in the MSIPSAP, one of the negative aspects of the renewable energies and their implementation and adoption is that they are very expensive. It implies that the framework in terms of finance provided by the FAA is not detailed and structured enough. The Action Plan (2011 to 2025) by the Ministry of Energy and Public Utilities has earmarked sustainable energy budget for the establishment of the EEMO by way of funds provided in the 2011 PBB and forecasted budget for 2012/2013 appearing in PBB. Nevertheless, the financing of renewable energy should be legislated by a distinct act of Parliament where the sources of funds and the obligations on the state are clearly spelt out.

The State Trading Corporation Act, 1982

The State Trading Corporation Act, 1982 (STCA) is responsible for the importation of petroleum products such as fuel oil, kerosene, Mogas, and gasoil. The objects of the Corporation are, inter alia, to negotiate the purchase of goods. The MSIPSAP evidently supports a move towards renewable energy. The STC should therefore be empowered to indulge in the manufacturing and processing of pollution-free energy through renewable sources.

Energy Strategy and Policy Documents

The National Long Term Energy Strategy, the Action Plan 2011–25 and the Outline of Energy Policy 2007–25 are the relevant documents in view of the energy projections done by the Ministry. They are very detailed documents with accurate information on all the fundamental components of the energy sector ranging from energy and equity, energy and environment, sources of renewable energy, and the issue of transport. For instance, the Long Term Energy Strategy recommends that a sustainable approach to the transport sector be as follows, encouraging the use of more efficient and lower emission vehicles and fuels, encouraging the use of bio-fuels, improving the efficiency of transport provision and use, and reducing the level and types of vehicles and fuel supply chain emissions.

A Critical Assessment of the Situation

Energy is irrefutably imperative in the unremitting development and growth of Mauritius whereby fossil fuels most precisely maintain their steady position as the prevailing source of primary energy involved in the creation of electricity and to drive the two prevalent consumers, the transportation and manufacturing sectors. It is absolutely pertinent to note at this stage that during the last decade, national energy requirement has escalated by an annual rate close to 5 per cent.[23] Presently, the Central Electricity Board generates around 45 per cent of the country's total power requirements from its four thermal power stations and nine hydroelectric plants while the remaining 55 per cent is procured from Independent Power Producers via a combination of bagasse and imported coal for generation.[24]

In 2011, 83.8 per cent of the total primary energy requirement was met by imported fossil fuels, while 16.2 per cent was procured from renewable energy sources. Bagasse brings about a contribution of 94 per cent of the renewable energy and 6 per cent is derived from hydro, wind, and fuel wood.[25] The importation of coal has substantially amplified over the years and made around 49.5 per cent of total fuel utilized for electricity production in 2011.[26] This dependence on external sources of fossil fuels and vulnerability to external shocks are an imperative challenge to energy security and a key promoter for change.

Furthermore, there is short penetration of renewable energy due to high initial investment costs while energy efficiency and conservation measures are also inadequate.

Today, the current energy challenge in Mauritius is to deliver unswerving reasonably priced energy while driving energy supply towards a more localized renewable source. Subsequently, the National Long Term Energy Strategy (2009–25) also intends to shrink the country's reliance on fossil fuels, proliferate the share of renewable energy, democratize energy supply, and stimulate energy efficiency and conservation. The prominent points accentuated on in this strategy, in the electricity sector are:

1. Intensification of the segment of renewable sources of energy in electricity supply (from 17.5 per cent presently to 35 per cent in 2025)
2. Expansion of energy efficiency and conservation in all sectors through demand-side management measures (with targeted energy efficiency gains of 10 per cent by 2025 over the 2008 baseline); and,[27]
3. The promotion of a financially sound and self-sustainable modern electricity sector, a translucent and impartial regulatory environment that fittingly balances the interests of consumers, shareholders and suppliers, conditions that can afford efficient supply of electricity to consumers and enhancement in customer services.

It goes without saying that while renewable energy will involve an escalating proportion of the energy mix, it will still be required to use conventional power such as Liquefied Natural Gas and Clean Coal. Subsequently, coal fired power stations will have to adhere to rigorous environmental safeguards. As far as the production of biofuel is concerned, the government has already established a framework for the use of locally produced ethanol.[28] The fact that cars driven by fossil fuel are decidedly polluting cannot be refuted at all. Mauritius has increasingly driven its steps towards a cleaner diesel from 2,500 parts per million (ppm) to 500 ppm and now 50 ppm.[29] Hence, Mauritius is gradually but surely moving towards greener technologies in the transportation sector. Moreover, while a fusion of cars is already accessible, electric cars have recently been introduced. These cars are charged by renewable energy like solar photovoltaic and wind can definitely be a solution towards greening further the transportation sector. So as to pave the way forward, the government incorporated in the budget exercise 2013, a deduction of the excise duty on such vehicles from 50 per cent to a flat rate of 25 per cent.[30]

The relevant legislation to regulate the electricity sector was passed in 2008 with the proclamation of the Utility Regulatory Authority (URA) Act. However, this now requires to be consolidated through the establishment of the URA. Moreover, the legal framework for energy efficiency has also been set up by the proclamation of the Energy Efficiency Act 2011 and an Energy Efficiency Management Office (EEMO) has been set up to encourage energy efficiency and conservation in all sectors. For instance, minimum energy performance standards for some domestic appliances and a wide-ranging framework to support sustainable buildings have been settled. Concerning the transport sector, revisions in vehicle taxation systems and measures to advance fuel efficiency have been hosted.

Transportation

Traffic congestion is a severe predicament and the total cost of congestion to the economy is estimated to be around 1.3 per cent of GDP. From 1990 to 2011, the total number of vehicles has escalated from 123,545 to 400,919, signifying an upsurge of 224.5 per cent with an annual increase of about 6 per cent.[31] The mass of vehicles has amplified substantially from 69 vehicles per km of road in 1990 to 190 vehicles per km of road in 2011.[32]

The swelling number of vehicles has been supplemented by an equivalent progression in fuel demand and carbon dioxide emission. Despite the relatively small size of the island, the average distances of commuting are high and time and energy consuming. Rationalization of land use could condense the number of trips and the commensurate average distance travelled, as well as promoting the use of public transport. This would necessitate interventions such as the densification of urban areas and the development of mixed use areas, in which employment, infrastructure, and houses are located together.

To make the transport sector more efficient, the Long Term Energy Strategy (2009–25) recommends a sustainable approach to transport comprising:[33] encouraging the usage of more efficient and lower emission vehicles and fuels, encouraging the habit of bio-fuels, cultivating the efficiency of transport provision and use, plummeting the level and types of vehicles and fuel supply chain emissions.

An integrated and sustainable approach to transport and land use should be also upheld. This could implicate the development of a Mass Rapid Transit (MRT) system with an efficient bus feeder network, which will deliver a reliable striking alternative to private vehicles. This approach will also boost walking and cycling through appropriate infrastructure.

Implementing the Vision of Sustainable Energy

Various MSI policies have been recommended to tackle the gaps, challenges, and opportunities, and henceforth backing the realization of the MSI vision for energy. To start with, the Long Term Energy Strategy aims at the creation of the enabling environment for the implementation of the Long Term Energy Strategy. The rationale behind being that this long-term plan for energy will enable demand to be fulfilled while also paving the way to a more sustainable energy production and consumption. Together with this, the aims behind Energy Efficiency Targets will be the usage and management of energy in an efficient way. In a more elaborate manner, the objectives will encompass the decline in energy consumption in non-residential buildings by 10 per cent by 2020 (taking 2010 figures as a baseline); plummeting public sector buildings energy consumption on average by 10 per cent by 2020 (in comparison to 2010); encouraging industry in shrinking their allied energy requirements, executing energy efficiency measures and linked influences of energy choices on Mauritius; refining the adeptness of transport by supplying modal shift (mass transit), inducing behaviours and promoting the adoption of more proficient technologies.

The rationale behind being that Energy targets will deliver a concrete basis for action, give a signal for innovation and acquaint potential suppliers with the prospective opportunities. They will construct the foundation of more detailed action planning. Moreover, Sustainable Public Transport is yet another policy put in place to cultivate an efficient reasonable, reachable, and cleaner transport system based on public transport efficient fuel use and good planning, resulting in plummeting the consumption of energy in the transport sector by 35 per cent by

2025, compared to 2010. The justification behind being that the public transport system presently has its reliance on old vehicles with high fuel consumption and air emissions. Had it been possible to modernize the transport fleet, it could speedily lessen the consumption of diesel. Additionally, the reintroduction of rail transport could be operative in dropping the number and length of car journeys.

Renewable Energy Targets is also another policy recommended to enable the Republic accomplish its energy needs in sustainable, locally compatible ways and to condense our reliance on fossil fuel sources of energy without driving exorbitant economic burdens on the nation. The goal is the ultimate intensification of the portion of contribution of sustainable renewable sources in electricity production of 35 per cent by 2025.[34] The validation of this recommendation comes from the idea that Energy targets will offer a tangible basis for action, give a signal for innovation, and familiarize potential suppliers about the probable opportunities while creating the basis of more detailed action planning. The next policy is that of Energy Security which will guarantee energy security through the application of MSI policies. The rationale being that MSI will enable the enhancement of energy security, by sponsoring both energy efficiency and the diversification of energy generation. Finally, the last policy will be that of the Power Sector Reform so as to ensure power sector reform and the introduction of independent economic regulation of electricity and launching the Utilities Regulatory Authority. It is to be observed that the introduction of the Utilities Regulatory Authority would be a speedy win that would progress both the clearness and proficiency of the utility sector.

Having discussed the policies and their rationales, various strategies have also been proposed to support the realization of the vision for energy. With regards to the Long Term Strategy Policy, the proposed strategy is to cultivate a short scenario based on energy plan for the coming thirty years to capitalize on the security of imminent energy supplies, building on the Long Term National Energy Strategy. Part of the strategy is also to recognize the developmental needs of the grid system to allow acceptance of new technologies in the future. Finally, the strategy also aims at confirming that transport and energy planning are core aspects of integrated planning within the Republic.

As far as the Energy Efficiency Targets policy is concerned, the strategy set is to back the activities and effective development of the Energy

Efficiency Management Office with lively partaking of the Central Electricity Board (CEB) and Ministry of Energy and Public Utilities (MEPU). Besides, it also aims to scrutinize the demand-side management to upkeep the control of peak energy loads. Finally, the strategy also encompasses the revision of Government Procurement Strategy to take account of energy efficiency as a key procurement principle and to carry on the work already embarked on to confirm sustained public awareness.

With regards to the sustainable public transport policy, the strategy is to encourage the implementation of the Mass Transit System, pedestrianization, and the use of low-energy modes of transport.[35] In the case of the Renewable Energy Targets policy, the strategy proposed is to create and maintain a pro-active energy supply plan so to deliver energy generation to fulfil future demand. Also, it intends to explore opportunities for furnishing additional economic incentives to stimulate renewable energy. Finally, while with regards to the Energy Security Policy, the strategy is to shrink dependence on external (international) fuel supplies, in the case of the Power Sector reform policy, the strategy proposed is to execute the actions in the Energy Regulator Terms of Reference.[36]

Undeniably, Energy has one of the most imperative involvements into our contemporary everyday lives. It fuels transport and agricultural machinery, helps the industry that produces export revenue to function, and is at the heart of the service sector that recruits the massive majority of the Mauritian workforce. The global competitiveness of Mauritius is terrifically inclined to the efficiency and cost of energy used in the production of its goods and services. As a matter of fact, energy outcomes are driven by three predominant intentions globally and these are, security of supply, affordability / efficiency, and minimal impacts on our natural environment.[37] At this stage, it is imperative to note that realizing all these objectives is difficult and necessitates a balance that can occasionally end in trade-offs against other objectives.

Moreover, with over 80 per cent of total primary energy coming from importation,[38] security of supply is indispensable for our small, isolated, and energy-resource poor nation. It should be noted that there is no local production of hydrocarbons or coal so as to be used in the production of electricity or fuel transport and machinery. This insufficiency of energy boils down to Mauritius being helpless to a huge level of import-dependence which gives birth to medium and long term risks to national welfare and sustainability. Refining security of supply will involve supply and demand-side intervention. As far as supply is concerned, up-front investment in our energy production and supply may be expensive and have implications for end-users. This could consist of new forms of base-load generation to arrange for firm power and/or renewable production that does not depend on any imported fuels. With regards to demand, pro-active intervention must happen through intensifying discipline of energy efficiency and demand side management.

As a matter of fact, affordability will be upgraded through the operation of a translucent, impartial, and commercially efficient energy sector that has an enhanced investment and operational framework to be channelled by an autonomous and consumer-focused Utility Regulatory Authority. This will be spotlighted on the development of a receptive and fiscally justifiable electricity utility and the incremental introduction of new renewable developments. The motivating objective will be to deliver value for household, commercial, and industrial consumers, while recognizing and meeting the objectives of a sheltered and quality supply of energy and environmental protection. Achieving environmental targets through the deployment of renewable energy can enhance security of supply through the fall of import reliance. Renewable energy, in the context of the principally oil-based stock of electricity generators, can have a constructive influence on the environment. Besides, in the context of ever-increasing and unpredictable nature of oil-based fuel sources, conventional forms of renewable energy are no-longer as economically exorbitant as they were formerly.

It remains a fact that future social and economic development of Mauritius is also much reliant on secure, affordable, and environmentally conscious energy. The international competitiveness of Mauritius depends on an efficient economy that is fuelled by reliable and competitive inputs to the goods and services produced. This is at a time when the global macroeconomic environment is volatile and unreliable and each country must make strenuous efforts to be remarkably resourceful and receptive. The MSI approach to energy and the development of the concept plan on energy is based on several vital key steps which are, a root-and-branch analysis of the energy profile of Mauritius which will imply the assessment of the current consumption and production of Mauritius and the feasibility of that trend in the short, medium, and long term; revising the past and present activities embarked on by public, private and NGO organizations; appraisal of the current energy policy, which is Long Term Energy Strategy and the extent to which the strategy is mirrored in present and future government programmes; bringing around the partaking of the general public, industry, academia, and NGOs in a consultative national process to solicit and secure their opinions and probable solutions to the future energy plan of Mauritius.

The goals of developing the energy sector in the spirit of sustainable energy are to attain the objectives of enhanced security of supply, afford-ability, and respect for the natural environment, the so-called energy policy 'tri-lemma' and to decouple the economy of Mauritius from world fuel markets, decline energy intensity, and upsurge the proportion of renewable energy reaching the grid. The targets to be achieved are as fol-lows, plummeting energy consumption in non-residential buildings by 10 per cent by 2020, reducing energy consumption in public sector buildings by 10 per cent by 2020, decreasing energy consumption in the transport sector by 35 per cent by 2025, and growing the segment of sustainable renewable sources in electricity production up to 35 per cent by 2025.[39]

Key challenges to achieving the above goals and targets will encom-pass the following, carrying out so in an economically practical way, where chosen solutions do not lead to long-term financial burdens on the state and energy consumers, reorganizing the electricity sector to demonstrate a fair and modern institutional architecture that mirrors the World Bank's classification of Mauritius as an upper middle-income economy, upholding the momentum of the complex and significant energy initiatives during a time of sustained global economic uncer-tainty, predominantly during 2013–14 when numerous enactments and institutions will be operationalized or consolidated, urgent action on energy use in transport, which is imposing significant pressure on petro-leum imports and restraining national productivity, all of which have repercussions for international competitiveness.

Scope of Action

Building Control Act

The Building Control Act, proclaimed in 2013, will implement desired regulations and standards across the Republic by law. In a more elaborate manner, the legislation is driven towards the promotion of the design, construction, and maintenance of buildings guaranteeing people's safety, society's well-being and the protection of the environment. Sustainability will be mirrored through a number of chief requirements including, water tightness of buildings and water management, waste management as part of the construction process, energy savings and optimum energy consumption for the proper operation of buildings, and decline in the heat island effect in urban areas.

It is to be noted that the spirit of the legislation will be supported by the Building Control Advisory Council (BCAC) with permanent members demonstrating the interests of energy efficiency, energy management, and environmental protection. BCAC shall bring forward actions comprising of advising the Minister on building regulatory issues and it will be accountable for the development of polices which stimulate safe, efficient, and sustainable construction of buildings.

Energy Efficiency

The Energy Efficiency Act provides for the EEMO to support the cause of energy efficiency, charged, as it will be, with the predominant goals to endorse efficient energy use and national awareness. This will be a main promoter of MSI objectives and a key contributor to the efforts of attaining the objectives and aims linked to the energy sector. The main areas of action to be driven by EEMO encompass, but are not restricted to the development and implementation of strategies, programmes, and action plans, comprising of pilot projects, for the efficient use of energy and the establishment procedures to observe energy efficiency and consumption. Issuance of guidelines for energy efficiency and conservation in all sectors of the economy are also part of the areas of action. The establishment of energy consumption standards, the collection, and maintenance of data on energy efficiency and consumption has been earmarked to provide for better information. The formulation and recommendation of innovative financing schemes for energy efficiency projects and the assistance in the preparation of educational courses and school curricula on the efficient use of energy form part of the initiative.[40]

Setting up of the Utilities Regulatory Authority

The Utilities Regulatory Authority (URA) is a fundamental element of the institutional architecture of developed countries. It is now time for our little island to establish this key institution which will take place over 2013–14. The URA will smooth the progress of the pillars supporting MSI—energy, environment, economy, education, and equity. The heart of setting up the URA is originated from the existing Long Term Energy Strategy and enacted legislation establishing its powers and responsibilities. It can be done in the prevailing electricity sector framework, whereby a single buyer will prevail for the anticipatable future in view of amplified involvement by the private sector in a more competitive process. Its capacities will need to be cautiously established and incrementally built up over time to make sure that the URA can realistically reinforce the delivery of its powers and responsibilities. This will imply that in order to promote an efficient, transparent, and investor-friendly electricity sector, the delivery of its mandate must be carried out in a new competitive environment.

Renewable Energy Procurement Framework/ Development Strategy

The Renewable Energy Procurement Framework (REPF) will improve the capacity of the public sector to acquire renewable sources of energy generation on a competitive basis, in view of commercially available and least-cost solutions. The REPF is the finest approach for the existing and near-term context of the energy sector in Mauritius. It will enable procurement of any form of renewable energy technology so as to accomplish policy objectives and be in line with the Renewable Energy Master Plan. The new framework will give elasticity over the timing, capacity, and technology to be sourced and provide the foundation for an internationally competitive process to draw reputable service providers for the advantage of the present and future consumers of Mauritius. The framework is intended to simplify the existing single buyer model and can function with or without the existence of an energy regulator. The REPF will back other existing national policies and MSI objectives such as plummeting our dependency on imported fossil fuel, expanding its energy mix in electricity, attaining the Republic's long term energy needs sustainably, and enabling the share of sustainable renewable sources in electricity production up to 35 per cent by 2025.

The MSI Process

It goes without saying that sustainable development is a worldwide aim, intending to meet the needs of current generations without compromising the ability of future generations to attain their own needs. The future of Small Island Developing States (SIDS) is predominantly at risk from intensifying costs of imported fuels and food, deteriorating natural environment and the influences of climate change on water, agriculture, and sea level rise. The 2005 Mauritius International Meeting on SIDS highlighted the pressing need for bigger commitment to diminish disparities, to cultivate sustainable consumption, and production patterns, to shield and manage natural resources sustainably for economic and social development, to safeguard health and integrate the objective of sustainable development. Maurice Ile Durable (MSI) has the vision of making Mauritius a model of sustainable development, mainly in the context of SIDS. It focuses on many aspects of sustainability, highlighted on the five Es, namely, Energy, Environment, Employment/Economy, Education, and Equity. The MSI Policy, Strategy, and Action Plan were accomplished in 2013, recognizing a range of activities that would help Mauritius to grasp the MSI vision. This Terms of Reference is to protect one of the activities identified in the Action Plan, for which the government of Mauritius necessitates the backing of technical specialists.

***

On 28 of August 2015, the Mauritius Renewable Energy Agency Bill was presented to the Parliament. According to the explanatory memorandum, the main object of this Bill is to provide for the establishment of the Mauritius Renewable Energy Agency for the purpose of promoting the development and use of renewable energy.[41] The proposed Agency's objects are the promotion of the adoption and use of renewable energy with a view to achieving sustainable development goals, advice on possible uses of liquid natural gas, create an enabling environment for the development of renewable energy, increase the share of renewable energy in the national energy mix, share information and experience on renewable energy research and technology and foster collaboration and networking, at regional and international levels, with institutions promoting renewable energy.[42]

In terms of its function, it is quite varied ranging from a duty of advice to the relevant ministries to the obligation of establishing mechanisms and framework to enhance the use of renewable energy in Mauritius. According to the Act, the Agency shall advise the Minister on all matters relating to renewable energy policy and strategy. Every five years, a renewable energy strategic plan is to be elaborated on and the necessary mechanism and framework to increase the use of renewable energy must be established. There is also the obligation to assess the feasibility and competiveness of renewable energy projects and make recommendations accordingly. The focus is equally on encouraging and supporting studies and research on the renewable energy technologies and ensuring their implementation. The Act also provides for the duty to compile and analyse data on use and benefits of renewable energy and develop guidelines and standards for renewable energy projects and for evaluation and approval of on-grid and off-grid renewable energy projects. The Act also defines a funding strategy for renewable energy projects and assesses the requirements for the improvement of skills for the implementation of renewable energy projects.[43]

The Agency is yet to come into force and it will only be able to operate effectively if it works in tandem with other regulatory bodies mentioned above. The success of it will equally be contingent on the competence of the people who would be administering it. They must be qualified personnel with a sound knowledge of energy and the environment. This is important as more often than not it is seen that political nominees would form part of such boards or agency without having any proper competence. The composition of the agency therefore is deemed to be decisive about the success of it.

One of the strengths of the energy sector lies in its institutional framework. As a matter of fact, the establishment of the Utility Regulatory Authority and Energy Efficiency Management Office will ease the process of attaining the objectives of the energy sector as set up by the MSIPSAP and hence driving faster towards sustainable development. Having reviewed all the major legislations above, it seems that the importance of efficient energy and sustainable ones have been indirectly embodied but there is still no clear and precise idea as to what the specific targets and achievements should be. At the same time, there may also be duplicity in the roles of the various authorities and boards. For instance, the URA and the CEB both have been entrusted with the obligation of producing electric energy in a sustainable manner and this is recognized in the respective acts of Parliament. Regrettably, the legislations are absolutely silent on the ways to achieve this objective. It can be contended that such technicalities are to be defined by the regulations. This indelibly cannot be challenged. However, the principles of sustainable energy and development as enshrined by international treaties under the UNEP should at least be underlined in all the core legislations namely the CEB Act, the URAA, and the Electricity Act while the Plug-ins can subsequently be added in subsidiary legislations. The threat of having details and information in only policy and strategy documents is that these documents do not cultivate an environment of accountability and answerability.

The two policies of the MSI, namely, Energy Efficiency Targets and Renewable Energy Targets need to be translated in a very clear and lucid Act of Parliament with specific time frame for achievement. The issue of sustainable development through efficient energy use is so pressing, principally with the effects of climate change that a document with a better binding force than simple policy papers from the relevant ministries will be a requisite. There should be a harmonizing exercise among ministries so that the targets for renewable energy and energy efficiency are identical and on the same wavelength. The Ministry of Energy and Public Utilities, the Ministry of Public Infrastructure, NDU, Land Transport and Shipping, the Ministry of Housing and Land, and the Ministry of Environment and Sustainable development must work towards a harmonized regulation to accomplish the above targets. The regulations should then be translated into practical guidelines that will inform any activities, actions, and undertakings by public or private parties that would fall under the realm of the above ministries. Energy law is indeed a crucial area in Mauritius which will welcome more additions and amendments.

Notes and References

1. Our Common Future: Report of the World Commission on Environment and Development, 'Our Common Future: Chapter 2: Towards Sustainable

Development', UN Doc A/42/427, available http://www.un-documents.net/a42-427.htm, accessed on 25 August 2016.

2. Steering Committee on Sustainable Development in the Public Sector established by the Procurement Policy Office of the Ministry of Finance and Economic Development (September 2009).

3. See, MSI 'Completed Projects', available http://www.gov.mu/portal/sites/MSI/MSIFComProj.htm, accessed on 1 March 2014.

4. SDI, 'Towards a National Policy for a Sustainable Mauritius' (April 2011), available http://www.stakeholderforum.org/fileadmin/files/Maurice%20 Ile%20Durable%20-%20Green%20Paper%20-%20Towards%20a%20 National%20Policy%20for%20a%20Sust....pdf., accessed on 25 May 2016.

5. MSI 'Consultative Workshops—Working Group on Energy Final Report', p. 11, available http://mid.govmu.org/portal/sites/mid/file/final% 20version%20WG1.pdf., accessed on 27 February 2016.

6. SDI, 'Working Group 4—Employment and Economy', p. 12, available http://mid.govmu.org/portal/sites/mid/file/wg4.pdf., accessed on 30 March 2013.

7. SDI, 'Consultative Workshop Working group 5—Education', p. ix, available http://mid.govmu.org/portal/sites/mid/file/final-WG5.pdf., accessed on 30 March 2016.

8. Ministry of Energy and Public Utilities 'Energy Resources', available http://publicutilities.govmu.org/English/Pages/Publications.aspx, accessed on 30 June 2016.

9. S. Deenapanray, 'Synthesis Report on Renewable Energy' (2006), p. 2.

10. Ministry of Energy and Public Utilities 'Energy Resources'.

11. Ministry of Renewable Energy and Public Utilities, 'Long-Term Energy Strategy', p. 33, available https://sustainabledevelopment.un.org/content/documents/1245mauritiusEnergy%20Strategy.pdf., accessed on 27 January 2016.

12. M. Romeela, S. Dinesh, and J. Pratima, 'Renewable Energy Potential in Mauritius and Technology Transfer through the DIREKT Project', p. 162, available http://psrcentre.org/images/extraimages/35%201012223.pdf., accessed on 30 January 2016.

13. CEB, 'Customer Service Charter', available http://ceb.intnet.mu/customer/Service%20Charter.pdf, accessed on 30 June 2016.

14. Romeela, Dinesh, and Pratima, 'Renewable Energy Potential in Mauritius and Technology Transfer through the DIREKT Project', pp. 160–2.

15. Ministry of Energy and Public Utilities, 'Energy Resources'.

16. Embassy of the United States, 'Mauritius—Economic Trends and Outlook', available http://mauritius.usembassy.gov/chapter_2.html, accessed on 30 June 2016.

17. Romeela, Dinesh, and Pratima, 'Renewable Energy Potential in Mauritius and Technology Transfer through the DIREKT Project', p. 160.

18. Consultancy work by R. Mahadew, et al.,—all lecturers from the University of Mauritius.

19. SDI, 'Consultative Workshop Working Group 1 Energy', p. 51, available http://mid.govmu.org/portal/sites/mid/file/final%20version%20WG1.pdf., accessed on 19 February 2016.

20. SDI, 'Consultative Workshop Working Group 1 Energy'.

21. Reducing energy consumption in non-residential buildings by 10 per cent by 2020 (taking 2010 figures as a baseline); Reducing public sector buildings energy consumption on average by 10 per cent by 2020 (in comparison to 2010); Supporting industry in reducing their associated energy requirements, implementing energy efficiency measures and associated impacts of energy choices on Mauritius; Improving the efficiency of transport by delivering modal shift (mass transit), influencing behaviours and supporting the adoption of more efficient technologies (targets are established under Policy A3).

22. Section 10 of the CEB Act.

23. See in general, J. Khadaroo, 'Peak Electricity Demand in Mauritius: Evolution and Forecasting' (July 2014), available http://web.uom.ac.mu/Staffdirectory1/FT001370/Research/Publications/Forecasting%20Peak%20Electricity%20Demand%202014-20.pdf., accessed on 24 April 2016.

24. MSI Policies Strategies and Action Plan (May 2013), pp. 10–13, available http://www.govmu.org/portal/sites/mid/file/4.%20Main%20Report.pdf., accessed on 24 April 2016.

25. SDI, 'Consultative Workshop Working Group 1 Energy', p. 55, available http://mid.govmu.org/portal/sites/mid/file/final%20version%20WG1.pdf., accessed on 19 February 2016.

26. MSI Policies Strategies and Action Plan (May 2013), p. 12.

27. SDI, 'Consultative Workshop Working Group 1 Energy'. p. 55.

28. Government of Mauritius, 'A Roadmap for the Mauritius Sugarcane Industry for the 21st Century', p. 5, available http://agriculture.govmu.org/English/Documents/Archives/Policy%20Documents%20and%20Reports/roadmap.pdf, accessed on 26 April 2016.

29. United Nations Environment Programme, 'Report on the Workshop on Global Fuel Economy Initiative', p. 3, available http://www.unep.org/Transport/PCFV/PDF/mauritius2013_Speech_Minister.pdf., accessed on 24 January 2016.

30. MSI Policies Strategies and Action Plan (2013), p. 37.

31. MSI Policies Strategies and Action Plan (May 2013), p. 41.

32. MSI Policies Strategies and Action Plan (May 2013), p. 41.

33. MSI Policies Strategies and Action Plan (May 2013), p. 41.

34. MSI Policies Strategies and Action Plan (May 2013), p. 64.

35. Ministry of Environment and Sustainable Energy 'Climate Change and Vehicle Emissions in Mauritius' (July 2013), p. 14, available http://www.unep.org/Transport/PCFV/PDF/mauritius2013_CCaVehicleEmissions.pdf., accessed on 25 February 2016.

36. MSI Policies Strategies and Action Plan (May 2013), p. 47.

37. O. Samantha, S. Ralph, and K. Nicolai, 'Contribution of Renewables to Energy Security (April 2007), p. 15, available https://www.iea.org/publications/freepublications/publication/so_contribution.pdf., accessed on 24 May 2016.

38. MSI Policies Strategies and Action Plan (May 2013), p. 41.

39. MSI Policies Strategies and Action Plan (May 2013), p. 67.

40. Republic of Mauritius, 'The Energy Efficiency Management Office', available http://eemo.govmu.org/English/Pages/default.aspx., accessed on 24 April 2016.

41. Explanatory Memorandum to the Mauritius Renewable Energy Agency (2015), available http://mauritiusassembly.govmu.org/English/bills/Documents/intro/2015/bill1115.pdf, accessed on 30 June 2016.

42. Section 4 of the Mauritius Renewable Energy Agency Bill, available http://mauritiusassembly.govmu.org/English/bills/Documents/intro/2015/bill1115.pdf., accessed on 26 May 2016.

43. Section 4 of the Mauritius Renewable Energy Agency Bill.

13

NEERAJ KUMAR GUPTA
PRATIBHA TANDON

Eco-Tax on Energy Resources

A Critical Appraisal with Special Reference to India

India is categorized as developing country as per the norms and indicators established for development.[1] In addition to this, India is the second largest country in terms of population and it is expected that it will very soon surpass China to become the largest.[2] To attain the status of a developed country, to maintain and feed such large population, to provide them with dignified life as mandated by the Constitution[3] requires that India increases the amount and pace of socio-economic development.

The need and right to develop has been recognized internationally[4] however, such right have certain limitations. The unsustainable pattern of development may adversely affect the human well-being thus, the developmental activities must be undertaken keeping in mind its environmental implications.[5] In the modern world, the developmental activities—economic or social—are dependent on the energy resources. A major resource of energy supply is fossil fuels such as coal, petroleum, and other allied resources. The use of these energy resources has resulted in greenhouse gas emission beyond the absorbing capacity of nature, resulting in environment pollution. Environment pollution has led to various adverse impacts such as climate change, deteriorating health of human and other species, extinction of various species, flooding, desertification, change in weather pattern, occurrence of natural disasters, and so on. In the light of these apparently conflicting needs of protecting and preserving environment and the right to develop, various measures have been suggested to reconcile the idea of development and maintaining wholesome environment.

These conflicting issues of environment and development are the core areas of environmental law. This also makes environmental law an interdisciplinary subject and it is confluence of various academic disciplines within law as well as non-legal branches of study.[6] Attempts are being made at national and international level to combat the issue of environment pollution. Various legal and non-legal measures have been suggested by international and national bodies and academicians. The chapter deals with the concept of environment tax, which is a branch of welfare economics, in the context of environmental law. Broadly speaking, inquiry in this chapter relates to the use of eco-tax as a measure to combat environment pollution and ensuring sustainable use keeping in mind the need of faster socio-economic growth.

The second part of the chapter is divided in various sub-headings. Firstly, it enquires theoretical basis of the eco-tax as a concept where we discuss the concept of externality. The concept was proposed in response to failure of market mechanism in fixing the correct price of the product and how this has led to unsustainable level of production and consumption of goods creating adverse impact on the environment. Secondly, it describes the solution to address the issue of externality and various criticisms levelled against the concept of eco-tax. It also describes how the concept of environmental tax entered in the realm of international environmental law. Lastly, it explains various types of eco-taxes and differentiates between eco-tax and environment related tax.

The third part of the chapter briefly sheds light on the magnitude of the consumption of conventional energy sources by India. The next part analyses Indian environment jurisprudence tracing the major developments which have taken place relating to environment protection. This part also analyses the legislations relating to eco-tax in India and judicial law-making in this filed. Fifth part critically analyses the concept of eco-tax, and then we conclude.

The Concept of Externality as Market Failure and Pigouvian Tax

Externality is an economic concept; the concept points towards the fact that an economic transaction between two persons may have some effect on the third person who is not involved in the transaction. It can also be termed as spill over effect of the transaction. An externality is a cost or benefit imposed on people other than those who buy or sell goods or services. The recipient of the externality is neither compensated for the cost imposed on him, nor does he pay for the benefit bestowed upon him. These costs and benefits are labelled 'externalities' because the people who experience them are outside of or *external* to the transaction to buy and sell the good or service. The externality can be of positive nature or negative.[7]

The concept of externality could be understood with the help of two examples. When a person invents a medicine and gets it patented he receives a limited monopoly over the invention. The patenting of the medicine allows him to reap the benefits of the investments made in the form of time and other goods which have resulted in such invention. This monopoly gives him the right to exploit the benefits of the invention. However, in the larger perspective, the invention also serves the society at large as it helps in maintaining better health of persons generally by curing the individuals ailing from the diseases. This benefit to the society at large is the externality, positive externality. Similarly, when a person produces a product using natural resources, in the process of production certain adverse impact on the environment may take place. The process may involve emission of hazardous chemical or toxic fumes or simply extinction of a natural resource. The society at large is the sufferer as the deteriorated environment may result in adverse health conditions. This is also an example of externalities, negative externality.[8]

Production of goods and services may result into positive as well as negative environmental externalities.[9] For example, when a machine is invented which increases the efficiency of fuel, it results in positive environmental externality, similarly, when coal or petroleum is burnt for production of goods or services it results in emission of various greenhouse gases, which has adverse effect on environment. In this chapter we will be restricting our inquiry to the issue of negative environmental externalities. These negative environmental externalities include air pollution, water pollution, and other types of pollution.[10]

Since the negative environmental externalities are not taken into account while fixing the price of the product, the cost of production is lower than the real cost as it does not include the cost borne by the society at large in the form of adverse effects of negative environmental externalities. These adverse effects may range from worsening health conditions to loss of vegetation and extinction of various species. This lower cost results in unsustainable pattern of production and consumption of the products.[11] The non-inclusion of the cost incurred due to the externalities is considered as market failure. It is argued that since the market has failed to take into account the costs of negative externalities, the state must intervene. Various methods of such correction have been suggested.[12] One of the mechanisms is levying tax on the externalities.

It was suggested by Arthur C. Pigou that the State should levy tax on such externalities. He suggested that the amount of tax must be such which neutralizes the social cost incurred in producing the goods or services.[13] In other words, the rate of tax must be such which is equivalent to the cost incurred by the neighbours who are affected by the pollution. This type of tax has been termed as Pigouvian Tax. The solution of taxation on externalities has been challenged on various grounds by academicians. It is pointed out that utility is not measurable; it is invalid to compare levels of utility between different people.[14] Different persons may place different utilities on the same commodity. It is also argued mathematical calculation of the value given to a product, service, or natural asset is highly misplaced. The mathematical calculations do not take into account preference given by different persons to different items.[15] When people demonstrate their preferences by exchanging, we can say that both parties felt that they would be better off trading goods than not. However, the solution of taxation is based on conjecture and surmises without taking into consideration the preference to a particular thing, the numbers arrived at will be a mere guesswork.[16]

Further, it was suggested that instead of taxation, defining the property rights in the resources will be more appropriate solution to the issue of externality.[17] It was argued that when the property rights are clearly defined and transaction costs are low, the individuals involved in situations of externalities can always negotiate a solution that internalizes any externality.[18] For example, a person running a factory in a village, which pollutes the river, vesting of proprietary right in the river in the villagers will lead to the negotiation between the factory owner and the villagers. Both parties can arrive at a solution best suited to them. Demarcation of property right theory is supported by the argument that it has the advantage of taking into consideration 'particular circumstances of time and place' by the parties, which can be perceived by them alone, to arrive at a solution. For instance, the factory owner may be aware of an alternate material which can be utilized for the purposes of production of goods that does not pollute the river, or the villagers might be aware that the river was stinky anyway and that it's best for them to move to other place. It is argued that these advantages cannot be taken into account either by the taxation regime or by command and control method of regulation.

Polluter Pays Principle and Eco-tax

The concept of negative environmental externalities has been recognized as an important issue at various international forums. This led to the emergence of the concept of 'Polluter Pays Principle'.[19] The polluter pays principle means that the polluter should bear the 'costs of pollution prevention and control measures', the latter being 'measures decided by public authorities to ensure that the environment is in an acceptable state'.[20] It requires that the polluter should bear the costs of pollution prevention and control measures, the latter being measures decided by public authorities to ensure that the environment is in an acceptable state.[21] In fact, the polluter pays principle is no more than an efficiency principle for allocating costs and does not involve bringing pollution down to an optimum level of any type, although it does not exclude the possibility of doing so.[22] The polluter pays principle was recognized as an important principle related to sustainable development in the Rio Declaration.[23] The ambit of the principle is expanding and various types of costs which relate to environment protection are being included under this principle.[24] Since the recognition of this principle, various other important conventions relating to environment have also recognized it as one of the important principles relating to environmental law.[25] It is suggested that the polluter pays principle can be implemented in various ways.[26] One of such ways is the use of economic instruments. The use of economic instruments was advocated by the report of the World Commission on Environment and Development.[27] Later the Rio Declaration also recognized it as one of the measures which can be utilized for implementing the polluter pays principle.[28]

Economic instrument could be understood as those decisions or actions of government that affect the behaviour of producers and consumers by causing changes in the prices to be paid for these activities. These instruments include effluent taxes or charges on pollutants and waste, deposit, refund systems, and tradable pollution permits. Environmental taxes also fall in the category of economic instruments.[29] There is no uniform definition of the term eco-tax. On the one hand Organization for Economic Co-operation and Development (OECD) defines environment tax as any compulsory, 'unrequited' payment to general government levied on tax-bases deemed to be of particular environmental relevance. The relevant tax-bases include energy products, motor vehicles, waste, measured, or estimated emissions, natural resources, and so on. Taxes are unrequited in the sense that benefits provided by government to taxpayers are not normally in proportion to their payments.[30] 'Requited' compulsory payments are levied more or less in proportion to services provided, for example fees and charges.[31] On the other hand, the regulation of the European Parliament and of the Council on European environmental economic accounts defines it as 'tax whose tax base is a physical unit (or a proxy of a physical unit) of something that has a proven, specific negative impact on the environment'.[32] It is to be noted that both definitions talk about tax-base. Various goods and services may be made the tax base and it may vary as per the policy of the state. Thus, if it appears to a state that any particular activity or any particular substance is creating an adverse impact on the environment, it can be regulated by levying the eco-tax on it.

Difference between Eco-tax and Environment Related Tax

Although the phrases eco tax and environment related tax appear similar, there is a huge difference. It is also possible that the environment related tax is actually working against the well-being of the environment. It is to be noted that sometimes the environment related taxes may not be able to reduce pollution.[33] This category of taxes may include taxes on energy resources which are levied for purposes other than the restoration of environment, such as for maintenance of roads or any similar purpose, other than the restoration or preservation of the environment.[34] Eco-tax, strictly speaking, is one of those taxes which simply internalize the externalities. In other words, only those taxes which are based on the calculation of the marginal social cost of an activity and are levied only to that extent which internalizes the marginal social cost can be categorized as environmental tax. Thus, strictly speaking, only the Pigouvian taxes qualify to be termed as eco-tax. On the other hand, environment related tax includes a broad area of taxes which relate to, or they are in some way or the other, connected to environment.[35]

Types of Environmental Taxes

Environmental taxes can be classified under various categories.[36]

1. Cost-Covering Charges: The object of this type of environmental tax is that the cost of regulation should be paid by those being regulated. It requires that those making use of the environment contribute to or cover the cost (cost covering charges) of monitoring or controlling that use in accordance with the polluter pays principle. Cost-covering charges can be further divided in two types of charges, (a) user charges, and (b) earmarked charges. The user charges are paid for a specific environmental service such as treating waste-water or disposing of waste. The revenue from earmarked charges is spent on related environmental purposes but not in the form of a specific service to the charge-payer, this can include activities like afforestation, refilling of the land after the mineral extraction, and so on.

2. Incentive Taxes are those taxes which are levied purely with the intention of changing environmentally damaging behaviour, and without any intention to raise revenues. The level of an incentive tax can be set according to externality cost of the environmental damage. Similarly, the incentive taxes can also signal expected behaviour from the consumers by raising the price of environmentally harmful activities.

3. Fiscal environmental taxes are the third category which is levied to change the behaviour, but yield substantial revenues over and above those required for related environmental regulation. The revenues

generated by these may be used in general budget deficits or shift taxes away from high income taxes, or high non-wage labour taxes, towards taxes on the consumption of resources and environmental pollution. These are also termed as 'Green Tax Reform'. These green tax reforms often consist of energy taxes and several other non-energy taxes like taxes on waste, waste water, pesticides, fertilisers, sulphur, and so on.

4. Environmental taxes designed mainly to raise significant revenues are called fiscal environmental taxes. Thus, we can say that taxes falling under category (3) and (4) are environment related tax and not eco-tax sensu stricto.

Environment Tax Reform

Eco-tax reform is a policy measure which is used by governments for better use of resources and to reduce pollution. For this purpose, economic instruments like environment taxes are utilized. Main objective of this reform is to increase the efficiency of resource use and improve the environment by levying taxes on the goods and activities which adversely affect the environment. In this policy, other taxes such as tax on income and labour are reduced gradually and that is why it is also called tax shift from 'goods' to 'bads'. To ensure that there is no additional burden on the taxpayers, normally this shift is revenue neutral. The idea behind this policy is that the tax base will shift from the good activities to bad activities which will not only reduce the harms on environment but it shall also create more capital and resources which can be utilized for employment generation and overall well-being of the residents. Due to the dual benefits expected from this type of policy, it is also termed as the double dividend tax reform. In addition, it is also expected that it will also affect the development of technology, particularly stimulating green innovation and green technology development.[37]

Magnitude of Consumption of Conventional Energy Sources in India

Though environment pollution cannot be attributed to conventional sources of energy alone, yet, the worldwide trend shows that whenever eco-taxes are proposed, the conventional sources of energy are targeted the most. Thus, keeping in mind the worldwide trend of giving attention to conventional sources of energy we try to analyse the magnitude of the consumption of conventional sources of energy in India. We shall also see the sector wise patterns of energy consumption in India.

The reports show that India's total consumption of energy is 2407 petajoules. The consumption has increased by 5.96 per cent if compared with the year of 2005–6. Coming to source wise consumption, coal and lignite consumption is 615.79 million tonnes, out of which, more than 78 per cent is consumed for electricity generation. On the other hand, Petroleum products account for 74,947 tonnes out of which more than 88 per cent is utilized in miscellaneous service sector and only 5 per cent is consumed in transport. The consumption of natural gas is 46.3 million tonnes of oil equivalent and here also the major share is equally divided between fertilizer industry (32.56 per cent) and power generation (31.02 per cent).[38] The data above gives a clear indication that the conventional sources of energy form a large portion of the total demand of energy. Further, the consumption is not limited to one aspect of life alone. While the high rate of coal consumption is due to electricity generation, on the other hand the petroleum products are being consumed by various service sectors. The use of natural gas is due to fertilizers and power generation.

The consumption level and the sectoral distribution of conventional sources of energy clearly indicate that greenhouse gas emission is not restricted to road transportation alone and there are various other segments where these resources are being consumed at high pace including electricity generation and service sectors. Thus, environment taxes on road transportation will prove only a small portion of the entire gamut of environment pollution being created from conventional sources of energy.

Environment Tax in India

In this part we shall be dealing with the environmental jurisprudence in India, in brief, and then we shall describe various legislations and judicial orders which can be categorized under the head of environment related tax. This part argues that the Indian environmental jurisprudence has grown from the infancy stages to more mature jurisprudence with the passage of time. It has taken inspiration from various sources to grow.

The environmental law regime in India has recognized various methods of environment protection at various stages.

Generally speaking, prior to the 42nd Amendment of the Constitution, the matters relating to environment were dealt by the common law system where courts provided relief to the litigants under the head of public nuisance for various types of environment pollution.[39] The Stockholm Declaration had massive impact on the Indian environmental jurisprudence. The 42nd Amendment of the Constitution, which took place in the year 1976, also contained provisions relating to protection, preservation, and improvement of environment in the form of Directive Principles of State Policies[40] and Fundamental Duties of Citizens.[41] Various legislations have also been enacted to protect and preserve the wholesome environment at national as well as regional level. Few of them are the Environment (Protection) Act 1986,[42] the Water (Prevention and Control of Pollution) Act 1974,[43] and the Air (Prevention and Control of Pollution) Act, 1981.[44] In addition to the legislative efforts relating to protection, preservation, and restoration of the environment, the role of Indian judiciary in environment protection deserves special mention. If seen in the long term perspective, we find a trend where the seriousness towards environment protection has grown on the part of higher judiciary and they have devised various mechanisms for ensuring the wholesome environment. Initially, the environment related matters were dealt under the common law concept of nuisance.[45] With the passage of time the courts started relating environmental issues with the issue of social justice.[46] Then came the period where the Supreme Court entertained a petition related to environment pollution under Article 32 of the Constitution and provided relief to the petitioner.[47] Though, the judgment nowhere mentions the wholesome environment as a fundamental right, but the relief was given under Article 32, which could be interpreted as the recognition of wholesome environment as fundamental right[48] and gradually the right to wholesome environment became established as part of right to life under Article 21 of the constitution.[49]

The Judiciary has also taken into consideration the international law related to environment and it has applied various principles relating to it. The Court recognized the concept of sustainable development as part of law of the land and observed that sustainable development has become part of the Customary International Law and 'there would be no difficultly in accepting them as part of the domestic law'. It was observed by the Supreme Court that 'it is almost accepted proposition of law that the rule of "customary international law" which are not contrary to the municipal law shall be deemed to have been incorporated in the domestic law and shall be followed by the Courts of Law'.[50] Further, the polluter pays principle was also described as one of the important principles relating to sustainable development by the judiciary.[51] The polluter pays principle has been applied in various cases since then.[52] Absolute liability and deep pocket theory are also recognized as forming law of the land on the issue of liability of the polluter.[53]

Another phase which is visible in the recent time relates to use of fiscal measures for protection, preservation, and restoration of environment. Here also, amount of legislative efforts and the judicial efforts appear to be on equal footing. In the next few paragraphs we shall be describing, in brief, the legislative and judicial efforts of protection, preservation, and restoration of environment with the help of fiscal measures.

Sikkim Ecology Fund and Environment Cess Act, 2005[54]

The Act aims to protect, preserve, and restore the environment and abate pollution.[55] Though the Act talks about other measures as well with regard to environment protection, the aim of the Act is to provide competence to levy the cess on the sale and use of the non-biodegradable—goods in the state and thereby discouraging their consumption. It is provided that the state is empowered to impose levy of cess or fee wherever the abuse of environment cannot be prevented.[56] The Act also empowers the imposition of restriction with regard to activities such as industry in specific areas, or the activities could be carried on only when the cess or fee is paid.[57] Further, state can also impose restrictions with regard to movement and entry of vehicles in the specified area or it can charge cess for allowing the entry or the movement of the vehicle.[58]

The cess collected under the Act is to be deposited in a separate non-lapsable fund called Sikkim Ecological Fund.[59] This Fund is meant to be utilized only for 'amelioration of the environment and the ecology' of the state.[60] The Fund is maintained as a current account in a nationalized bank.[61] The State may appoint a person to be the prescribed authority and also such other persons to assist him as may be required for the proper execution of the collection work.[62]

Every dealer or person including the person engaged in the business of works contract bringing non-biodegradable materials or goods for sale, in Sikkim from outside Sikkim is bound to pay the cess as prescribed on the sales turnover. Similarly, every person who brings non-biodegradable materials in Sikkim from outside Sikkim for own use and consumption or for whatsoever purpose other than making sales in Sikkim, is also bound to pay the cess on the purchase invoice value. In addition to that, every hotel, resort, lodge, or motel operating or carrying on business within Sikkim are also liable to pay cess as per the turnover of their business. On entry in the state of various vehicles as prescribed by the state can also levy cess at such rates as prescribed. Apart from the above categories the state is also empowered to prescribe and levy cess on such items having impact on environment directly or indirectly.[63] The rates are also prescribed in the Act with regard to cess to be levied on various items.[64] The list of non-biodegradable items is provided in the Schedule attached to the Act. The list includes items like cement, iron and steel, parts of motor vehicles, paints, varnishes, glass and glassware, wireless reception instruments, and so on.

The Act also prescribes the mechanism as to how the Act has to be implemented and the authorities which shall function under the Act. It also provides punishments in case of non-compliance of the provisions contained in the Act. It is required by the Act that every dealer that is liable to pay the cess has to register under the prescribed authority within 45 days of becoming liable. If dealers are still unable to comply, they are liable to pay Rs 500/day starting from the expiry of the 30 day notice. The state government of Sikkim has the responsibility for proper management of the ecology fund.

Goa Green Cess Act[65]

Goa Green Cess Act on products and substances causing pollution, namely, the Goa Cess on Products and Substances Causing Pollution (Green Cess) Act, 2013[66] is enacted by the Legislature of Goa with the aim to levy and collect cess on the products and substances including hazardous substances, which upon their handling or consumption, or utilization, or combustion, or movement, or transportation causes pollution of the lithosphere, atmosphere, biosphere, hydrosphere, and other environmental resources of the state of Goa, under the concept of polluter pays principle. Further, it also aims to provide for measures to reduce the carbon footprint left due to such activities and for matters connected therewith or incidental thereto.[67] The Act provides for levying of the cess at the rate of two per cent on the sale value of the products and/or substances, the handling, utilization, consumption, combustion, transportation, or movement, of which, by any means, causes pollution within the State of Goa, from every person carrying out any of the above activities.[68] The word 'product' is defined in the broadest possible manner. It means those products which upon their handling, consumption, utilization, combustion, or movement, or transportation cause pollution of the lithosphere, atmosphere, biosphere, hydrosphere, and other environmental resources and causes emission of carbon dioxide and other greenhouse gases or discharge other types of effluents and includes asphalts, automotive gasoline, fuel oils, kerosene, lubricants, naphtha, waxes, other hydrocarbon compounds including mixtures, and products obtained from crude oil and natural gas processing, and such other products which the government notifies as such for the purposes of the Act.[69] Similarly, the word substances are also defined in the same line to include coke, coal, chemicals and chemical products, hazardous substances, and such other substances as the government notifies.[70]

The Act provides that the cess collected under the Act shall be utilized for undertaking the measures to reduce the carbon footprint, 'by means of such programmes or schemes as may be decided by the government'.[71] The Act also provides for the establishment of Environmental and Energy Audit Bureau to identify sensitive areas of energy and environmental conservation and to recommend appropriate measures and solutions for reducing carbon footprint, and to suggest measures for deriving benefits under carbon credit trading and related matters in the state.[72] The Act also envisages the machinery for the implementation of the Act.[73] The Act also provides for imposition of penalties in case of failure to observe the provisions of the Act.[74]

Cess on Coal

In the year 2010,[75] the Government of India levied the 'Clean Energy Cess',[76] as duty of excise, on the goods produced in India as specified in the Tenth Schedule (the schedule contains the entries coal, briquettes, ovoids, and similar solid fuels manufactured from coal, lignite, whether or not Agglomerated excluding jet, Peat including peat litter, whether or not agglomerated). The purpose of this cess was financing and promoting clean energy initiatives, finding research in the area of clean energy, or for any other purpose relating thereto. The rate of clean energy cess was initially fixed at the rate of Rs 100 per tonne however, exemption was granted with regard to Rs 50 and effective rate was notified as Rs 50 per tonne.[77] In the Finance Act, 2016, this rate has been enhanced to Rs 400 per tonne. The Finance Act has also changed the nomenclature from 'Clean Energy Cess' to 'Clean Environment Cess'.[78]

Swachh Bharat Cess

The Finance Act 2015[79] has levied a cess to be called the Swachh Bharat Cess, as service tax on all or any of the taxable services at the rate of two per cent on the value of such services for the purposes of financing and promoting Swachh Bharat initiatives, or for any other purpose relating thereto.[80] This Cess is in addition to any cess or service tax leviable on such taxable services.[81] The proceeds of the cess levied shall first be credited to the Consolidated Fund of India and the central government may, after due appropriation made by Parliament by law in this behalf, utilize such sums of money of the Swachh Bharat Cess for the purposes of financing and promoting Swachh Bharat initiatives or for any other purpose relating thereto.[82] The manner of collection, assessment, exemption, and so on, is to be in accordance with the other service taxes.[83]

Water Cess Act[84]

The Act is intended to provide for the levy and collection of a cess on water consumed by persons carrying on certain industries and by local authorities, with a view to augment the resources of the Central Board and the State Boards for the prevention and control of water pollution. The Act extends to all states to which the Water Pollution Act, 1974 applies and also to the Union territories. According to Section 3 of the Water Cess Act, 1977, a cess is to be levied and collected for the purposes of the main Act and for utilization thereunder from every person carrying on any specified industry and from every local authority. The specified industry is listed in the First Schedule to the Act.[85] The cess is levied on the basis of the water consumed for the purposes specified in the Second Schedule, column 1. The maximum rate of cess is laid down in the Second Schedule, column 2. The Act empowers the central government to specify the rate from time to time within the maximum rate as provided in the Schedule. It is provided in the Act that for the purposes of measuring and recording the quantity of water consumed, meters are to be affixed.[86] The Act provide for various administrative matters relating to the implementation of the Act under section 5 to 10. Under section 11 of the Act it is provided that where the cess as levied under the Act is in arrears, the prescribed authority may impose a penalty not exceeding the amount of cess in arrears. Section 14(1) of the Act provides punishments in case of furnishing of a false return, while section 14(2) of the same Act punishes wilful evasion of cess. Section 16 of the Water Cess Act empowers the central government by notification to add industries in the First Schedule.

Gujarat Green Cess Act, 2011[87]

The Act provides that cess shall be levied on generation of electricity except on generation of renewable energy at the generating company at the generating station or at the captive generating plant, or the stand by generating plant.[88] The cess is payable irrespective of the fact whether such electricity is consumed within the state or not.[89] Rate of cess is twenty paise per unit of the electricity generated.[90] However, the state is empowered to exempt from payment of the cess, having aggregate installed capacity of not more than one thousand kilowatts.[91] It is provided in the Act that the proceeds of the cess, interest, and penalty recovered under it shall first be credited to the Consolidated Fund of the State, and after deduction of the expenses of collection and recovery there from shall, under appropriation duly made by law in this behalf, be entered in and transferred to a separate fund called the Green Energy Fund, for being utilized exclusively for the purposes of this Act.[92] The cess collected and fine generated is to be utilized for promoting the generation of electricity through renewable energy, purchase of non-conventional energy, and taking initiatives for protecting environment in the state.[93] This Act also creates the authorities for the enforcement of the Act and provides for offences and penalties in case of breach of the provisions contained therein.[94]

The Act was challenged in the High court of Gujarat[95] and the Act was held to ultra vires. The main contention of the industrialists in the petition presented to the court was that the state legislature lacks competence to levy any tax on generation of electricity as it is the field reserved for the Parliament and it is already occupied by the Parliament. On the other hand, it was contended on behalf of the state that the cess levied under the Act is for health, and so on, which falls in the exclusive domain of the state. In the alternative, it was argued that the cess is not tax per se and it is a type of 'fee'. The court came to a conclusion that the cess was levied on the generation of electricity and therefore in breach of the constitutional mandate. Further, the utilization of the cess was for renewable energy and other laudable activities yet it was not being utilized directly or indirectly for the benefit of the payer of such cess, which is the nature of a 'fee' levied. Appeal has been filed against the judgment of the High Court which is pending till date. However, the Supreme Court stayed the Gujarat High Court judgment quashing the state legislation as unconstitutional.[96] The matter is pending in the Supreme Court.[97]

Judicial Approach on Environment Tax

National Green Tribunal imposed an environment cess of Rs 1,000 and Rs 2,500 per visit respectively for every petrol and diesel tourist vehicle from Manali going to Rohtang Pass, along with the cap of one thousand vehicles on daily basis on the ground that the tourist surge to the hill station posed a threat to its fragile ecology.[98] However, cess was stayed by the Tribunal after some time.[99]

Supreme Court imposed the Environment Compensatory Charge of Rs 700 and 1,300 on commercial vehicles entering National Capital for the purposes of transit. The rate of tax is Rs 700 for one light duty vehicle and two-axle trucks and Rs 1,300 for three- to four-axle trucks. The toll tax operators are also required to install radio frequency identity system at their own cost. The Green Cess levied has exemptions to vehicles carrying essential commodities like foodstuff, passengers, and emergency including ambulances. As per the directions of the Court, the toll tax would be collected by the toll operators and passed to the Delhi government.[100]

In another order related to Delhi's high level pollution, the Supreme Court passed an order banning the registration of cars above the engine capacity of 2000 CC and above[101] temporarily, later on the registration was lifted on the undertaking given by the car manufacturers that they are willing to pay one per cent of the ex-showroom price as environment protection charge on the SUVs running on diesel.[102] The court prescribed in this order that Central Pollution Control Board shall open separate account for the deposit of this amount. The court observed that the fixation of amount at the rate of one per cent is an interim and it can be modified but not retrospectively.[103] In the same order the court also allowed the registration of vehicles to be used by the Tihar Jail for ferrying inmates without payment of any cess.[104] Another interesting order on the registration of diesel vehicle was passed by the Apex Court where the Delhi Police asked for the registration of diesel vehicles. In this order the court asked Delhi Police to pay thirty per cent of the purchase value before registration.[105]

Critical Analysis

The concept of environment tax was mooted to address the issue of negative environmental externality. It was provided by Pigou that the rate of tax must be such which is equivalent to the cost incurred by the society at large. However, the criticism on the issue of quantifying the losses in numbers and giving it a monetary value is a difficult task. On the other hand, Coase has argued that the market can decide the cost of externality and there is no need of any intervention by the state. However, it should be noted here that the market is a composition of existing persons and there is no place in the market for the non-existent persons or future generations, the solution provided by Coase does not take into account the future generations. The environmental law as developed at international and national level provides for intergenerational equity. Thus, it can be said that state intervention appears to be necessary.

It is to be noted that the environmental taxes have certain advantages if compared to command and control mechanism of environment regulation. The taxation regime provides incentives to the polluter to reduce pollution to the maximum possible extent in comparison to command and control system where the polluters need not worry once he is polluting within the prescribed amount. Another important benefit is environment tax regime incentivises the innovation in the sense that the polluters are always looking for better technology so that they have to pay minimum amount of tax. Researchers have also given the double dividend theory which argues that the environment tax not only abates pollution, but it also provides additional resources for society which leads to betterment of the society by funding the employment generation and other welfare activities, apart from green tax shift.

However, the problem lies where the environment tax has to be introduced for the first time. Resistance is possible, as the taxpayers would not be inclined to pay more and more taxes. Thus, politically, it is a difficult task for any government to introduce such taxes in the democratic setup.

Another difficult question which arises is that how to utilize the revenue generated from the environment tax. It is argued that the taxpayer will be willing to pay environment taxes only when he has this confidence that his payments will be used exclusively for environment protection and restoration. However, in the current scenario, where the concept of sustainable development is a dominating principle for protection of environment, it becomes difficult to demarcate between the activities which are exclusively termed as relating to environment protection. The Sustainable Development Goals as envisaged by international community require the state to ensure reduction in poverty, elimination of hunger, good health and well-being, quality education, gender equality, clean water and sanitation, affordable and clean energy, decent work and economic growth, industry innovation and infrastructure, reduced inequalities, sustainable cities and communities, responsible consumption and production, climate action, protection of life below water and life on land, peace, justice and strong institutions, and global partnership, are the goals which in itself cannot be exclusively compartmentalize as relating to environment protection.[106]

Coming to Indian scenario, it is clear that efforts in the field of environment tax are visible in legislative as well as judicial law making. Various legislations relating to environment tax definitely adhere to the concept of externality. However, it is very difficult to predict as to what is the parameter to provide the amount of tax to be levied. We find that various legislations provide rate of cess from 20 paise per unit electricity production to 2 per cent purchase value of the goods of non-biodegradable nature. Another issue which can be highlighted here relates to utilization of the fund generated. The legislations as enacted in India use the broadest possible terminology where they provide for utilization of the fund.[107] Thus, even if the fund is earmarked for environment protection, still there is ample amount of discretion which can be exercised by the government. Secondly, the challenge to the legislation of Gujarat highlights the issue of federal structure and taxation. Any law made by the legislature need to adhere to Constitution. However, it is expected that the matter will be taken up by the Parliament and now the Goods and Services tax will be accommodating this issue.[108]

Similarly, the Supreme Court also appears to be recognizing the concept of externality when it observed that the trucks enter Delhi only to save toll taxes. However, the problem relating to rate of charge to be levied is visible in the orders of Supreme Court, where the court agrees to 1 per cent environment compensation charge in case of diesel SUVs and on the other hand it asks the jail authorities to pay 30 per cent of purchase value of the vehicle. Another important issue that needs to be addressed is that what goods or services are to be given exemption and what not. We find in the order relating to trucks' entry in Delhi that certain trucks which carry foodstuffs and essential commodity were exempted from the charges. It is purely a policy matter whether to exempt certain goods or not or whether a class of people must be taxed or exempted, which requires sufficient data to arrive at a decision.

***

On the above analysis, it can be said that the idea of environment tax as a measure to protect and preserve environment has its own merits and demerits. It needs to be understood that the socio-economic development of a country like India is still required at a fast pace, any legislation in the field of environment tax is bound to face stiff resistance from various segments of society. However, it also needs to be understood that we have a duty to protect and preserve the environment. The state intervention is needed and that is why we find that various states have started acting towards this direction.

It needs to be emphasized here that in India we find that there is huge population which is still reeling under poverty, putting them under the burden of environment tax might add to their poverty. Thus, any policy relating to environment tax must be framed in such a manner which takes into account the problems of the lowest strata of society.

Further, taxation may have various benefits in comparison to command and control mechanism, yet, it cannot act independently, as it is clear from the orders of the Supreme Court in the case of entry of commercial vehicles in Delhi. In other words, any mechanism relating to environment tax has to act in tandem with other modes of laws relating to environment.

Notes and References

1. World Economic Situation and Prospects 2015, available http://www.un.org/en/development/desa/policy/wesp/wesp_archive/2015wesp_full_en.pdf, accessed on 24 October 2016.

2. United Nations, Department of Economic and Social Affairs, Population Division (2015), 'World Population Prospects: The 2015 Revision, Key Findings and Advance Tables', Working Paper No. ESA/P/WP.241: p. 10, available https://esa.un.org/unpd/wpp/Publications/Files/Key_Findings_WPP_2015.pdf, accessed on 27 October 2016.

3. See generally, Constitution of India, 1950, Part IV, Directive Principles of State Policy, Articles 36 to 51. See also the case of *Maneka Gandhi v. Union of India* where it was held that the right to life as contained in article 21 of the Constitution of India is not mere life like a living corpse; it means a life with human dignity. The concept of human dignity has been reiterated in various following judgments. For more details on this subject see generally, Ruma Pal and Samaraditya Pal (Rev.) M.P. Jain, *Indian Constitutional Law* (Gurgaon: LexisNexis, 2010), Chapter XXVI, Part E.

4. Principle 8 of the Stockholm Declaration provides that economic and social development is essential for ensuring a favourable living and working environment for man and for creating conditions on earth that are necessary for the improvement of the quality of life. Principle 21 of the same Declaration provides that states have sovereign right over the natural resources. This recognition impliedly confers the right to develop on the states. Similarly, principle 2 of the Rio Declaration also confers this sovereign right over the natural resources in the territory.

5. See generally, Stockholm Declaration and particularly, Principles 4, 5, 6, 11, and 12.

6. P. Leelakrishnan, *Environmental Law in India*, Third Edition (Gurgaon: LexisNexis, 2015 [2008]), p. 3.

7. See generally, W.J. Baumol and W.E. Oates, *The Theory of Environmental Policy* (Cambridge, UK: Cambridge University Press, 1988). For a critique on the subject, see, B.P. Simpson, 'An Economic, Political, and Philosophical Analysis of Externalities', available https://reasonpapers.com/pdf/29/rp_29_8.pdf, accessed on 24 October 2016.

8. The concept of externalities was propounded by economist A. Marshall and it was developed by A.C. Pigou. For more details see, Arthur Cecil Pigou, *The Economics of Welfare*, 4th Edition (London: Macmillan and Co. Limited, 1932), p. 106–12, available http://oll.libertyfund.org/EBooks/Pigou_0316.pdf, accessed on 25 October 2016.

9. B. Caplan, 'Externalities', the Concise Encyclopaedia of Economics, available http://www.econlib.org/library/Enc/Externalities.html, accessed on 25 October 2016.

10. Baumol and Oates, *The Theory of Environmental Policy*, p. 12.

11. Pigou, *The Economics of Welfare*, pp. 173–4.

12. R.H. Coase argues that creation of property rights is one of them, another example is setting the norm of maximum allowable limits of emissions of toxic fumes and chemicals.

13. Pigou, *The Economics of Welfare*, pp. 174–5.

14. J.M. Buchanan, 'Cost and Choice: An Inquiry in Economic Theory', Liberty Fund, *Collected Works of James Buchanan*, Vol. 6 (1963), p. 66, available http://oll.libertyfund.org/titles/buchanan-cost-and-choice-an-inquiry-in-economic-theory-vol-6-of-the-collected-works?q=Cost+and+choice#Buchanan_0102-06_12, accessed on 25 October 2016.

15. Buchanan, 'Cost and Choice'.

16. Buchanan, 'Cost and Choice'.

17. R.H. Coase, 'The Problem of Social Cost', *The Journal of Law and Economics*, III (1960), pp. 1–44, 4–5, available at http://econ.ucsb.edu/~tedb/Courses/UCSBpf/readings/coase.pdf, accessed on 25 October 2016.

18. Coase, 'The Problem of Social Cost'.

19. The principle was recognized firstly by the Organization for Economic Co-operation and Development (OECD) countries. For more details on the subject as existing in OECD countries see, 'Environment Monographs: The Polluter-Pays Principle', OECD Analyses and Recommendations, Environment Directorate Organisation For Economic Co-Operation and Development, Paris (1992), available http://www.oecd.org/officialdocuments/publicdispla ydocumentpdf/?cote=OCDE/GD(92)81&docLanguage=En, accessed on 15 September 2016.

20. OECD Recommendation of the Council on Guiding Principles concerning International Economic Aspects of Environmental Policies, 26 May 1972—C(72)128, available http://acts.oecd.org/Instruments/ShowInstrumentView.aspx?InstrumentID=4&Lang=en ; Recommendation

of the Council on the Implementation of the Polluter- Pays Principle, 14 November 1974 - C(74)223, available http://acts.oecd.org/Instruments/ShowInstrumentView.aspx?InstrumentID=11&Lang=en&Book=False, accessed on 25 October 2016.

21. OECD Recommendation of the Council on Guiding Principles concerning International Economic Aspects of Environmental Policies.

22. OECD Recommendation of the Council on Guiding Principles concerning International Economic Aspects of Environmental Policies.

23. Principle 16, Rio Declaration. A/CONF.151/26 (Vol. I) Available http://www.un.org/documents/ga/conf151/aconf15126-1annex1.htm.

24. Apart from the head of Pollution prevention and control costs, other heads are being recognized such as, Inclusion of the costs of administrative measures, Inclusion of the cost of damage, Inclusion of accidental pollution and many more. For more details on this, see, 'Environment Monographs: The Polluter-Pays Principle'.

25. For Example, International Convention on Oil Pollution Preparedness, Response and Cooperation, 1990, Concluded at London on 30 November 1990, No. 32194, 1891-I-32194-English, Preamble, available https://treaties.un.org/doc/Publication/UNTS/Volume%201891/volume-1891-I-32194-English.pdf, accessed on 22 October 2016; Preamble, Convention on the Transboundary Effects of Industrial Accidents, 2105, I-36605, available http://www.unece.org/fileadmin/DAM/env/documents/2006/teia/Convention%20E%20no%20annex%20I.pdf, accessed on 22 October 2016; Convention For The Protection Of The Marine Environment of The North-East Atlantic, available http://www.ospar.org/site/assets/files/1290/ospar_convention_e_updated_text_in_2007_no_revs.pdf] and Stockholm Convention On Persistent Organic Pollutants [http://www.pops.int/documents/convtext/convtext_en.pdf, accessed on 25 September 2016.

26. Damages, if pollution occurs and its adverse impact harms the human or environment, prescribing remedial measures in the form of better technology to be used in the process of production and supply of goods and services respectively. For more details see, 'Environment Monographs: The Polluter-Pays Principle'.

27. See, *Our Common Future*, Report of the World Commission on Environment and Development, p. 152.

28. Principle 16, Rio Declaration.

29. 'Glossary of Statistical Terms, Economic Instruments', available https://stats.oecd.org/glossary/detail.asp?ID=6408, accessed on 25 October 2016.

30. OECD, *The Political Economy of Environmentally Related Taxes*, (2006), p. 28.

31. OECD, *The Political Economy of Environmentally Related Taxes*.

32. Regulation (Eu) No. 691/2011 of the European Parliament and of the Council of 6 July 2011 on European environmental economic accounts, available

http://eur-lex.europa.eu/legal-content/EN/TXT/PDF/?uri=CELEX: 02011R0691-20140616&from=EN, accessed on 25 October 2016.

33. Regulation (Eu) No. 691/2011.

34. Regulation (Eu) No. 691/2011.

35. E.M. Naess and T. Smith, 'Environmentally Related Taxes in Norway: Totals and Divided by Industry', Statistics Norway/Division for Environmental Statistics, available https://circabc.europa.eu/sd/d/.../NO%20470%20env% 20taxes.pdf, accessed on 27 October 2016.

36. European Environment Agency, 'Environmental Taxes: Implementation and Environmental Effectiveness', Environmental Issues Series, no. 1 (1996) Copenhagen, available http://www.geota.pt/rfa/docs/gt.pdf], accessed on 25 October 2016; [http://www.ecosmes.net/cm/navContents?l=EN&navID=ec oTaxes&subNavID=1&pagID=10&flag=1, accessed on 25 October 2016.

37. P. Ekins, 'Resource Productivity, Environmental Tax Reform and Sustainable Growth in Europe', Anglo-German Foundation for the Study of Industrial Society 2009, available http://www.petre.org.uk/pdf/FinRepFin. pdf, accessed on 27 October 2016.

38. All the figures are taken from the *Energy Statistics 2015*, Central Statistics Office Ministry of Statistics and Programme Implementation Government of India, New Delhi, available mospi.nic.in/mospi_new/upload/Energy_ stats_2015_26mar15.pdf, accessed on 27 August 2016.

39. Leelakrishnan, *Environmental Law in India*, p. 18. See also, B.N. Kirpal, 'Developments in India Relating to Environmental Justice', available http:// www.unep.org/delc/Portals/119/publications/Speeches/INDIA%20.pdf, accessed on 5 October 2016.

40. Article 48A Protection and Improvement of Environment and Safeguarding of Forests and Wild Life—The State shall endeavour to protect and improve the environment and to safeguard the forests and wild life of the country.

41. Article 51A (g) Fundamental Duties—It shall be the duty of every citizen of India to protect and improve the natural environment including forests, lakes, rivers, and wild life, and to have compassion for living creatures.

42. Act 29 of 1986.

43. No. 6 of 1974.

44. Act No. 14 of 1981. All these laws were based on the command and control mechanism of regulation of environmentally harmful activities. In other words, it provided that particular activities relating to environmental pollution shall not be carried on without the permission of the State and it also provided the standards in regards to the activities to be carried on. The breach of the provisions contained in these Acts was made punishable.

45. For more details on this area see, Leelakrishnan, *Environmental Law in India*, pp. 15–34.

46. The *Municipal Council, Ratlam v. Vardichand*, (1980) 4 SCC 162 decision interpreted the statutory provision of Section 133 Cr.P.C in a manner which might have never been thought of ever before.

47. *Rural Litigation and Entitlement Kendra v. State of UP* (1985) 2 SCC 431, Section 133.

48. Leelakrishnan, *Environmental Law in India*, pp. 223–51, 224.

49. *Subhash Kumar v. State of Bihar* AIR 1991 SC420.

50. *Vellore Citizens' Welfare Forum v. Union of India* AIR 1996 SC 2715, para 15.

51. *Vellore Citizens' Welfare Forum v. Union of India* AIR 1996 SC 2715, para 15.

52. See for example, *Indian Council for Enviro-Legal Action v. Union of India* AIR 1996 SC 1446; *M.C. Mehta v. Kamal Nath* AIR 1997 SC 734.

53. *M.C. Mehta v. Union of India*, AIR1987 SC 965.

54. Sikkim Act No. 1 of 2005.

55. Section 3(1).

56. Section 3(1)(c).

57. Section 3(1)(e).

58. Section 3(1)(f).

59. Section 13(1).

60. Section 14.

61. Section 13(3).

62. Section 4.

63. Section 5.

64. Section 6.

65. The Goa Cess on Products and Substances Causing Pollution (Green Cess) Act, 2013 (Goa Act 15 of 2013).

66. Goa Act 15 of 2013.

67. As per the long title of the Act.

68. Section 4.

69. Section 2(e).

70. Section 2(g).

71. Section 5.

72. Section 6.

73. Section 3.

74. Section 7.

75. The Finance Act, 2010 No. 14 of 2010, available http://lawmin.nic.in/legislative/textofcentralacts/2010.pdf, accessed on 29 September 2016.

76. Section 83(2).

77. For more details, see *e-Book on Clean Environment (energy) Cess,* available www.nacenkanpur.gov.in/download3.inc.php?rid=239, accessed on 24 October 2016.

78. Section 235 of the Finance Act 2016, available http://www.cbec.gov.in/resources/htdocs-cbec/fin-act2016.pdf., accessed on 24 October 2016. For more details see, e-Book on Clean Environment (energy) Cess.

79. The Finance Act, 2015 No. 20 of 2015, available http://www.cbec.gov.in/Cbec_Revamp_new/resources/htdocs-cbec/finact2015.pdf, accessed on 29 September 2016.

80. Section 119(2).

81. Section 119(3).

82. Section 119(4).

83. Section 119(5).

84. The Water (Prevention and Control of Pollution) Cess Act, 1977 (No. 36 of 1977).

85. These industries are ferrous metallurgical industry, Non-ferrous metallurgical industry, Mining industry, Ore processing industry, Petroleum industry, Petro-chemical industry, Chemical industry, Ceramic industry, Cement industry, Textile industry including cotton, synthetic and semi-synthetic fibres manufactured from these fibres, Paper industry, Fertilizer industry, Coal (including coke) industry, Power (thermal, diesel) and Hydel generating industry, Processing of animal or vegetable products industry including processing of milk, meat, hides, and skins, all agricultural products and their wastes, and Engineering industry.

86. Section 4.

87. The Gujarat Act No. 3 of 2011.

88. Section 3(1).

89. Section 3(2).

90. Section 3(3).

91. Section 3(4).

92. Section 4.

93. Section 6 (1).

94. See generally, the Act.

95. It is an oral order, *Reliance v. State*, available at http://indiankanoon.org/doc/56617404/, accessed on 27 October 2016.

96. 'SC Admits Gujarat Govt's Petition for Green Cess on Conventional Power Firms', available http://www.financialexpress.com/archive/sc-admits-gujarat-govts-petition-for-green-cess-on-conventional-power-firms/1137782/, accessed on 25 October 2016.

97. Civil Appeal no. 5135 -5157 of 2013.

98. 2015 SCC OnLine NGT 327.

99. 'Tribunal Stays Environmental Cess on Vehicles at Rohtang Pass', available http://www.financialexpress.com/archive/sc-admits-gujarat-govts-

petition-for-green-cess-on-conventional-power-firms/1137782/, accessed on 25 October 2016.

100. *M.C. Mehta v. Union of India*, I.A. 365 of 2015 in No. 345 in WP (C) No 13029 of 1985 order dated 9 October 2015, available at 2015 SCC Online SC 968.

101. 2015 SCC Online SC 1327, para 10.

102. I.A. No. 376-377, 80-81, 82, 85, 86, 90 in I.A. Nos. 380-381, 423, 445-446 in WP(c) 13029 of 1985, available at 2016 SCC Online SC 1130.

103. Para 6.

104. In I.A. No.467–8.

105. In I.A. No. 447–8.

106. For more details see, the official website at http://www.un.org/sustainabledevelopment/sustainable-development-goals/, accessed on 25 October 2016.

107. See, the Sikkim Act and Act of Goa.

108. A. Mohan, 'GST Rate for Polluting Products Could Be Higher', *Business Standard* (15 October 2016), available http://www.business-standard.com/article/economy-policy/gst-rate-for-polluting-products-could-be-higher-116101401430_1.html, accessed on 25 October 2016.

14

M. AFZAL WANI

Humanity, Energy, and Law
Urgencies and Challenges

Anything that occupies space and has weight is called 'matter'. Anything that gives 'life' to that matter is 'energy'.[1] Nonexistence of energy is death, excess of energy is a threat, and low energy is a crisis. Excessive production of energy is gluttony, excessive use of energy is lunacy, and misuse of energy is a peril. In other words, to sustain life there should be an appropriate and careful production, preservation, and utilization of energy. Human beings are one of the varieties of life on the planet earth. They need energy to survive and sustain their activities. They have a thinking mind, a punching hand, and erect posture. That gives them an advantage on all other forms of creation including animals and plants. They can feel, experience, ponder, measure, make, and manage. They can produce more energy on need and regulate when in excess, therefore, owing a responsibility to themselves for their sustenance and much more to other forms of matter and life, whether animate or inanimate.

There are multiple types of energy which can change form and each kind is put to use as per requirement.[2] There are sources and techniques for production of energy which have to be used cautiously and carefully to the advantage of everything around. Human wisdom, that is, ability and capacity to perceive, ponder, and prepare has to control and regulate the process of energy production, proliferation, and utilization to its best. It should not allow anything that can bring in misuse, overuse, and devastation.

Historically, human beings have the experience of using all forms of energy from conventional sources of wind, wood, water, and waste to mineral reserves and nuclear backup. Met with, besides situations of harmony and peace, with disharmony, hate and war, the unwanted production, misuse and overuse of energy has been a part of human history. That has lessons to teach enough and experience to learn much. Time is to wake up, realize, rationalize, respond, and assure a safe environment to sustain.

Quantum of Energy Required

The quantum of energy that humanity requires and its production by the use of energy producing technologies are being estimated by relevant quarters along with the assessment of greenhouse gas emissions taking place during the process of production and use of energy. The purpose is to look for fulfilment of energy requirements out of sustainable resources and evaluation of future intensity of greenhouse emissions. The yearly energy demand of the world at present is 409 EJ (9,741 Mt_{oe}) of commercially traded primary energy.[3] Primary energy encompasses not only the final use of the energy but also energy used in energy transformation industries such as electricity generation, that is, the energy contained in the coal that was burnt to produce electricity. The current level of primary energy use releases 29 Gt of carbon dioxide into the atmosphere each year, 4.4 t per person. In terms of just carbon mass this is 7.9 Gt of carbon (GtC), or 1.2 t per person.[4] This level will increase during the present century itself unless the fractional share of each fuel changes.[5] EIA (Energy Information Administration termed International Energy Outlook or IEO) predicts rise in energy use up to 2020 by 60 per cent to 650 EJ. Use of oil is likely to continue as preferred fuel by 40 per cent as now for its consumption in transportation system. Though there will be more generation of electricity from natural gas, water, and alternative fuels the share of oil in the energy production would continue to be so.[6]

Natural gas consumption till 2020 shall rise to 177 EJ (162 trillion cubic feet, 4,200 Mt_{oe}) to represent 28 per cent of energy market by 2020. Nuclear power is expected to be 359 GW in 2020 with greatest expansion in China, South Korea, India, and Taiwan.[7] Renewables would represent eight per cent market share mostly from large scale hydro projects especially in China, India, Malaysia, and other developing nations.[8] The electricity consumption over the period is expected to increase to 22 TkWh (79 EJ).[9] Growth in transportation system especially in personal transport (car ownership) is obviously notable. All that use of energy would increase carbon emissions from it till 2020 to 9.9 GtC_{eq} from 7.9 GtC_{eq} in 1999.[10]

For policy makers, also, IPCC (Intergovernmental Panel on Climate Change), established by World Meteorological Organization and the United Nations Environment Programme, has commissioned a Special Report on Emissions Scenarios (SRES), predicting emissions of greenhouse gases. It assesses the scientific, technological, and socio-economic information relevant to an understanding of anthropogenic climate change and considers, inter alia, the anthropogenic emissions as carbon dioxide (CO_2), methane (CH_4), nitrous oxide (N_2O); hydrofluorocarbons (HFC_S), perfluorocarbons (PFC_S), sulphur hexafluoride (SF_6), hydrochloro-fluorocarbons ($HCFC_S$), chlorofluorocarbons (CFC_S), sulphur dioxide (SO_2), carbon monoxide (CO), nitrogen oxide (NO_X), and, non-methane volatile organic compounds ($NMVOC_S$).[11] The emission scenarios in SRES are being presented in four settings. One, the world would be of very rapid economic growth, highest population growth in mid-century and decline thereafter, and the introduction of new and more efficient technologies. Major underlying themes are convergence among regions, capacity building, and increased cultural and social interactions with reduction in regional differences in per capita income. The technological emphasis would be in alternative on, fossil fuel intensive, non-fossil fuel energy sources, or a balance across all sources. Two, in the offing is a very heterogeneous world with the underlying currents of self-reliance and local identities, continuously increasing global population, and a slower and fragmented technological change. Three, as in 'one' above with respect to population, the world would witness rapid changes in economic structure towards a service and information economy, with reduction in material intensity, and introduction of clean and resource efficient technologies. Global perspective would be stressed for economic, social, and environmental sustainability with enhanced equity and impartiality. Four, move towards emphasis on local solutions to economic, social, and environmental sustainability, a continuous low rate increase in population, intermediate levels of economic development, and a slower and diverse technological change. All the four descriptions point to a world of higher emissions and an unnatural and pretentious climate with cumulative carbon emissions in 2,100 of 770 GtC to 2,540 GtC.[12]

Threat from the Situation

The quantum of energy the world is to use in coming times from now involving the process of production and emissions with use of resources is one of the biggest challenges before humanity. Devastating indicators are the ensuing change in temperature, melting of ice sheets, rise in sea level, diminishing fauna and flora, decreasing food supply, reduction of water resources, drought, disease, tornadoes, storms (with or without rain, hail and snow), barren soil, poisonous air, geopolitical disasters, dismay and disappointment. As per WHO, the per annum loss of human life presently due to these factors is 150,000 which will double in the coming two decades.[13] Increase in malaria, heat stroke, dehydration, malnutrition, is the obvious outcome and apparent impact of these changing conditions. In the given scenario, can law play the role of a redeemer and contribute in constructing an alternative safe future for all is a million dollar question. The question becomes more relevant when it is being claimed that much of the observed recent adverse global climate change is due to human activities, that is, it is anthropogenic in nature and character. In the contemporary world, human capacity has immeasurably multiplied which in many dimensions proves more supportive for violation of norms rather than proving supportive to them. Thus, the issues need enough attention to shape comprehensive policies and legislations resulting in better hope for the world.

Identified Activities Needing Legislative Intervention

The role of law in prevention of excessive production, excessive use, and misuse of energy against an unruly gluttony and lunacy causing peril becomes inevitable and most crucial. Given the emerging legal culture at the national and global levels, the expectations from law are unimaginably voluminous. Cultivation of a good governance mechanism, required ethics, and proper quality standards would be greatly supportive to the cause. At the international level all governments real-ize a need for regulated industrial activities which have a direct relation with use and over use of energy. A kind of dialogue among common people, academics, and the policy makers is informally and formally going on in the whole world discussing issues of energy security, impact of unguided production and use of energy, and failure of nations to forge consensus on pressing issues. At the national level, energy security and depletion of resources are prime concerns to be addressed. It has been rightly observed, 'In the years ahead, the active intervention of government in the economic and social affairs of nations is certain to increase greatly. This is a matter of necessity rather than of ideology. The most industrialised, urbanized, and congested nations will be in the need of the strongest and the most ubiquitous forms of government control.'[14]

Regulation of Fuel and Transport

In India, oil was stuck at Makum in 1867 and Digboi was the only refin-ery till the country came out of British governance.[15] It was only in 1974 that oil was stuck in Bombay High. Till 1950s India's petroleum industry was controlled by some Anglo American companies. For exploration and production of petroleum and natural gases, the Oil and Natural Gas Commission was created in 1956.[16] Indian Oil Corporation was then created to refine and market the products. India's challenge is apprehended exhaustion of meagre oil and gas resources and enormous increase in imports. The challenges as per the Government of India's Integrated Energy Policy document include maintenance of crude oil and gas supplies in a constrained world market amidst rising prices; demand management and rational pricing of petroleum products and natural gas; creation of space for private players and real market compe-tition; proper regulation of LPG and Kerosene subsidies; and protection of environment through product up-gradation.[17]

To meet the concerns direct and indirect methods are at hand for use to prolong the existing resources and ensure required supplies with sustainability. Use of energy efficient automobiles like hybrid cars and trucks combining electric motors with petroleum engines to reduce fuel consumption is one such significant measure. Promotion of public transport in urban areas to ensure restricting over use of private vehicles and development of mass transport as underground trains, dedicated bus lanes and caring metros need due attention. These ideas have become a common expression of all those understanding and facing the problems of environmental degradation impacting their health and survival, especially in bigger cities. Conversion of coal into oil; import of gas through pipelines, import of LNG through long term contracts, in-situ coal gasification through technological development, tapping and utilization of coal-bed methane, exploration of new gas resources and of gas hydrates through research and development must ensure better sustainability.[18]

Hybrid cars which use two or more engines (besides a conventional petrol or diesel engine, an electric motor) are becoming more and more popular day by day. They not only conserve fuel but also produce less CO_2 emissions though the electric engine powers the car at a lower speed. Still the number of people actually using hybrid cars is not as expected due to lack of knowledge about their working and good results. After the development of this new technology in early 1990s, it is only now that its use is being advocated with due force. There are also more government incentive programmes that use credits and special discounts to support the purchase and use of hybrid vehicles. Many cities are switching their public transportation and service vehicles over to hybrid cars and buses as a part of the programme to become more environmentally responsible.[19] Also, there are electric cars that use rechargeable batteries.

One of the biggest advantages of hybrid cars over gasoline powered car is that it runs cleaner and has better gas mileage which makes it environment friendly.

Although most of the persons relate hybrid vehicles with the vehicles that utilize electricity as their principal alternate fuel, more options are on hand now. Certain cars use hybrid technologies with propane and natural gas as well. Any hybrid car is now known as one that has an engine which can knob between a fossil fuel and an alternate fuel source.[20] Hybrid cars are supported by lower annual tax bills and exemption from congestion charges. A hybrid car is to a large extent un-contaminating requiring less fuel causing less emission and less dependence on fossil fuels. Ultimately it helps in reducing the price of fuel in domestic market. It has an advantage that applying brake each time while driving a hybrid vehicle recharges the battery a little. 'An internal mechanism kicks in that capture the energy released and uses it to charge the battery which in turn eliminates the amount of time and need for stopping to recharge the battery periodically.'[21] Such vehicles are made up of lighter materials and the engine is also smaller calling for less energy to run, thus save energy.

By using a hybrid car one can have control over his budget as gas prices carry on going higher. To reiterate, the best advantage owning and driving a hybrid car is healthy impact on environment. It reduces the dependence on fossil fuels and lowers your carbon imprint on the environment.

It is considered as a disadvantage that hybrid cars are, because of low power as compared to gas powered engine, suited for city driving only. Hybrid cars are also comparatively expensive than the petrol cars but the extra amount can be counterbalanced with lower running cost and tax exemptions.[22]

It is, however, notable that the high voltage present inside the batteries can prove lethal in case of any accident. Experts shall be doing more to find technological answers to this question.

Being conscious about assuring minimum use of energy and avoiding its adverse effects many car trips can be replaced, as we every day keep commenting during traffic jams in Delhi, with use of public transport, carpooling, and riding bicycles. At home, facilities should be made available by government to avoid multiple trips to avail services at places needing use of a vehicle. Community life should be engineered to make possible natural life style in preference to more mechanical energy dependent fashion. Metro feeder electric bus services may be made available in all localities with reasonable charges.

Congestion and parking charges should be levied to discourage use of private cars. The concept of congestion charges, although relatively unknown in India, has been effectively implemented in cities like London, New York, Milan, and Singapore.[23] Levying of this charge would mean that vehicles driven into congested areas of a city would have to pay an entry fee or heavy parking charges. The congestion charge would deter people from taking private vehicles to congested areas of cities and encourage them to use public transport. This would result in lesser number of vehicles on roads.

Discarding of vehicles running on oil which are aged over 15 years should be done. Reduction of travelling on roads is to be facilitated by appropriately allowing staff to work at home using communication and information technology and paying least visits to offices, not visiting outlets by cars for purchasing air or railway tickets or paying local taxes, and so on, as is generally done by people by taking out their personal vehicles. State may take measures for providing such facilities in different localities with fixed calendars. Arrangement for schooling of children with least travelling and least use of personal cars needs to be religiously thought about.

Promotion of Alternative Energy Sources

There should be development of alternative energy sources—solar, wind, energy plantations, bio-mass municipal waste.[24] Biodiesel and ethanol can substitute diesel and petrol. Jatropha bushes can be planted in railway lands. It is a drought-resistant perennial, growing well in marginal/poor soil, relatively quickly, and lives producing seeds for 50 years. This wonder plant produces seeds with an oil content of 37 per cent, usable as fuel without being refined—burns with clear smoke-free flame while tested successfully as fuel for simple diesel engine.[25] The by-products are press-cake (a good organic fertilizer) and the insecticide (contained in oil).

A National Mission on Jatropha Biodiesel has been adopted in India under which the committee on development of bio-fuel, under the patronage of the Planning Commission of India, presented a report in April 2003 recommending a major multi-dimensional programme to replace 20 per cent of India's diesel consumption. The Ministries of Petroleum, Rural Development, Poverty Alleviation, and the Environment and others have been integrated to blend petro-diesel with a planned 13 million tonne of bio-diesel using 11 million hectares of presently unused land to cultivate Jatropha.[26] The programme has defined a number of research and development needs, especially to look for genetically improved tree species, to produce better quality and quantity of oil. This includes tree improvement programmes, identification of contender plus trees, standardization of nursery techniques (vegetative/seed/tissue culture), scientific data for planting density, fertilization practices, plantation actions, research on inter-cropping for agriculture and agro-forestry, techniques of storage and transport of bio-diesel, engine development, and modification, marketing and trade.[27] Watering techniques, water and irrigation needs, and wastewater use are not part of the programme.

The 'micro-missions' planned under the programme require the Ministry of Forestry to carry plantations on forest lands, the Novod to plant on non-forest lands, Ministry of Rural Development to provide for other land implementation and the Khadi Village and Industries Commission to strive for procurement of seeds and oil extraction.[28]

Some of the institutions presently involved in R&D activities in the area are, Punjab Agricultural University (PAU), Coimbatore Horticultural University with 250 I/day bio-diesel production facility, Indian Institute of Petroleum (IIP), Indian Institute of Chemical Technology (IICT), Indian Institute of Technology (Delhi, Madras), Indian Oil Corporation (IOCL) with 60 kg/day bio-diesel production facility at Faridabad, Mahindra & Màhindra (works on tractors from Karanji bio-diesel; pilot plant in Mumbai).[29] Negative reports, from wherever received, will have to be examined by these institutions to find out ways for attaining better. One such report provides,

> Jatropha has not been able to meet the expectations of those involved in the Jatropha programme in southern India. 85 per cent of the interviewed farmers have discontinued cultivation of Jatropha due to poor performance. Jatropha biodiesel production was advocated based on the idea that Jatropha could be cultivated on degraded or barren land, that demand for inputs was low, and that the crop was resistant to drought and pests. Experiences in the field show that Jatropha has failed to survive and/or grow on poor soils and that a majority of the farmers planted Jatropha on cropland. The plantations have not been able to tolerate drought as well as expected, and pest attacks have occurred in several cases. Farmers have experienced that the crop requires inputs for survival and growth and have used irrigation, fertilizers, manure, and pesticides. Even when planted on fertile land and provided inputs, Jatropha did not produce a harvest or else not a sufficient yield.[30]

Fuel Storage

Cavernous storage of oil as a buffer stock can be helpful in meeting the concerns of environmental degradation and use and production of energy. Strategic reserves of crude oil and petroleum products were first recognized as policy tool in the aftermath of the first oil shock in

1973. Currently, the International Energy Agency countries hold strategic stocks of about 90 days of net imports.[31]

India has already the facility at certain places. Very recently the United Arab Emirates' national oil company, Adnoc, showed keenness for storing crude oil in India's maiden storage facility. The Arab realm has also offered two-thirds of the oil to India for free.

India is building underground storage facilities at Visakhapatnam in Andhra Pradesh, and Mangalore and Padur in Karnataka, with a capacity to store about 5.33 million tonnes of crude oil to guard against global price shocks and supply disruptions. Abu Dhabi National Oil Company, or Adnoc, has decided to use half of Mangalore's 1.5 million-tonne facility, which would make up six million barrels of oil. The fact, however, is that tax issues need to be sorted out before Adnoc can start using the storage facility.[32]

Augmentation of Railways

There should be a move to augment railways in freight haulage for saving of diesel oil. By this the freight lost to roadways can also be won back. According to the World Bank, Transportation, Water, and Information and Communications Technologies Department, Transport Division, goods transported by railways in India in 1998 were 158,474 million tonne-km which in 2014 has arisen to 665,810 million tonne-km.[33] The trend depicts an obvious increase in transportation of goods by railway and must fetch more turn over.

Electrification of Railway can be an alternative to replace diesel locomotives. Indian government is getting to electrification of half of its rail lines (35,000 km) in the next three years seeking to save about Rs 16,000 crore forex per annum on fuel import. Electrifying remaining 35,000 km of the track may be done by some company under the Power Ministry.[34] The project can be launched instantaneously after necessary discussions among relevant bodies. The electrification of whole rail track in India will result in to power consumption mounting by seven billion units. The investment can be predictably recovered in three years. The electrifying company can get 75 per cent of the savings which may be made available to it for next ten years leaving the remaining amount to be enjoyed by Railways. Out of 75 per cent savings 50 per cent may be used to shell out costs and interest and residual 25 per cent will be a profit.[35]

Use of Hydrogen Based Vehicles

Dependence on oil imports can be reduced by use of hydrogen based vehicles run with fuel produced using locally available energy sources. The fuel cell vehicles use hydrogen gas to power an electric motor and like all other electric cars and trucks produce zero damaging tailpipe emissions. The pattern by which hydrogen fuel is prepared and the way it is delivered can affect cleanliness of the hydrogen fuel cell vehicles. It is being believed that the fuel 'coming from the dirtiest sources of hydrogen, natural gas, today's early Fuel Cell Electric Vehicles (FCEVs) can cut greenhouse gas emissions by over 30 per cent when compared with their gasoline-powered counterparts. Under California's renewable hydrogen requirements, reductions are over 50 per cent. Future fuel cell vehicles will likely be even cleaner.'[36]

At present crude oil is the main source of transportation fuels. Its demand increases every year. Crude oil is a finite resource and its demand may exceed its supply. Technology is to be adopted gradually to replace crude oil as feed before its supply becomes constraining. Indirect liquefaction can be used to convert natural gas into normally liquid products as are presently produced from crude oil. Natural gas being the most hydrogen-rich carbon source, even with significant displacement of carbon-based energy carriers like transportation fuels, not all applications can be carbon-free like petrochemicals. During constrained crude oil supply, indirect liquefaction is a more C-efficient route for alternative C-based products. The effective feed H:C ratio determines the CO_2 footprint. In case of meaningful price difference between natural gas and crude oil there is an economic incentive for gas-to-liquid conversion.[37]

In indirect coal liquefaction, first the coal is gasified into a syngas in two major steps, (a) gasification to produce a synthesis gas (syngas); and, (b) conversion of the carbon monoxide (CO) and hydrogen (H_2) into the syngas to a range of hydrocarbon fuels/products such as gasoline, diesel, methanol and chemicals). Most frequently, Fischer-Tropsch (FT) synthesis followed by subsequent liquids product refining is used to convert syngas to fuels; alternately, methanol formed from syngas can be converted to gasoline via ExxonMobil's MTG process. Direct coal liquefaction requires an external source of hydrogen, which may have to be provided by gasifying additional coal feed and/or the heavy residue produced from the DCL reactor. Many argue that indirect liquefaction with the current state-of-the-art technologies is more competitive than direct liquefaction. ICL has been demonstrated commercially by Sasol since the 1950s, and the ICL process is more amenable to carbon dioxide (CO_2) capture.[38]

Nuclear Power

Electricity generation through nuclear energy makes it a process of less fuel offering more energy, significant save on raw materials, better transport, and more experience in handling and extraction of nuclear fuel. The cost of nuclear fuel may be only 20 per cent of the cost of energy generated.[39] The production of electric energy is continuous for almost 90 per cent of annual time. It restricts the price unpredictability of other fuels as petrol. The stability benefits the electrical planning.[40] Nuclear energy does not depend on other natural aids or resources. Instead, it is a solution for the main disadvantage of renewable energy, like solar energy or eolic energy,[41] because the period of sun or wind may not coincide with the hours with more energy demand.[42]

Whereas nuclear energy is an alternative to fossil fuels, it reduces the consumption of fuels such as coal or oil is benefiting the situation of global warming and global climate change. By that we also improve the quality of air improving health and quality of life.[43]

There are disadvantages of nuclear power also which are to be seen. In spite of the high level of sophistication of the safety systems of nuclear power plants the human aspect has always to be taken due care. Facing an unexpected event or managing a nuclear accident, as in Chernobyl and Fukushima, there is no guarantee of safety and security. Different incorrect decisions during the management of the nuclear plant can cause distressing expositions and detonations. At the Fukushima nuclear accident, the operations of the staff were remarkably dubious. One significant issue is the complexity in the management of nuclear waste. Ordinarily it takes numerous years to remove its radioactivity and hazards. The nuclear reactors, wherever constructed, have an expiration time. On completion of that time they have to be dismantled, but at the same time nuclear energy producing countries have to maintain a regular number of operating reactors. [44]

As the nuclear plants have a limited life and the investment for the construction of a nuclear plant is very high, there must be an attempt to recover its cost as shortly as possible. This can raise the cost of electricity generated thereby. This may be so that the energy generated is low-priced compared to the price of fuel, but the recuperation of its construction may be much more high-priced.

Nuclear power plants are to be maintained with high security due to emergence of anti-establishment sentiments in different forms in different parts of the world as they can be considered as key targets by such forces. Since nuclear power plants generate external dependence that will have to be managed with due diplomatic advertence and care. Countries not having uranium mines and many have not even nuclear technology, they have to hire both. Prevention of radioactive explosions is a herculean task to be ensured at any cost to save loss of life and property. That can otherwise prove devastating.[45]

The most alarming concern is the use of the nuclear power in the military operations. Several countries have signed the Nuclear Non-Proliferation Treaty, but the risk that nuclear weapons could be used in the future does always exist.

All these concerns shall need thorough investigation and policy making resulting into a comprehensive set of legislations and an effective implementation and enforcement mechanism.

Cradling and Controlling of Indian Energy Programmes

Thinking for Safe Energy

Here some account of Indian situation is given to reflect the trends so far in the area of nuclear energy use and law. In India, the Fukushima disaster of Japan led the country to rethink about effective safety measures for its energy programmes. It was more so because the International Atomic Energy Agency did through a global expert fact-finding mission ascertain and identify initial lessons to be learned for a safe nuclear global community.[46] The mission's report stated in clear terms that 'there were insufficient defences for tsunami hazards. Tsunami hazards that were considered in 2002 were underestimated. Additional protective measures were not reviewed and approved by the regulatory authority. Severe accident management provisions were not adequate to cope with multiple plant failures'.[47] The report states, 'Japan has a well-organized emergency preparedness and response system as demonstrated by the handling of the Fukushima accident. Nevertheless, complicated structures and organizations can result in delays in urgent decision making.'[48] The incapacity to anticipate such severe state of affairs is a premonition to nations that are mounting nuclear capacity at a hyperactive rapidity. India has to find in it a call for required care and take an exceptional opportunity to reassess the makeup of the DAE (Department of Atomic Energy), and establish more dependable and reliable structures through transparent method and modus operandi. DAE management should not classify the related audit reports as 'top secret' and shelve them.[49] The Three Mile Island incident (1979) and Chernobyl accident (1986) should not be put into oblivion. The reports and action-taken reports should be published and made accessible. By this the DAE can gain considerable public faith in its performance and give a noteworthy poise to its engaging with stakeholders.

Safety Evaluation of the Energy Programme

After Fukushima, the Nuclear Power Corporation of India Ltd. under-took safety evaluation of various operating power plants and under con-struction nuclear power plants. Accident at Fukushima Nuclear Power Plants (NPP) in Japan occurred on 11 March 2011 due to earthquake followed by Tsunami. On 15 March 2011, CMD NPCIL constituted four task forces to review consequences of occurrences of similar situations in INDIAN NPPs.

> [O]n 11th March 2011, an Earthquake of magnitude 9.0 struck near Fukushima, Japan. It was followed by Tsunami of ~15 meter high waves after an hour of earthquake. Magnitude of the earthquake and the tsu-nami wave height were more than considered in the design. There were total 13 NPPs located in the affected zone, out of which 10 were operat-ing and 3 were under maintenance outage. All 10 operating plants at the affected area did automatically shut down on sensing the earthquake. Out of 13 NPPs in the affected zone, 4 NPPs at Fukushimac Daiichi got affected. Remaining 9 plants were safe. All the 6 plants located in Fukushima Daiichi were of BWR type.[50]

The report titled as *Safety Evaluation of Indian Nuclear Power Plants Post Fukushima Incident* proposed many safety measures for incorporation in all audited nuclear power plants in a time-bound manner. The proposed measures are, inter alia, about strengthening technical and power systems, shutdown of automatic reactor on sensing seismic activity, expansion, and rise of tsunami bunds at all coastal stations.[51] Most of the people, however, do not accept the claim in the report that 'adequate provisions exist at Indian nuclear power plants to handle station blackout situations and maintain continuous cooling of reactor cores for decay heat removal'.[52]

However, in the same breath, the report provides assurance by stating that, 'adequate provisions exist at Indian nuclear power plants to handle station blackout situations and maintain continuous cooling of reactor cores for decay heat removal'. The report reminds, 'the incidents at Indian nuclear power plants, like prolonged loss of power supplies at Narora plant in 1993, flood incident at Kakrapar plant in 1994 and tsunami at Madras (Chennai) plant in 2004 were managed successfully with existing provisions.'[53]

The report under the sub-head 'Continued Monitoring and Periodic Safety Assessment' says, 'Safety is a moving target. Continued monitoring, periodic safety assessment and improvement of Indian nuclear power stations including national and international operating experience, are performed by Nuclear Power Corporation of India Limited (NPCIL) as well as by the Atomic Energy Regulatory Board (AERB)'.[54]

The Government of India at present is ardent about the promotion of Indian nuclear energy programme in a big way. There seems no looking back on nuclear energy expansion but it has to be assured that the use of nuclear energy meets the highest safety standards. This is a matter on which there can be no compromise. With the memory of Bhopal accident, as the legal, administrative, and political failure, these assurances should be real. [55]

There are numerous domestic and international power projects in the pipeline. DAE has projected a multi-fold increase in the current production of electricity and hopes to achieve targets exceeding 30,000 MWe by 2020 and 60,000 MWe by 2032.[56] Carrying out of these higher targets is to be achieved by importing high-capacity reactors and through native initiatives. The trust deficit on safety measures needs to be addressed with equal concern as there are doubts in the minds of the people. It has to be manifested that from the very early days of the nuclear power, safety has been the prime concern of India and provisions are built in the nuclear reactors to avoid any accident.

Issues like land acquisition are also a major concern for infrastructure. The main apprehension in this expansion is to convince the public of the safety and security of nuclear power plants and also arrive at a comprehensive information and communication package for states in whose territory projects are being executed. Because of combined existence of civilian and military nuclear programmes in India the nation may not be in a position to accomplish the required regulatory autonomy, process, and engagement that is ordinarily said to have been adopted by many European countries and in the US. [57]

Many countries in Europe moved on to peaceful civilian use of nuclear energy in preference to military programmes which enabled them to establish credible, autonomous, and competent regulatory mechanism with strong public and stakeholder engagement.[58] Italy shows its own peculiar trend. It had four operating nuclear power reactors but shut the two down after the Chernobyl accident. The 10 per cent of its electricity now from nuclear power is all imported. The government intended to have 25 per cent of electricity supplied by nuclear power by 2030, but this prospect was rejected at a referendum in June 2011.[59] Italy is the only G8 country without its own nuclear power plants, having closed its last reactors in 1990. In 2013, its gross electricity generation was 288 billion kWh. Of this, 110 billion kWh was from gas-fired generation, 50 billion kWh from coal, 18 billion kWh from oil, 53 billion kWh hydro, 22 billion kWh from solar, and 15 billion kWh from wind. Net imports were 42 TWh, mostly from France and Switzerland. Per capita electricity consumption in 2012 was 4900 kWh.[60]

In 2009, Croatia concluded an agreement with Albania for the construction of a joint nuclear facility near the Montenegrin border. They formed a working group of five experts to focus on the technical implementation of the project. In January 2010 the government approved the creation of the Agjencia Kombetare Berthamore (National Atomic Agency) to supervise the development of nuclear projects and set up the legal infrastructure. A proposed site for a 1500 MWe plant is in the Shkoder region, on a lake of that name, bordering Montenegro, or at Drac on the north coast, or Durres.[61]

In 2010, the Serbian government announced that it would take an equity stake, possibly of five per cent, in Bulgaria's Belene nuclear plant. In November Bulgaria invited Serbia, Croatia, and Macedonia to take equity of one per cent to two per cent in the Belene plant. Croatia is co-owner with Slovenia of the Krsko nuclear plant close to the border. It is considering joining with Slovenia in building a new reactor at Krsko, joining with Hungary in building new nuclear capacity at Paks, or building a plant of its own at Dalj or Prevlaka by 2020. A decision is expected late in 2012. In 2010 Croatia rejected an offer to invest in Bulgaria's Belene plant.[62]

Portugal's electricity production is about 47 billion kWh/yr gross and in 2007 this came 27 per cent from coal, 27 per cent from gas, and 22 per cent from hydro. Net imports are about 5 TWh/yr from Spain. Its electricity grid is closely linked with Spain's, so that a large nuclear plant on the Atlantic coast could serve both countries. In 2004 the government rejected a proposal to introduce nuclear power but this is now being reviewed.[63]

The bifurcation of nuclear establishment into civilian and military, subsequent to commitment under India-US civil nuclear cooperation provides the prospect of an empowered regulatory system.[64]

Legal Framework and Policy

Safety should be accorded prime precedence in all the activities. The existence of a liability regime is no panacea for operation of nuclear power plants and is an add-on to a necessary well structured legal, regulatory, and institutional environment. Just after Independence of India, the Constituent Assembly adopted the Atomic Energy Act of 1948 providing for the establishment of an Atomic Energy Commission. In 1954 Department of Atomic Energy was set up and later 'to provide for the development, control, and use of atomic energy for the welfare of the people of India' a new Atomic Energy Act of 1962 was passed repealing the 1948 Act. The Act covers the identification, positioning, setting up, operation, mining, and safety of the atomic reactors. Now in the institutional hierarchy for the Indian Atomic Energy is included the AERB which works for the safety of the atomic energy sector while the Department of Atomic Energy performs the executive functions with research and development. For making the policies at the top is the Atomic Energy Commission.

Main Responsibilities of the Central Government

Section 3 of the 1962 Act empowers the central government to, (a) produce, develop, use, and dispose of atomic energy either by itself or through any authority or corporation established by it, or a government company and carry out research into any matters connected therewith, and (b) manufacture or otherwise produce any prescribed or radioactive substance and any articles which in its opinion are or are likely to be, required for or in connection with the production, development, or use of atomic energy or such research as aforesaid and to dispose of such described, or radioactive substance or any articles manufactured or otherwise produced, (a) to buy or otherwise acquire, store, and transport any prescribed or radioactive substance and any articles which in its opinion are or are likely to be, required for or in connection with the production, development or use of atomic energy and (b) to dispose of such prescribed or radioactive substance or any articles bought or otherwise acquired by it either by itself or through any authority or corporation established by it or by a government company.

The responsibility does mainly remain with the central government to look to the safe operations concerning production and use of nuclear energy and all other related matters. The nuclear facilities should, in any case, be located, designed, built, specially made, and managed in accordance with strict quality and safety standards. The regulatory framework is expected to be robust. Under section 27 of the Atomic Energy Act, 1962 the AERB is to carry out regulatory and safety functions as required under Section 16, 17, and 23 of the Act. The Board consists of a full-time Chairman and a full-time Member-Secretary with part time or full-time members not exceeding five, including the Chairman and Member-Secretary. The Board is responsible to the Atomic Energy Commission.

Section 16 provides for control over radioactive substances empowering the central government to prohibit the manufacture, possession, use and transfer by sale or otherwise, export and import and in an emergency, transport and disposal of any radioactive substances without its written consent. Section 17 provides for making special provisions for safety and authorizes the central government, as regards any class or description of premises or places in which radioactive substances are manufactured, produced, mined, treated, stored or used or any radiation generating plant, equipment or appliance is used, make necessary rules, (a) to prevent injury being caused to the health of persons employed at such premises or places or other persons either by radiations or by the ingestion of any radioactive substance, (b) to secure that any radioactive waste products resulting from such manufacture, production, mining, treatment, storage, or use as aforesaid are disposed of safely, and (c) to prescribe qualifications of the persons for employment at such premises or places and the regulation of their hours of employment, minimum leave, and periodical medical examination. And the rules may in particular and without prejudice to the generality of this sub-section provide for imposing requirements as to the erection or structural alterations of buildings or the carrying out of works.

Regarding the transportation of any radioactive substance or any other such substance which has been declared as dangerous to health under this Act, the central government can make necessary rules to prevent causing of any injury by such haulage. The rules can be made in the form of prescribing prerequisites for such transportation or imposing of particular kinds of prohibitions and restrictions on employers, employees and other persons falling within the area of risk.

In the event of any contravention of the rules made under this section, the central government shall have the right to take such measures as it may deem necessary to prevent further injury to persons or damage to property arising from radiation or contamination by radioactive substances including, without prejudice to the generality of the foregoing provisions and to the right to take further action for the enforcement of prescribed penalties, the sealing of premises, vehicle, vessel or aircraft, and the seizure of radioactive substances and contaminated equipment. Under Section 24 the contravention of any rules made under Section 17 or any requirement, prohibition, or restriction imposed under any such rules punishable with imprisonment for a term which may extend to five years or with fine or both. Any person who fails to comply with any notice served on him regarding control over mining or concentration of substances containing uranium under Section 5 or with any terms and conditions that may be imposed on him under that section or fails to comply with any notice served on him to obtain information regarding materials, plant, or processes under Section 7 or knowingly makes any untrue statement in any return or statement made in pursuance of any such notice or obstructs any person or authority in the exercise of powers of power of entry and inspection under Section 8 or power to do work for discovering minerals under 9 or contravenes any other provision or any order made under the Act is punishable with imprisonment to the extent of one year or with fine, or with both.

In relation to any factory owned by the central government or any authority or corporation established by it or a government company and engaged in carrying out the activities covered by the provisions of this Act, Section 23 gives to the central government the power of administration of the Factories Act, 1948 authorizing it to do all things for the enforcement of its provisions, including the appointment of inspecting staff and the making of necessary rules.

Role of AERB

As specified by the central government under the Atomic Energy Act, 1962 the functions of the AERB are to: [65]

1. Develop safety codes, guides, and standards for siting, design, construction, commissioning, operation, and decommissioning of the different types of plants, keeping in view the international recommendations and local requirements and develop safety policies in both radiation and industrial safety areas;
2. Ensure compliance by DAE and non-DAE installations of safety codes and standards during construction commissioning stages;
3. Advise AEC/DAE on technical matters that may specifically be referred to it in connection with the siting, design, construction, commissioning, operation, and decommissioning of the plants under DAE; and,
4. Review from the safety angle requests for authorising/ commissioning/ operation of DAE Projects/plants.

The AERB must continue to lay down safety standards and requirements and monitor and enforce all the safety provisions with due autonomy and professionalism. Nuclear power generation as governed by the Atomic Energy Act, 1962 should meet the above challenges while the Act encompasses all the activities concerned with atomic energy, including electricity generation.[66]

Employee Safety

Employees need to be protected from radiological damage to person and property. For their radiological protection legislative control is given to central government under the Atomic Energy Act, 1962 to ensure proper design, appropriate plant layout, and adequately protective and restricted air contamination levels in different zones of a plant.[67] There should be source control in the form of preventive measures by proper selection of methods and materials/components minimizing possibilities of exposures to radiation.[68]

The AERB has prescribed dose limits for exposures to ionizing radiations for occupational workers. Investigation limits are also prescribed at which investigation of exposure cases exceeding these limits are carried out by its committee.[69] The practices followed are:

1. The gaseous wastes from reactor buildings are filtered using prefilters and high-efficiency particulate air filters and released after monitoring through a stack of 100 m height.
2. The release rate and integrated releases of different radio nuclides are monitored and accounted for to demonstrate that the releases are within the prescribed limits.
3. The radioactive liquid wastes generated are segregated, filtered, and conditioned as per procedure and after adequate dilution to comply with the limits of discharges, disposed to the environment water body.
4. The activity discharged is monitored at the point of discharge and accounted on a daily basis.
5. The limits on the annual volume and activity of discharge, daily discharges, and activity concentration, which are site specific.
6. The radioactive solid wastes are disposed off in brick-lined earthen trenches, reinforced cement concrete (RCC) vaults, or tile holes, depending on the radioactivity content and the radiation levels.[70]

Regulation and Management of Energy Sites

The discharge of radioactive waste is governed by the Atomic Energy (Safe Disposal of Radioactive Wastes) Rules, 1987, issued under the Atomic Energy Act, 1962.[71] Under these rules a wide range of 'persons' can be authorized to dispose of nuclear waste, such as, any individual, cooperation, association of persons whether incorporated or not, partnership, estate, trust, private or public institution, group, government agency, or any state or any political sub-division thereof or any political entity within the state, any foreign government or nation or any political sub-division of any such government or nation or other entity; and, any legal successor, representative, or agent of each of them. The rules mention necessary requirements for the safe and proper disposal of radioactive wastes. For disposal of radioactive wastes it is mandatory to obtain authorization under these rules from the competent authority.[72] The disposal of the waste can be allowed in accordance with the conditions related to the procedure, materials, and equipment generating radioactive waste in the installation; environment and surrounding nature around the installation; safety devices and other equipment in the installation for conditioning treatment and disposal of radioactive wastes; estimates of annual releases, discharges, and leakages in normal conditions and its anticipated environment impact; potential accidents, design features, and monitoring equipment to control the release of radioactivity; and, procedure to be followed in the safe collection of radioactive wastes.

The Environment (Protection) Act, 1986 and rules made under that are applicable to such ventures. Air (Prevention and Control of Pollution) Act, 1981 makes provisions of Atomic Energy Act, 1962 applicable for radioactive air pollution.[73] The Atomic Energy (Working of the Mines, Minerals, and Handling of Prescribed Substances) Rules, 1984 also standardize related mining, milling, processing, and handling of radioactive materials. The Atomic Energy (Radiation Protection) Rules of 2004 allow handling of radioactive materials and operation of any radiation generating equipment only by authorized persons and in accordance with the terms and conditions of a licence.[74] Compliance with safety measures and training are strictly to be complied with. Express licence is required for nuclear fuel cycle facilities, land based high intensity gamma irradiators other than gamma irradiation chambers, neutron generators.[75] Any custodian of the radiation sources has to ensure the physical security of the sources at all times.[76]

Nuclear Civil Liability

The Civil Nuclear Liability Act, 2010 is an important legislation on protection to participants and victims of nuclear industry and accidents.

It embodies the established international principles of nuclear liability but does not maintain parity in operators' liability being higher in the US, Japan, and Germany. It limits the operators' liability to Rs 1,500 crores as compared to $10.2 billion cap in the US, $6.2 billion in Japan and no limit in Germany.[77] In case of gross negligence the Act allows the operator to seek damages from the supplier in contradiction to the principles of law followed by many countries engaged in nuclear energy.[78] The damages recoverable from supplier are to the extent of the operators' liability or the contract price whichever is lesser with a limitation period of five years.[79] The Act allows pursuing claims under any other law in the country opening door for criminal liability as well as claims under law of torts.[80] For enforcement of claims the Act provides for the establishment of a Nuclear Damages Claim Commission for quick access to justice.[81]

An environmental survey programme is carried out by the HPUs (Health Physics Units) and the Environmental Survey Laboratories (ESLs) of BARC (Bhabha Atomic Research Centre). The basic objective of the monitoring and surveillance is to assess the radiological impact and demonstrate compliance with the radiation exposure limits set for the members of the public. The radiological surveillance of the environment is carried out by professionals of the ESLs.[82]

The Disaster Management Act, 2005, provides for the effective management of disasters, including accidents involving nuclear power plants. Under the provisions of this Act, the National Disaster Management Authority (NDMA) has been established at the national level, whose chairperson is the Prime Minister.[83] The NDMA has the responsibility for laying down policies, plans, and guidelines for disaster management for ensuring timely and effective response to any disaster.[84] National plan, state plans, and district plans are prepared by the respective authorities constituted for the purpose.[85]

In India, a nuclear power plant is generally located in a relatively low-population zone, with the basic objective of limiting the dose-received members of the public and population as a whole under normal and accident conditions to ALARA (as low as reasonably achievable) levels. As defined in Title 10, Section 20.1003, of the *Code of Federal Regulations* (10 CFR 20.1003), ALARA is an acronym for 'as low as (is) reasonably achievable,' which means making every reasonable effort to maintain exposures to ionizing radiation as far below the dose limits as practical, consistent with the purpose for which the licensed activity is under-taken, taking into account the state of technology, the economics of improvements in relation to state of technology, the economics of improvements in relation to benefits to the public health and safety, and other societal and socioeconomic considerations, and in relation to utilization of nuclear energy and licensed materials in the public interest.[86] In order to achieve the above objective, the area around the plant is divided into, Exclusion zone; Sterilized zone; and Emergency planning zone (EPZ). An exclusion zone of 1.5 km radius around the plant is established, which is under the exclusive control of the operating organization, and no public habitation is permitted in the area. A sterilized zone is established in 5 km radius around the plant.[87] This has the potential for extensive contamination in case of a severe accident.[88] Development activities within this area are controlled so as to check an uncontrolled increase in the population. In this area, only natural growth of the population is permitted. In this sterilized zone, the total population should be small, preferably less than 20,000. So that this may continue to remain the case, 'only natural growth of population is permitted'. The figure of 1.5 km results from 'rounding off' the 1.6 km (1 mile) radius and in some DAE documents both figures are used. There are, in addition, some 'desirable characteristics' for any reactor site, including population density of less than two-thirds the state average in an area of a 10-km radius around the site; distance to population centres with a population of more than 10,000 to be more than a 10-km radius from the site; population centres with a population of more than 100,000 to be at a distance of more than 30 km from the site.[89]

The Emergency Planning Zone (EPZ) is defined around the plant up to a 16-km radius and provides for the basic geographical framework for decision making on implementing measures as part of a graded response in the event of an emergency. It is divided into 16 equal sectors to optimize the emergency response mechanism and to provide the maximum attention and relief to the region's most affected during an offsite emergency.[90]

The emergency measures required for the purpose consist of notification, alerting, assessment of situation, corrective actions, mitigation, protection, and control of contamination. It is essential to ensure that the measures would reduce the overall impact to the public to a level significantly lower than what they would be in the absence of such measures. The emergency response manual gives details of the protective measures and the intervention levels approved for initiating protective measures to limit radiation exposures.[91] Evacuation is a very effective countermeasure but is very carefully considered before a decision to implement is taken. The benefits and risks of this countermeasure are carefully assessed in terms of averted dose. If radiation levels in the affected zone continue to exist beyond acceptable levels, then relocating the affected population is resorted to.[92]

National Electricity Policy

National Electricity Policy is also a matter to be examined under the theme of energy and law. Section 3(1) of the Electricity Act, 2003 required the central government to prepare the National Electricity Policy for development of the power system based on optimal utilization of resources. The policy as framed is providing guidance to the Electricity Regulatory Commissions in discharging their functions and to the Central Electricity Authority for preparation of the National Electricity Plan.[93] The Policy aims at accelerated development of the power sector, providing supply of electricity to all areas and protecting interests of consumers and other stakeholders keeping in view availability of energy resources, technology available to exploit these resources, economics of generation using different resources, and energy security issues.[94] The policy provided for access for electricity to all households in the country. For sustained supplies of power the National Electricity Policy aims at achieving the objectives which include[95] access to electricity for all households, power-demand to be fully met—Energy and peaking shortages to be overcome and adequate spinning reserve to be available, supply of Reliable and Quality Power of specified standards in an efficient manner and at reasonable rates, per capita availability of electricity to be increased to over 1,000 units, minimum lifeline consumption of 1 unit/household/day, financial Turnaround and Commercial Viability of Electricity Sector, and protection of consumers' interests.

One main thrust of the policy is developing rural electrification distribution backbone and village electrification. The policy envisages fiscal support in terms of capital subsidy to States for rural electrification. Special preference is sought in this regard to be given to *dalit bastis*, tribal areas, and localities of other weaker sections.[96] The policy pursues full development of hydro potential.[97] Choice of fuel for thermal generation is to be based on economics of generation and supply of electricity.[98] Appropriately, the policy envisages exploitation of non-conventional energy sources including water, solar, biomass, and wind for additional power generation capacity.[99] Development of a National Grid is an important feature of the Policy.[100]

About nuclear power, clause 5.2.19 of the Electricity Policy provides,

> Nuclear power is an established source of energy to meet base load demand. Nuclear power plants are being set up at locations away from coalmines. Share of nuclear power in the overall capacity profile will need to be increased significantly. Economics of generation and resultant tariff will be, among others, important considerations. Public sector investments to create nuclear generation capacity will need to be stepped up. Private sector partnership would also be facilitated to see that not only targets are achieved but exceeded.

The AERB is, as discussed above, laying down safety standards and requirements and it is expected that nuclear power generation as governed by the Atomic Energy Act, 1962 meets the challenges and concerns associated with production of nuclear energy, including electricity generation.[101]

★★★

Without availability of energy and proper use of that the human life is an 'uncertainty'. Having come far ahead of the stone-age it is now dependant on too much of technology and artificiality. The process of management of the resources and fulfilment of wants is far more complex. Regulation has to be made largely by law to ensure saner and safer use of resources and careful production of energy. Legislations so far are more or less skeletal, symbolic, and inadequate. Hence, much has to be done by the legislature depending upon reliable research done in this regard. This can be possible only by adopting a proper agenda for research and development for the country.

In order to properly plan to meet the energy requirement of the country, it is essential to accurately forecast the energy demand and survey the availability of resources. The conventional approach of energy planning tends to overestimate and project high-energy demands and fails to account for internal energy conservation mechanisms. The government should ponder about sustainable advancement of energy sector and intervene in the energy market when necessary.

Notes and References

1. Life can be animate and 'inanimate'.

2. Potential energy, kinetic energy, and so on, are the examples. Heat, sound, electricity, and light are also forms of energy.

3. D.A. Coley, *Energy and Climate Change* (West Sussex, England: John Wiley & Sons, 2008), pp. 312–22.

4. Coley, *Energy and Climate Change*, pp. 312–22.

5. Coley, *Energy and Climate Change*, pp. 312–22.

6. Data with text, pictorial, and graphic representations can be seen from Energy Information Administration, *International Energy Outlook* 2004, EIA04..

7. See Coley, *Energy and Climate Change*, pp. 314–15.

8. Coley, *Energy and Climate Change*, pp. 314–15.

9. In Asia demand for air conditioners, refrigerators, and space heating is fast growing.

10. Coley, *Energy and Climate Change*, pp. 317–20.

11. Coley, *Energy and Climate Change*, pp. 317–20.

12. Coley, *Energy and Climate Change*, The International Energy Agency publishes predictions in their annual *World Energy Outlook (WEO)* available at http://www.iea.org/weo/.

13. See A.J. McMichael, *et. al.*, *Climate Change and Human Health*, p. 276 (World Health Organization: Geneva, 2003): World Health Organization (WHO), The World Health Report 2002. Geneva, Switzerland.

14. W. Friedmann, *Law in a Changing Society*, Fifth Indian Reprint (New Delhi: Universal Law Publishing Co. Pvt. Ltd.), p. 521.

15. B. Styanarayan, *Energy Law and Policy in India* (Vishakhapatnam: GK Print House, 2014), p. 83.

16. Styanarayan, *Energy Law and Policy in India*, p. 83.

17. Styanarayan, *Energy Law and Policy in India*, p. 84.

18. 'What Is Hybrid Car? Its Advantages and Disadvantages', see http://www.conserve-energy-future.com/advantages-and-disadvantages-of-hybrid-cars.php, accessed on 7 September 2016.

19. 'What is Hybrid Car? Its Advantages and Disadvantages'.

20. 'What is Hybrid Car? Its Advantages and Disadvantages'.

21. 'What is Hybrid Car? Its Advantages and Disadvantages'.

22. 'What is Hybrid Car? Its Advantages and Disadvantages'.

23. S. Bhatia, 'Congestion Charging in India, the Government Initiative and the Way Forward', available http://www.indiantollways.com/category/congestion-charging/, accessed on 15 September 2016.

24. S. Bhatia, 'Congestion Charging in India, the Government Initiative and the Way Forward'.

25. See the website at http://www.jatrophabiodiesel.org/aboutJatrophaPlant.php?_divid=menu1, accessed on 12 September 2016.

26. For more details on the subject, see the website abovementioned.

27. For more details on the subject, see the website abovementioned.

28. For more details on the subject, see the website abovementioned.

29. (Courtesy 'case study' jatropha curcas by GFU), available http://www.jatrophabiodiesel.org/Indian Programs.php?_divid=menu5, accessed on 12 September 2016.

30. L. Axelsson and M. Franzen, 'Performance of Jatropha Biodiesel Production and Its Environmental and Socio-economic Impacts—A Case Study in Southern India', Department of Energy and Environment Chalmers University of Technology (Göteborg, Sweden, 2010), available http://www.focali.se/filer/FRT%20201006%20-%20Jatropha.pdf, accessed on 15 September 2016.

31. See the official website at http://www.indiaenergy.gov.in/default.php, accessed on 15 September 2016.

32. The facts presented here are taken from the official website http://www.indiaenergy.gov.in/default.php.

33. 'Railways, Goods Transported (million ton-km)', available http://data.worldbank.org/indicator/IS.RRS.GOOD.MT.K6?locations=IN, accessed on 5 September 2016.

34. Rail Budget 2016: Outlay for Railway Electrification Increased by Almost 50%, *Economictimes.Com*, 25 February 2016, available http://economictimes.indiatimes.com, accessed on 5 September 2016.

35. Rail Budget 2016: Outlay for Railway Electrification Increased by Almost 50%.

36. 'How Clean Are Hydrogen Fuel Cell Vehicles?', available http://www.ucsusa.org/clean-vehicles/electric-vehicles/how-clean-are-hydrogen-fuel-cell-vehicles#.V1fe2tKqqko, accessed on 5 September 2016.

37. 'How Clean Are Hydrogen Fuel Cell Vehicles?'

38. 'Indirect Liquefaction Processes', available https://www.netl.doe.gov/research/coal/energy-systems/gasification/gasifipedia/indirect-liquefaction, accessed on 15 September 2016.

39. For railway electrification see map titled 'Status of Railway Electrification on Indian Railways' at http://www.indianrailways.gov.in/railwayboard/

uploads/directorate/rail_elec/downloads/MAP_221214.pdf, accessed on 15 September 2016.

40. 'Status of Railway Electrification on Indian Railways'.

41. P. Sparrow, 'Wind Energy: Advantages and Disadvantages', available http://www.ianswer4u.com/2012/02/wind-energy-advantages-and.html#axzz4KJAxWQgl, accessed on 15 September 2016.

42. Sparrow, 'Wind Energy: Advantages and Disadvantages'.

43. See http://www.indianuclearenergy.net/#, accessed on 17 September 2016.

44. 'Cheaper Renewable Energy Soars Past Nuclear Power In India', available http://scroll.in/article/805316/cheaper-renewable-energy-soars-past-nuclear-power-in-india, accessed on 17 September 2016.

45. 'Can the World's Biggest Electricity Company Restart India's Nuclear Power Programme?', available http://qz.com/602992/can-the-worlds-biggest-electricity-company-restart-indias-nuclear-power-programme/, accessed on 17 September 2016.

46. M.P. Ram Mohan, 'How Safe Is India's Nuclear Energy Programme?' *Mint*, 23 August 2011, available http://www.livemint.com/2011/08/22202845/How-safe-is-India8217s-nucl.html, accessed on 14 September 2016.

47. Mohan, 'How Safe Is India's Nuclear Energy Programme?'.

48. Mohan, 'How Safe Is India's Nuclear Energy Programme?'.

49. Mohan, 'How Safe Is India's Nuclear Energy Programme?'.

50. 'Review of Indian NPPs—Post Fukushima Event', available http://www.npcil.nic.in/pdf/ presentation.pdf, accessed on 16 September 2016.

51. 'Safety of Indian Nuclear Power Plant', available www.aerb.gov.in/AERBPortal/pages/English/t/.../SAFETYOFINDIANNPPs.pdf..., accessed on 16 September 2016.

52. 'Safety of Indian Nuclear Power Plant'.

53. 'Safety of Indian Nuclear Power Plant'.

54. Details are covered in Section2 of Report, 'Safety Evaluation of Indian Nuclear Power Plants, Post Fukushima Incident'. Unit Commercial Operation Periodic safety review (PSR) Remark, available http://www.npcilnic.in/pdf/presentation.pdf, accessed on 16 September 2016.

55. See R. Rajaraman (ed.), *India's Nuclear Programme—Future Plans, Prospects and Concerns* (New Delhi: Indian National Science Academy, 2013), accessed pp. 1–278

56. 'Addressing Safety of Indian Nuclear Power Plants - Atomic Energy', available www.aerb.gov.in/AERBPortal/pages/English/t/.../SAFETYOFINDIAN NPPs. Pdf, accessed on 16 September 2016.

57. See Mohan, 'How Safe Is India's Nuclear Energy Programme?'.

58. Mohan, 'How Safe Is India's Nuclear Energy Programme?'.

59. 'Nuclear Power in Italy', available http://www.world-nuclear.org/information-library/country-profiles/countries-g-n/italy.aspx, accessed on 22 September 2016.

60. 'Nuclear Power in Italy'.

61. 'Emerging Nuclear Energy Countries', available http://www.world-nuclear.org/information-library/country-profiles/others/emerging-nuclear-energy-countries.aspx, accessed on 22 September 2016.

62. 'Emerging Nuclear Energy Countries'.

63. 'Emerging Nuclear Energy Countries'.

64. Mohan, 'How Safe Is India's Nuclear Energy Programme?'.

65. See, S-5180 The Gazette of India: December 31,1983/Pausa 10,1905 [Part Ii-Sec.3(Ii) Department of Atomic Energy New Delhi, 15 November, 1983(S.O.4772 --)][No. 25/2/83-ER]

66. Section 22.

67. Sections 16, 17, and 23.

68. Section17.

69. See M.V. Ramana, 'The Indian Nuclear Industry: Status and Prospects', available www.cigionline.org, accessed on 10 October 2016.

70. Ramana, 'The Indian Nuclear Industry: Status and Prospects'.

71. Rules made under section 30 (ii) (i) of Atomic Energy Act, 1962.

72. Rules made under section 30 (ii) (i) of Atomic Energy Act, 1962.

73. Section 25, the Air (Prevention and Control of Pollution) Act, 1981.

74. Clause 3, The Atomic Energy (Radiation Protection) Rules of 2004.

75. Clause 3, The Atomic Energy (Radiation Protection) Rules of 2004.

76. Clause 22, The Atomic Energy (Radiation Protection) Rules of 2004.

77. Section 6, Civil Nuclear Liability Act, 2010.

78. Section 17, Civil Nuclear Liability Act, 2010.

79. Section 24, Civil Nuclear Liability Act, 2010.

80. Section 46, Civil Nuclear Liability Act, 2010.

81. Section 35, Civil Nuclear Liability Act, 2010.

82. See the website of Bhabha Atomic Research Centre at http://www.barc.gov.in/pubaware/enr_n1.html, accessed on 10 October 2016.

83. Section 3, Disaster Management Act, 2005.

84. Section 6, Disaster Management Act, 2005.

85. Section 18, Disaster Management Act, 2005.

86. For additional details, see 'Dose Limits for Radiation Workers' available http://www.nrc.gov/reading-rm/basic-ref/glossary/alara.html, accessed on 10 October 2016.

87. See https://public-blog.nrc-gateway.gov/2011/03/22/all-about-epzs/, accessed on 10 October 2016.

88. 'Zoning Concept and Emergency Planning', available https://www.ncbi.nlm.nih.gov/pmc/articles/, accessed on 10 October 2016.

89. 'Protecting Health and Safety', available http://www.safesecurevital.com/emergency-preparedness/emergency-planning.html, accessed on 10 October 2016.

90. See also 'Nuclear Power in India', available http://en.wikipedia.org/wiki/Nuclear_power_in_India, accessed on 10 October 2016.

91. See https://www.ncbi.nlm.nih.gov/pmc/articles/PMC2796747/, accessed on 10 October 2016.

92. See https://www.ncbi.nlm.nih.gov/pmc/articles/PMC2796747/, accessed on 10 October 2016.

93. Clause 3, National Electricity Policy, 2005.

94. Clause 2, National Electricity Policy, 2005.

95. Clause 2.0 Aims and Objectives, National Electricity Policy, available http://pib.nic.in/archieve/others/2005/nep20050209.pdf, accessed on 10 October 2016.

96. Clause 5.1.3, National Electricity Policy.

97. Clauses 5.2.2; 5.2.5 to 5.2.11, National Electricity Policy.

98. Clauses 5.2.12 to 5.2.18, National Electricity Policy.

99. Clauses 5.2.20, National Electricity Policy.

100. Clause 5.3.2, National Electricity Policy.

101. Report on Energy Policy, available http://planningcommission.nic.in/reports/genrep/rep_intengy.pdf, accessed on 10 October 2016.

15

MAANSI VERMA

Achieving Redistributive Energy Justice

A Critical Analysis of Energy Policies of India

As per a 2007 report of the Government of India, 'Energy is the sine qua non of development'.[1] On comparing the per capita energy consumption and Human Development Index (HDI) across various countries, the report concluded that an essential component of national development is access to energy. There is a mutually reinforcing relationship between access to reliable energy and development. Availability of energy ensures increased activity and productivity which contributes to a higher GDP which in turn ensures availability of capital and resources for investment in infrastructure to ensure continued availability of energy. The relationship however is not linear and is dependent on a number of factors. It is indeed because of the presence of interconnected and interdependent factors that the World Energy Council looks at the problem of energy access as 'Energy Trilemma'—a challenge to simultaneously deliver 'secure, affordable, and environmentally sustainable energy' to all.[2] In a similar vein, The Energy and Resources Institute (TERI) highlights in the National Energy Map for India that on one hand growth of an economy hinges on 'cost effective and environmentally benign energy sources' and on the other hand, energy demand drives economic growth.[3]

In this context, energy justice becomes a conversation among multiple factors—poverty (income poverty causing energy poverty leading to dependence on inefficient and polluting energy sources), energy security (systemic features including fossil fuel subsidies, loss making utilities, centralized grid, transmission and distribution, losses and unavailability of capital and government support for electrification in remote, low density rural areas), environmental concerns (reducing energy intensity of economy, increasing share of renewables, decentralized, and off-grid solutions, green subsidies including feed-in tariffs, improving energy efficiency), and justice (adverse health effects, continued deprivation). It is nigh impossible to achieve one objective without interventions across all the related variables.

This paper thus endeavours to knit together many of these themes. The first section of the paper is devoted to identifying the magnitude of the problem where facts and narratives are given to identify the various dimensions of injustice. The second section of the paper delves deeper into the interconnection between energy justice and other essential variables viz. economic development, energy security, and environmental concerns. The third and the final section of the paper deals with the current policy scenario and evaluates it on the energy justice parameter.

Multiple Facets of Injustice

There is no set parameter to measure injustice. Facts, figures, and narratives presented in this section, however, paint a picture of a general sense of inequity, deprivation, and negative fallouts of the present development paradigm. All of this points to the fact that 'energy justice' needs to be at the centre stage of all energy related policy decisions and not merely be one of the outcomes which may result from a decision taken in isolation. In this section we look at three broad sources of injustice—energy poverty, negative effects of iniquitous energy deprivation, and adverse costs of present development paradigm.

Energy Poverty

Eliminating energy poverty is essentially a quest for access. The International Energy Agency considers availability of modern energy services, that is, electricity and clean cooking fuels as the indicator for energy access.[4] It further identifies that globally close to 1.3 billion people are without electricity and around 2.6 billion people rely on unclean cooking fuels. As per India's Intended Nationally Determined Contribution (INDC) submitted just ahead of the meeting of Conference of Parties (CoP) 21 in December 2015, India alone is house to 24 per cent of the global population without electricity and 30 per cent of the global population relying on solid biomass for cooking.[5]

In India, various attempts have been made through different instruments to measure the general well-being of people, and one of the thrust areas is access to modern energy services. Take for instance the Socio-Economic and Caste Census (SECC) conducted in 2011 which measured 'deprivation' of households according to certain factors. The methodology involved 'excluding' and 'including' households based on set parameters and then measuring their deprivation.[6] It is interesting to note that if any of the following is true for a household, it will not be considered deprived—motorized two/three/four wheeler/fishing boat, mechanized three/four wheeler agricultural equipment, owns a refrigerator, owns landline phone, irrigation equipment (with corresponding amounts of land). All of these indicators require access to either fuel or electricity. After excluding on the basis of any of these factors and including based on a criteria, in rural India, 10.72 crore households were considered for deprivation, that is, they did not have access to modern energy services apart from other services.

But the picture is not complete as the dependence on solid fuels for cooking was not captured. For this, we look at the 68th Round of the National Sample Survey (NSS).[7] The survey reports that among rural households surveyed across the country, 67.3 per cent relied on firewood and chips as primary energy source for cooking, whereas 15 per cent relied on LPG.[8] The element of injustice surfaces as the data reveals that as many as 87 per cent of the Scheduled Tribe (ST) households and 69.8 per cent of the Scheduled Caste (SC) households in rural India relied on firewood and chips for cooking.[9] Again in terms of use of kerosene for lighting, it was reported that the proportion of households in rural areas relying on kerosene among the poorest 5 per cent of the households was as high as 57.1 per cent and it comes down to 3.7 per cent for the highest income fractile.[10]

Energy poverty is exacerbated by the fact that the fuel the poor rely on is also inefficient. As per a report based on data collected under India

Human development Survey carried out in 2004–5, kerosene is not just a poor lighting fuel—it provides less electricity than a 40 watt bulb—but it is also more expensive.[11] Energy poverty is also about poverty of choice. While kerosene is readily available in open market and across Public Distribution Stores, it is not that easy to get an LPG connection. Another dimension of energy poverty which is rarely captured is the time and effort expended in gathering fuelwood or biomass for producing energy. This time and energy, which usually falls in the share of a woman's or child's household labour has an opportunity cost—time which could have been used by the woman in engaging in some productive and income bearing enterprise or the time which could have been utilized by the child for education.[12] Thus, it can be argued that energy poverty is a multi-dimensional concept and a careful understanding of its many aspects is needed to tackle it fully.

Inequity and Negative Health Effects of Energy Deprivation

Inequity at the global level often plays out in the tussle between developed v. developing countries. It is in this backdrop that India, conscious of and leveraging its status of a developing economy, states in its INDC that per capita emissions in India in 2010 were about 1.56 metric tonnes, which is much lower than the per capita emissions of developed countries which hovers between 7 to 15 metric tonnes.[13] But in its 2007 report, Greenpeace advocated for a common but differentiated responsibility within the nation also.[14] The report found through a survey of households that the richest income class within the survey (earning more than Rs 30,000 per month) was producing 4.5 times more CO_2 than the poorest income class (earning less than Rs 3,000 per month) and almost three times the average Indian.[15] Interestingly, the report also notes that the bulk of this more than average emission among the richest households comes from modern equipment like computers, kitchen appliances, phones, personalized transport, and so on, which are exclusive to few, and not from the energy services needed at the very subsistence level like electricity for lighting which is widely available and used.[16] Calculation of averages thus inflicts injustice of its own kind where subsistence consumption of a poor household is used as an illusion to hide the unsustainable consumption patterns of the richest few.

Deprivation from modern energy services and dependence on inefficient and unclean sources of fuel for cooking and lighting also has its attendant negative health consequences discriminately affecting the poor the most. As per World Health Organization (WHO), inefficient cooking, and heating practices in poorly ventilated smoke filled homes can increase the acceptable levels of fine particulate matters 100 times and disproportionately affect women and children who spend the most time near domestic hearths.[17] As far as India is concerned, in 2004 WHO estimated that indoor pollution is likely to cause 488,200 deaths per year as compared to 119,900 deaths caused annually by outdoor pollution.[18] Despite acknowledging and often even quoting these findings by WHO, the Government of India has not undertaken any research to study pollution related mortality in the country[19] or to study the effect of pollution in rural areas.[20]

Adverse Costs of Present Development Paradigm

It has been argued that environmental justice is also closely linked to energy production—by using geographic and regression analysis,[21] a study concluded that disproportionately high siting of coal fired power plants in the poor and minority communities in the state of Illinois presented an environment justice issue.[22] Similar claims have been made from other US states as well. In a study of twelve coal plants across six US States carried out by National Association for the Advancement of Colored People, it was found that proportion of coloured people living around these coal plants was as high as 76 per cent with an average income much lower than the national average.[23] The argument of injustice can be stretched to allege blatant discrimination because the victim is a poor, voiceless, marginalized person whose only contribution to the development story would be living in the vicinity of polluting coal plants in silence.

Closer to home, in a spatial analysis of poverty across India, a study found that high poverty is unusually found near certain economic zones, such as mining areas and is characterized by unstable working conditions, low literacy, and a high proportion of deprived population.[24] As far as the mineral rich Chhattisgarh state is concerned, the spatial analysis revealed that a large number of poor people are settled near mining areas which attract low skilled cheap labour.[25] These areas are also tribal dominated and are reported to have a sizeable number of marginal, forced, and bonded labour. Using the Tendulkar poverty line,[26] the study also found that some of the richest coal bearing districts also bear the burden of high poverty. Korea district, for instance, with 45.5 per cent of the population below poverty line, has vast reserves of high grade coal.[27] The same story is repeated in Jharkhand, one of the richest states in terms of mineral reserves, but poorest in terms of economic indicators. Even in Jharkhand, high percentage of poverty was found near mining areas.[28] The spatial analysis also indicates that these high poverty areas around rich mines are further characterized by low energy usage, sparse road connectivity, dense forests, and poor water and sanitation facilities. Who benefits from the rich minerals is a question we must ask in our quest to find justice.

The isolation and the rough terrain which prevents these people from accessing basic facilities, however, are no obstacle in the exploitation of coal mines and establishment of thermal plants or hydel dams. Consider for instance, the plight of Sarojini and Rama Naik, interviewed by Greenpeace for their study, belonging to Vatehalla Hamlet in Mahime Village, HonavarTaluk in Uttara Kannada district in Karnataka who have to traverse a 10 kilometre journey to procure kerosene for their lighting needs despite living in the vicinity of two large hydel dams—55 MW Linganalaki and 240 MW Gerusoppa.[29] Consider also for instance, the case study of Parli Thermal Power Station in Beed District of Maharashtra which witnessed protests from residents of nearby villages over the issue of fly ash which caused allergy and rashes.[30] As per government's own admission, it has stipulated norms for consumption of fly ash, but it has conducted no research to study the correlation between sickness and fly ash generation.[31] The issue of fly ash generation is just one of the issues within the larger debate of pollution from thermal power plants and the government has blatantly stated that no study has been conducted by Central Pollution Control Board (CPCB) on rising temperature, health fatalities, and other ailments due to pollution from thermal power plants.[32] How is one to account for the ailments that could have been caused by all the pollution if one does not even attempt to enumerate it? Or maybe the situation is not that emergent, or maybe we would rather not know. Lack of research on issues of injustice is itself an injustice.

Another element of injustice would be the effect of climate change on agriculture. Energy generation through fossil fuels leads to increase of Green House gas emissions which in turn contribute to climate change which affects rainfall patterns causing flood or draught like conditions. The causation may not be direct, but a correlation is evident. In India, in 2015, due to deficient rainfall, ten states declared draught and as many as 897 farmers committed suicide for agrarian causes, the most being in Maharashtra which witnessed severe draught for many months in three consecutive years.[33]

The theme repeats itself, a top down and centralized process of economic development uniformly applied from government offices distant from ground realities fuelling unsustainable consumption patterns among the privileged few which in turns reinforces the existing development cycle. The invariable victims of the present paradigm are the poor, the marginalized, and the disempowered and the injustice perpetuates.

Thus in the first section an attempt has been made to elucidate the various manifestations of energy injustice and their magnitude. The lens of justice is needed to make sense of the numbers–of marginalized people being deprived of clean cooking fuels; of backward states lacking reliable energy supply; of deaths caused due to indoor pollution; of poor tribals near coal rich mines. To say that energy justice is merely about lack of access is oversimplifying a multi-layered issue, one which needs a comprehensive and integrated response. But before that, we look at other parallel aspects which have a bearing on the struggle against injustice.

Interconnections, Correlations, and Combinations

If poverty exacerbates energy injustice or if lack of access to energy leads to poverty, then not much analysis is needed to anticipate a strong correlation between energy availability and economic development. If injustice is caused because of erratic supply or inefficient use of power or dependence on large scale power projects, then it can be argued that the concerns of energy justice are interlinked with concerns of energy security as well. And if pollution caused by large scale thermal plants cannot be an oversight, then the question of environment sustainability cannot be divorced from energy pursuits. In this section, three broad themes will be investigated—the energy-economic development nexus, the energy security dilemma, and the sustainability question.

Energy–Economic Development Nexus

In its 2006 report, 'Fuel for Life', World Health Organization argues that household energy and development are 'inextricably linked'—with increasing prosperity and development, households tend to move from more inefficient and polluting solid fuels like crop waste, wood, charcoal to cleaner and more efficient fuels like kerosene, LPG, and electricity.[34] Further explaining the link between income poverty and energy poverty, the report highlights that, according to 2003 figures for India, more than 90 per cent rural households in the poorest income quintile used solid fuels as compared to less than 40 per cent rural households in the richest income quintile.[35] Interestingly, it is the poorest households who end up spending the most on energy as well, mostly because they rely on inefficient fuels. The report finds that in India, households belonging to the poorest income quintiles ended up spending more than 8 per cent of the household income on energy as compared to the less than 5 per cent of income that the households belonging to the richest income quintile were spending.[36]

The link between energy and economic development, however, does not just play out at the household level, but also mirrors the economy wide development trend. It has been argued that economies which are at the lowest levels of income and social development tend to see a higher dependence on biological sources like wood, dung, and so on, which elevates to more processed biofuels and some commercial fossil energy in the intermediate stage and finally with industrialization and advanced development taps into commercial fuels and electricity.[37] The fact that overall development of a state affects the choice of energy source is reflected in the findings of the 68th Round of NSS where higher incidence for use of LPG in households was found in more developed states of Tamil Nadu, Kerala, and Punjab and the least use of LPG in household was found in the states of Chhattisgarh, Jharkhand, and Odisha[38] Similarly, percentage of households relying on kerosene for lighting is found to be disproportionately high in states of Bihar, Uttar Pradesh, Assam, Jharkhand, Odisha, and West Bengal.[39]

This is understood as the interaction of energy with other development indicators. For instance, construction of a road connecting a remote village is likely to quicken the pace of its electrification which in turn could enable setting up of small or medium enterprises, schools, hospitals. In a study on effect of rural electrification in India, the impact of lighting and irrigation from motor pumps on education was studied. It was concluded that with pump irrigation, farm income increased and with lighting there was a marked percentage increase in consumer surplus. The cumulative effect of benefits from electrification was seen in the improved rate at which 'household income rises with years of schooling' and improved education in turn increased farm and nonfarm income.[40]

As per the World Bank's and UNDP's Energy Sector Management Assistance Program (ESMAP) 2002 study, this relation between energy and income works the other way round as well. Unless the household income and productivity are increased, people will not be in a position to purchase energy services and energy services will not become viable.[41] But the India Human Development Survey (IHDS) data finds that poor households living in poor states are doubly deprived as not only can they not pay for the electricity, the supply itself is either unavailable or erratic.[42] Thus, mere increase in household expenditure on energy may not necessarily mean that a higher quantity of energy is being consumed, as that will depend also on quality and reliability of supply, energy price, efficiency of the equipment used, and so on.[43] And this is the reason energy poverty cannot be addressed without also addressing the structural issues affecting the energy security of the country.

Issues of Energy Security

The National Institute for Transformation of India (NITI) Aayog has launched an open source web based tool—India Energy Security Scenario 2047—to estimate various energy scenarios and assist policy makers in arriving at crucial decisions to support the government's vision of achieving power for all, rural electrification, 175 GW generation from renewables, and reduction in import efficiency.[44] Energy security, thus, requires inputs from various sectors, produces outputs on different variables, and can lead to multiple outcomes. A report prepared by the erstwhile Planning Commission on Integrated Energy Policy uses the following words to define energy security:

> We are energy secure when we can supply lifeline energy to all our citizens irrespective of their ability to pay for it as well as meet their effective demand for safe and convenient energy to satisfy their various needs at competitive prices, at all times and with a prescribed confidence level considering shocks and disruptions that can be reasonably expected.[45]

The definition further explains that 'lifeline energy' is a basic necessity which will be procured by people in environmentally unsustainable ways if the government does not provide it. Across studies, it has been highlighted that energy access is an issue of energy security, because if all the unmet needs are added up, the demand for energy in a growing population will be considerable and will have to be factored in. For instance, in a study by TERI, it has been argued that with sustained economic growth and rising incomes demand for modern fuels and technology will increase, electrification will spread, and people are bound to shift to cleaner fuels for cooking purposes like LPG.[46] A scenario like this will have its own policy implications—the most crucial being—how much investment is to be made in exploitation of which energy resource and in what format.

As per the estimated scenario in the India Energy Outlook, a 2015 report by International Energy Agency (IEA), by 2040 India is poised to become the 'largest source of growth in global coal use' and even oil demand increases 'by more than any other country'.[47] Even in its INDC, India admits that in the future coal will continue to be the dominant source of power supply in order to ensure 'reliable, adequate, and efficient' electricity.[48] Not just this, the import of oil, gas, and coal are also likely to increase in the future indicating a growing energy bill.[49] As far as the source of energy generation is concerned, the future of secure energy seems to be dependent on fossil fuels. But the demand side of energy security requires as much, maybe even more consideration than the supply side. The National Energy Map of India prepared by TERI estimates that in the Business As Usual[50] (BAU) Scenario, end use consumption of energy in India is likely to increase eight times between 2001 and 2031.[51] This demand is largely driven by growth in energy intensive passenger and freight transport, increased infrastructural demand in the industry sector, and growth in residential sector. In the face of this fact, energy demand management assumes great importance.

It has been suggested that for effective demand management, it is essential to increase efficiency and reduce energy intensity of India's growth. The Integrated Energy Policy points that India's energy intensity of GDP growth had been continuously falling since the seventies, and suggests that with increasing efficiency it will become possible to reduce intensity even further.[52] But even with improvements in efficiency, unsustainable energy generation is likely to have its consequences. In a report prepared by McKinsey & Company, it was estimated that, anticipating a GDP growth of 6–9 per cent per annum and growth in demand in key sectors like power, industry and transport, with reasonable assumptions about improvement in energy efficiency, greenhouse gas emissions are set to increase from 1.6 billion tonne of carbon dioxide equivalent in 2005 to 5–6.5 billion tonne of carbon dioxide equivalent.[53] And with growing consciousness towards anthropogenic effects on climate, the concept of environmental sustainability now has a bearing on energy security as well as energy justice concerns.

The Sustainability Question

'It must be understood that poverty is a big polluter; so is the extravagant way of life and a profligate pattern of consumerism a grave threat to environment.'[54]

This statement is an admission of the fact that energy security policy cannot be divorced from climate policy. Poverty forces inefficient use of scarce resources but it is no solution to replace this with power generated from other polluting fossil fuels. It has been recommended in one study after the other that the only way to sustainably address the lack of energy access is to tap the renewable energy generation potential.[55] In the Ninth Session of the Commission on Sustainable Development, it was observed that for sustainable development it was essential to make available a 'cost effective mix of energy resources' in which there is a greater share of renewable energy along with improvement in technology and efficiency.[56] This is a reiteration of the argument that developing countries must not traverse the unsustainable path undertaken by the developed countries, because the world cannot simply afford that anymore and because today alternatives are available. There is a need to break from the past and push the boundaries of human imagination in finding sustainable solution to the development dilemma.

Following a sustainable development paradigm has its attendant benefits as well. As per India's INDC submission, an Asian Development Bank report estimated that the economic damage and losses in India from climate change will be 1.8 per cent of the GDP annually by 2050 and mitigation expenses, as per a NITI Aayog estimation will be USD 834 billion by 2030 at 2011 prices.[57] Proposing stricter environmental and social safeguards in development projects, an opinion piece by the Director General of Independent Evaluation at Asian Development Bank argues, that from economic perspective a nominal expenditure on setting safeguards makes sense as it saves the huge expenditure that may have to be incurred in the future to mitigate negative externalities.[58] These negative externalities also encompass the huge social costs involved with displacement of people due to large scale development projects and the unrest caused. The loss of biodiversity, if any, would inflict an irreparable environmental damage, the cost of which will be a drain on the overall cost of development.

Thus, in the second section we have explored the essential linkages which must be addressed in an integrated and comprehensive manner if energy injustice is to be tackled fully. India cannot be energy secure if a major proportion of its population is deprived of energy or dependent on unclean energy. India also cannot be either energy secure or equitable if it continues to depend heavily on fossil fuels or if it does not manage the future demand of energy. But ensuring energy access to all requires exploiting new and renewable resources which calls for a major overhaul in the energy policy with systemic, financial, and infrastructural changes. With this we move to the third and final section of the paper which concerns itself with the policy scenario.

Energy Justice and Policy

Any study looking into the question of energy poverty, access, or security ends with policy recommendations. Because it is essentially a matter of policy, not even much of choice. In the absence of enabling policy, people are devoid of the capabilities needed to make sustainable choices. But the policy must be rooted in an understanding of the core issue it hopes to address, which in the case of this paper is energy justice. Thus this section of the paper speaks of three broad themes—the conceptual underpinnings of redistributive energy justice, an overview, and critique of select current energy policies and a collection of policy recommendations.

Redistributive Energy Justice: The Concept

This paper so far has claimed that, issues of energy poverty and production, its attendant ills, the disproportionate deprivation and health consequences are various facets of injustice which are further interlinked with concerns of economic growth, energy security, and sustainable development. Addressing energy justice thus requires, as has been argued elsewhere as well, a 'whole systems approach',[59] where a set of subsystems must coexist and coordinate to accomplish a set goal.[60] It has further been argued that the energy justice movement which finds its roots in the environment justice movement considers 'distribution, procedural, and recognition tenets'.[61] The environment justice movement encompasses within itself such social justice concepts like 'self-determination, sovereignty, human rights, inequity, access to natural resources, and disproportionate impacts of environmental hazards'.[62] Again, the movement rests on distributional justice and procedural justice—who gets what and how. Consider for instance the Principles of Environment Justice adopted at the People of Colour Environmental Leadership Summit in 1991 which affirm, among others, the declaration that public policy must be based on mutual respect and justice and must be free from any discrimination or bias and provide fair access for all to full range of resources with a right to participate as equal partners at all levels of decision making and enforce a right to ethical, balanced, and ecological use of renewable resources.[63] Distribution or rather redistribution becomes an essential cog in the wheel of justice.

John Rawl's *A Theory of Justice*, as has been proposed in another paper, provides us with the necessary philosophical understanding of redistribution—he argues for legitimate principles of justice which have been arrived at through a fair negotiation, in which each person has equal rights to most basic liberties and social and economic inequities are rearranged so that they may bestow greatest benefit to the most disadvantaged.[64] Extrapolating this principle in the concept of energy justice, it can be argued that not only is it essential that everybody has equal access to lifeline energy but also that the current energy systems be reworked to eliminate the disproportionate effects on the marginalized and deprived. Looking at energy justice from this perspective, a study proposes that new energy infrastructure development must not simply be efficient, but must lead to allocation and distribution of costs and benefits in a way which appear 'just and equitable'.[65]

To this end, another study on decentralized generation of electricity through micro grids elucidates that the current energy systems suffer from 'path dependency' as their existence is shaped by the historical roots in which there were different conditions, the system addressed different needs and hence is not suited to 'further the integration of new energy sources'.[66] The study further proposes that any change in the system will require 'empowerment of new actors and hence disempowerment of current incumbents', and a regime of good governance which fosters cooperation in 'use, monitoring, maintenance, rules for extraction and contribution'.[67] The new system which must emerge from this churning of ideas of justice, redistribution, capacities, and economic efficiencies will have to look at policies from a different perspective. For instance, in the Integrated Energy Policy, the Expert Committee of the Planning Commission argues that taxes and subsidies on different fuels should be rationalized so that consumer choices are not distorted but at the same time environmental taxes and subsidies will be justified as they are meant to affect choices in order to address 'environmental externalities'.[68] This exemplifies the redistributions required in the system to ensure that costs and benefits flow to the recipients in an efficient yet just and equitable manner.

Adding the human element to the strictly economic considerations of energy is the essence of energy justice approach which recognizes that social considerations can no longer be ignored while designing strategies for sustainable development. Polices will have to be redesigned to address the historic injustice in a manner which does not reinforce the divide. And this brings us to the segment where an overview of current energy policies is attempted along with their critique.

Select Current Energy Policies

It would be impossible to review all the energy policies of the country within the scope of this paper. Since the aim of this paper is to assess the policy framework from a redistributive energy justice perspective, the focus will be on fundamental shifts in the policy or radical changes in the regime in the recent past which can have a bearing on issues related to energy justice.

The Statistics Related to Climate Change—India, a 2015 report by Government of India presents some interesting statistics.[69] The relevant statistics are summarized below:

1. Between 1994 and 2007, emission from electricity sector grew by a compounded annual growth rate (CAGR) of 5.6 per cent, of transport sector by 4.5 per cent, of residential sector by 4.4 per cent and of other energy by 1.9 per cent.[70]

2. As of total installed capacity of all power utilities across India, as on 31 March 2015, 70.57 per cent was generated from Thermal (Coal, Oil and Diesel), 15.4 per cent was generated from Hydro, 11.8 per cent from renewable resources and rest from Nuclear.[71]

3. Estimated potential of renewable sources for electricity generation as on 31 March 2014 was 147615 MW, with 69.6 per cent of it coming from wind power, 13.3% coming from small hydro, 11.88 per cent coming from biomass and rest from other sources (this estimate did not include potential from solar).[72] Out of this estimated potential, grid connected installed capacity of renewable resources was a mere 31692.18 MW.[73]

4. Similarly, as against the estimated potential of distributing 12,339,300 family type biogas plants, 4,753,085 were distributed.[74] As against an estimated potential of 19749.44 MW of small hydro projects, 4,055.355 MW had been installed and 683.21 MW were under implementation as on 31 March, 2015.[75]

5. As far as Solar (Photovoltaic) PV systems are concerned, 42,157.61 kWp (Kilowatt peak) of stand-alone power plants were installed and 30,668 kWp of grid connected power plants were installed up till March 2013.[76]

It is clear from these facts that dependence on fossil fuels is still considerable and efforts towards realizing the estimated potential of energy generation from renewable sources needs to be reinvigorated. In a recent op-ed it was pointed out that India is not fully exploiting the advancement in technology available which has made solar energy as competitive as coal.[77] The author of the editorial appreciated the auction of solar power in Andhra Pradesh recently which had fetched as low a price as Rs 4.63 per unit making it fully competitive with coal based power especially for meeting peak loads but he also lamented the fact that this was for Solar PV power plants which can only work during the day and cannot store solar energy, making it imperative to guarantee it with thermal power thereby defeating the purpose. He further pointed to the advancements made in construction of Concentrated Solar Thermal Power plants which can even store energy and hence only need a 15 per cent backup of natural gas based energy. The initial cost of setting up such a plant will be higher but will be offset by savings in efficiency and transmission as compared to Solar PV Plants.

Following the 'precautionary principle' India must make a technological leap of faith and lap up the best available technology because India is ramping up its efforts on the solar front and plans to add as much as 2,000 MW of solar capacity in the first quarter of 2016 itself and with falling solar prices, India is becoming an attractive destination for foreign investors as well.[78] India needs to find innovative ways of making investment in renewable energy attractive just as it continues to do for thermal power. As per the 2015 budget announcement by Finance Minister, India is poised to set up 5 coal-based Ultra Mega Power Plants, each of 4,000 MW capacity in a 'plug and play' mode where all clearances and linkages will be in place before the project is awarded through a bidding process, which as per government's estimate is likely to fetch investment of Rs 90,000 crore.[79] As an abatement mechanism, the technology proposed to be used will be supercritical and the government proposes to increase the cess on coal which will then be used to fund renewable energy projects.[80] But having said that, it must be noted that out of the 87 GW of new coal capacity added by India since 2010, 61 GW has been subcritical indicating a hesitation to move to supercritical and eventually ultra-supercritical technology which is greener but much more expensive.[81]

It has further been argued that simply investing in energy creation is not sufficient and India needs to find ways to connect this to the grid and balance the shortage in supply from renewable resources through multiple sources including coal, but also gas and hydel, across states which would require better load forecasting and improvement in technology to tap the potential of renewable sources.[82] To this end the

Draft National Renewable Energy Act, 2015 speaks of 'decentralized renewable energy' and aims to provide for 'facilitating and coordination related aspects for grid connected renewables'.[83] Distributed or decentralized renewable energy, however, has not been defined in the draft, but it does call for 'off grid' solutions defined as those which generate electricity from renewable sources to a specified area but are not connected to the main distributing grid. The Act specifically links decentralized renewable energy to the issue of energy access and calls upon all state governments to identify those villages and hamlets which still have to be electrified and for that purpose, explore the possibilities of using off grid, mini or community grids, and decentralized renewable energy where grid extension is technically or economically unfeasible. For this purpose, the Act also urges the government to provide the necessary incentives, facilitating framework, and technology development. The government has not taken any action on the draft and it does not seem likely that the Act will be pushed through in the near future.

It is interesting to note here that under the erstwhile Rajiv Gandhi Rural Electrification Scheme (RGGVY), there was a specific component for Decentralized Distributed Generation (DDG), which provided for electrification of villages and hamlets with a population of at least hundred people through solar PV, small hydro, Biomass gasification, biofuels, and so on.[84] The scheme guidelines specifically provided for using the decentralized technology for villages where grid connectivity was not foreseen in the next five-seven years but the technology should be set up in a manner that it is grid compatible. In 2013, the scheme guidelines were amended to include also those villages/hamlets where power was supplied for only six hours a day, with the aim of supplementing the existing grid network.[85] This was a welcome step as now even villages which were grid connected but poorly supplied could be covered through DDG. In 2015, this scheme was subsumed under the Deen Dayal Upadhyay Gram Jyoti Yojana which has the larger objective of rural electrification through feeder separation of agricultural and non-agricultural consumer in rural areas, strengthening the sub transmission and distribution in rural areas and metering.[86] The scheme, however, proposes to implement the projects sanctioned through outlay approved under the RGGVY including the DDG scheme. Just about Rs 900 crores as subsidy has been earmarked for this purpose as compared to the overall budgetary support Rs 33,453 crore for other components of the scheme. Other decentralized distribution schemes and off grid solutions are managed by the Ministry of New and Renewable Energy[87] which spent a total of Rs 1,101.97 crore in 2014–15 and Rs 469.15 crore in 2015–16.

There is a clear thrust towards universal electrification not in a redesigned redistributive manner but in a conventional centralized grid connected mechanism. To feed this grid, reliable energy supply of a particular voltage is needed which mostly comes from coal based power plants. Moreover, centralized grid distribution is also plagued by failing health of power distribution companies which are bogged down by huge debts, electricity thefts, and unpaid dues. Interestingly, some of the most debt ridden power utilities also belong to the more rural states like Bihar, Jharkhand, Uttar Pradesh, Haryana, Telangana, and so on.[88] Despite this, decentralized distributed energy generation through renewable energy is seen only as feeding the grid even when there is emerging technology to establish renewable energy generation as complete in itself through various hybrid models.[89] The government has recently launched the Ujjwal Discom Assurance Yojana to revitalize the power discoms by the state government undertaking to manage their debt which runs into lakhs of crores.[90] As against this, the only clear policy announcement of a dedicated investment in renewable energy was when India mentioned in its INDC, that it proposes to introduce tax free infrastructure bonds to the tune of Rs 50 billion for investment in renewable energy projects in 2015–16.[91] This has now been followed by the issue of Rs 1,716 crore tax free bonds by Indian Renewable Energy Development Agency in January 2016.[92]

As far as the petroleum sector is concerned, green push has come mainly from cut in subsidies. In its INDC, India announced that it has cut its subsidy on petroleum by 26 per cent in the fiscal year 2014–15 as compared to 2013–14.[93] Much of it has been possible due to fall in global oil prices. Using this opportunity, India has hiked the excise on petrol and diesel resulting in an implicit tax on petroleum products. Another controversial and bold step undertaken by the Ministry of Road Transport is to fast track the implementation of Bharat Standard VI fuel norms for vehicles which has been lambasted by the automobile industry for being too ambitious as even the present norms of BS IV are not available throughout the country and the validation of current technology takes more time.[94] But as has been argued in an op-ed, we are moving towards a 'reconfigured energy system' where oil prices are on a downward spiral and tremendous work is being done on developing and scaling up alternative fuel technologies to make them commercially viable.[95] In this backdrop, any step taken by the government to promote better technologies is welcome as long as it is supported by the necessary research and development and financial support required. For instance, the government recently launched the FAME (Faster Adoption and Manufacturing of Hybrid and Electric Vehicles in India) scheme under the National Electric Mobility Mission Plan wherein the Ministry of Heavy Industries will subsidize the cost of an electric vehicle over and above the diesel/petrol cost.[96] This is a good example of redistributing costs and benefits as such subsidy is meant to push the demand for electric vehicles and thereby reduce demand for oil, a polluting fossil fuel. Another laudable redistributive effort of the government is the scheme for voluntary giving up the LPG subsidy.[97] By redirecting the subsidy to those, especially in rural areas, who would otherwise not be able to procure it, government takes a step towards securing energy justice.

To conclude, in this section, myriad policy initiatives have been overviewed and critiqued from the lens of energy justice. The direction in which funds are directed, the promotion of technology, market reforms, and selection of beneficiaries, all of these have a bearing on energy justice and will either reinforce the injustice or abate it. Having reviewed the existing policies, the last and the final theme of the paper, deals with some policy recommendations to promote energy justice.

Policy Recommendations

In February 2015, NITI Aayog submitted a report to the Government of India laying down a roadmap for renewable electricity.[98] The report categorically recommends that Renewable Energy needs to be seen as a resource of national and strategic importance and to this end it calls for a National Renewable Energy law or policy which lays down binding targets like an enforceable Renewable Purchase Obligation (RPO) and provides for integrated energy resource planning. For the purpose of making investments in renewable energy attractive, the report urges the government to reduce the risk perception by de-risking, make finance for such projects easily accessible and reduce administrative hassles by providing for standardized contracting and improve grid connectedness. Some of these recommendations have found their way into the National Tariff Policy which is being amended to give a push to clean energy.[99]

Intermittency is a big concern in energy generation from renewable sources and hence there is a tendency to complement it with guaranteed supply from thermal power plants. However, it can be argued that no system will be 100 per cent reliable and with advancement in technology it will be more and more possible to improve the transmission. A study suggests that it is possible to rely completely on wind, water, and solar energy if geographically dispersed naturally variable energy sources are interconnected; a non-variable energy source like hydropower is used to fill temporary gaps; if there is smart weather forecasting and smart demand and supply management in the grid; and if electric power can be stored at the site of generation.[100]

Improvement in technology, at all levels, from a more efficient and cleaner cooking stove to an ultra-supercritical thermal plant to an electric vehicle is needed. The much publicized 'Make in India' scheme can be put to good use for the purpose of advancement of energy justice if the government gives the much needed push to research and development especially with a renewed focus on more sustainable energy. For instance, within the 'Swachh Bharat Mission', a major component of which is Solid Municipal Waste Management, waste to energy creation has been given a major push with the Ministry of New and Renewable Energy promoting a dedicated scheme for this purpose.[101] To similar effect would be efforts directed towards smaller decentralized and distributed generation which will not only be a panacea for access to energy especially for marginalized and deprived, but it will also lead to co-creation and empowerment with the community owning the process of energy creation through locally available resources.

Finally, it is absolutely imperative that energy justice becomes the driving force behind all decisions. Every cost and benefit analysis of a policy decision must weigh the impact it would have on those who will be most adversely affected by it or those who will continue to remain deprived despite that decision. Such a cost and benefit analysis could also reveal the gains from taking a more conservative step today than spending money on mitigating the negative externalities tomorrow. India can take inspiration from the historic Executive Order issued by the president of US which made every federal agency accountable towards ensuring environment justice in all its missions 'by identifying and addressing, as appropriate, disproportionately high, and adverse human health or environmental effects of its programmes, policies, and activities on minority populations and low-income populations'.[102] To this end, there needs to be more rigorous data collection and analysis and the government cannot simply get away from its responsibilities by saying that it has no data to prove injustice when it is plainly evident.

***

In this paper an attempt has been made to explore the various facets of energy justice which essentially stems from our sense of equity and fairness and hence needs no convincing. The data presented in the paper points to injustice subtly existing in the details. The interlinkages show that it is indeed impossible to achieve economic growth unless issues of energy poverty, security, and generation are addressed. But to address these issues, one cannot simply be dictated by the present development paradigm which has resulted in a lopsided growth.

The central argument of this paper is that not only the common thread determining the success or failure of current policy decisions is energy justice which may be elucidated in concerns for universal electricity access, clean cooking, and sustainable development, but that energy justice itself is the ultimate goal and not the gross investments or the gross capacity created. Numbers exist to measure the progress but the principle shows the way and unless energy justice is explicitly made the goal of all energy related policy, the poor, voiceless, out of sight, and hence out of mind will continue to languish in oblivion. It is not to say that no progress has been made so far. Indeed, much has been achieved but more still needs to be achieved and it is for this future path yet to be taken that redistributive energy justice must become the pivot around which all decisions revolve.

Notes and References

1. Ministry of Environment and Forests and Bureau of Energy Efficiency under Ministry of Power, Government of India, 'India: Addressing Energy

Security And Climate Change' (2007), p. 1, available http://www.ceeindia.org/cee/pr/publication.pdf, accessed on 13 January 2016.

2. World Energy Council, '2015 World Energy Issues Monitor, Energy Price Volatility: The New Normal' (2015), p. 6, available https://www.worldenergy.org/wp-content/uploads/2015/01/2015-World-Energy-Issues-Monitor.pdf, accessed on 25 August 2016.

3. The Energy and Resources Institute (TERI) and Office of the Principal Scientific Adviser, Government of India, 'National Energy Map for India: Technology Vision 2030' (2006), p. 1, available http://www.teriin.org/div/psa-fullreport.pdf, accessed on 25 August 2016.

4. International Energy Agency, 'Energy Poverty', available http://www.iea.org/topics/energypoverty, accessed on 25 August 2016.

5. Government of India, 'India's Intended Nationally Determined Contribution: Working Towards Climate Justice' (2015), p. 5, available http://www4.unfccc.int/submissions/INDC/Published%20Documents/India/1/INDIA%20INDC%20TO%20UNFCCC.pdf, accessed on 25 August 2016.

6. Ministry of Rural Development, Government of India, 'Socio-Economic and Caste Census 2011', available http://www.secc.gov.in/reportlistContent, accessed on 25 August 2016.

7. Ministry of Statistics and Programme Implementation (MOSPI), Government of India, 'Energy Sources of the Indian Households for Cooking and Lighting, 2011–12', July 2015, available http://mospi.nic.in/Mospi_New/upload/nss_report_567.pdf, accessed on 25 August 2016.

8. MOSPI, 'Energy Sources of the Indian Households for Cooking and Lighting', p. 25.

9. MOSPI, 'Energy Sources of the Indian Households for Cooking and Lighting', p. 9.

10. MOSPI, 'Energy Sources of the Indian Households for Cooking and Lighting', p. 10.

11. S.B. Desai, et al., 'Households Assets and Amenities', in *Human Development In India—Challenges For A Society In Transition* (New Delhi: Oxford University Press, 2010), p. 64.

12. S. Pachauri, et al., 'On Measuring Energy Poverty in Indian Households', *World Development*, 32(XXII) (2004), pp. 2083–104, doi:10.1016/j.worlddev.2004.08.005.

13. World Energy Council, '2015 World Energy Issues Monitor', p. 2.

14. G. Ananthapadmanabhan, K. Srinivas, and V. Gopal, 'Hiding Behind the Poor', Greenpeace India Society (2007), p. 3, available http://www.green-peace.org/india/Global/india/report/2007/11/hiding-behind-the-poor.pdf, accessed on 25 August 2016.

15. Ananthapadmanabhan, Srinivas, and Gopal, 'Hiding behind the Poor', p. 9.

16. Ananthapadmanabhan, Srinivas, and Gopal, 'Hiding behind the Poor', p. 11.

17. World Health Organization, 'Household (indoor) Air Pollution', available http://www.who.int/indoorair/en/, accessed on 25 August 2016.

18. World Health Organization, 'India: Country Profiles of Environmental Burden of Disease' (2004), p. 1, available http://www.who.int/quantifying_ ehimpacts/national/countryprofile/india.pdf?ua=1, accessed on 25 August 2016.

19. Ministry of Environment and Forests, 'Increasing Pollution Responsible for High Child Mortality', Unstarred Question No-1425 answered in Rajya Sabha on 22 August 2013, available http://rajyasabha.nic.in/, accessed on 13 January 2016.

20. Ministry of Environment And Forests, 'Air Pollution in Rural Areas', Starred Question No-323 answered in Rajya Sabha on 13 February 2014, available http://rajyasabha.nic.in/, accessed on 13 January 2016.

21. Regression analysis is a statistical process for estimating the relationships among variables.

22. C. Clark, 'Environmental Justice and Energy Production: Coal-Fired Power Plants in Illinois' (MPP Research Paper, Oregon State University, 2008), p. 9.

23. A. Wilson, 'Coal Blooded- Putting Profits before People', National Association for the Advancement of Colored People (2011), p. 30, available http://www.naacp.org/page/-/Climate/CoalBlooded.pdf], accessed on 15 January 2016.

24. L. Bhandari and M. Chakraborty, 'Spatial Poverty Across India', *Livemint* (20 April 2015), available http://www.livemint.com/Home-Page/ BH4Yupc3zGc2gs067s6eKN/Spatial-poverty-across-India.html, accessed on 25 August 2016.

25. Bhandari and Chakraborty, 'Spatial Poverty in Chhattisgarh', *Livemint* (09 March 2015), available http://www.livemint.com/Opinion/ TelfQDSiZ1IPufDbwZ526K/Spatial-poverty-in-Chhattisgarh.html, accessed on 25 August 2016.

26. An expert group constituted by Planning Commission under Suresh Tendulkar in 2005, came out with a new poverty line which marked a shift from calorie based poverty line to income based poverty line drawn after taking into account consumption patterns and price fluctuations.

27. Official Website of District Korea, Chhattisgarh, available http://korea. gov.in/resources.htm, accessed on 15 January 2016.

28. L. Bhandari and M. Chakraborty, 'Spatial Poverty in Jharkhand', *Livemint*, 6 October 2014, available http://www.livemint.com/Specials/bqVly6xj4usB3DiTibS3DK/Spatial-poverty-in-Jharkhand.html, accessed on 25 August 2016.

29. S. Sharma, 'Still Waiting', Greenpeace India Society (2009), p. 8 [http://www.greenpeace.org/india/Global/india/report/2009/11/stillwaiting.pdf], accessed on 25 August 2016.

30. Sharma, 'Still Waiting', p. 16.

31. Ministry of Environment, Forests and Climate Change, 'Sickness of people due to fly ashes', Unstarred Question No-1946 answered in Rajya Sabha on 06 August 2015, available http://rajyasabha.nic.in/, accessed on 13 January 2016.

32. Ministry of Environment, Forests and Climate Change, 'Pollution caused by thermal power plants', Unstarred Question No-1941 answered in Rajya Sabha on 06 August 2015, available http://rajyasabha.nic.in/.accessed on 13 January 2016.

33. Ministry of Agriculture and Farmers Welfare, 'Failure of Crops Due to Unseasonal Rains and Monsoon Failure', Unstarred Question No. 2257 answered in Rajya Sabha on 18 December 2015, available http://rajyasabha.nic.in/. accessed on 14 January 2016.

34. World Health Organization, 'Fuel for life: Household Energy and Health' (2006): p. 8, available http://www.who.int/indoorair/publications/fuelforlife.pdf, accessed on 17 January 2016.

35. World Health Organization, 'Fuel for life', p. 16.

36. World Health Organization, 'Fuel for life', p. 18.

37. M.T. Toman and B. Jemelkova, 'Energy and Economic Development: An Assessment of the State of Knowledge', *The Energy Journal*, 24(IV) (2003):pp. 93–112 doi: 10.5547/ISSN0195-6574-EJ-Vol24-No4-5

38. MOSPI, 'Energy Sources of the Indian Households for Cooking and Lighting', p. 27.

39. MOSPI, 'Energy Sources of the Indian Households for Cooking and Lighting', p. 37.

40. D.F. Barnes, K.B. Fitzgerald, and H.M. Peskin, 'The Benefits of Rural Electrification in India: Implications for Education, Household Lighting, and Irrigation', World Bank, Washington, DC (2002), as in Toman with Jemelkova, p. 18.

41. World Bank and UNDP, 'Energy Sector Management Assistance Program, Annual Report' (2002), p. 2 available https://www.esmap.org/sites/esmap.org/files/AR_2002.pdf, accessed on 13 January 2016.

42. Desai, 'Households Assets and Amenities', p. 7.

43. S.R. Khandker, D.F. Barnes, and H.A. Samad, 'Energy Poverty in Rural and Urban India—Are the Energy Poor Also Income Poor?' (Policy Research

Working Paper 5463, The World Bank, Development Research Group, Agriculture and Rural Development Team, 2010), p. 8.

44. NITI Aayog, 'NITI Aayog launches India Energy Security Scenarios 2047—An Interactive Energy Platform', Press Information Bureau, Government of India (28 August 2015), available http://pib.nic.in/newsite/PrintRelease. aspx?relid=126412, accessed on 16 January 2016.

45. Report of the Expert Committee, Planning Commission, Government of India, 'Integrated Energy Policy' (2006), p. 54, available http://planningcommission.nic.in/reports/genrep/rep_intengy.pdf, accessed on 25 August 2016.

46. The Energy and Resources Institute (TERI), 'India's Energy Security: New Opportunities for a Sustainable Future'. (2009), p. 6, available http://www.teriin.org/events/CoP16/India_Energy_Security.pdf, accessed on 10 January 2016.

47. International Energy Agency, World Energy Outlook Series 'India Energy Outlook' (2015), p. 11, available www.worldenergyoutlook.org/india/, accessed on 12 January 2016.

48. Government of India, 'India's Intended Nationally Determined Contribution: Working Towards Climate Justice', p. 38.

49. TERI, 'National Energy Map for India: Technology Vision 2030', p. 7.

50. BAU scenario assumes that with normal rainfall and an investment rate of 27.6 per cent of the GDP, the GDP is projected to grow at a rate of 6.1 per cent over the modelling time frame.

51. TERI, 'National Energy Map for India: Technology Vision 2030', p. 184.

52. Planning Commission, 'Integrated Energy Policy', p. 21.

53. McKinsey and Company, 'Environmental and Energy Sustainability: An Approach for India' (2009), p. 11, available http://www.mckinsey.com/business-functions/sustainability-and-resource-productivity/our-insights/environmental-and-energy-sustainability-an-approach-for-india, accessed on 10 January 2016.

54. Government of India, 'India's Intended Nationally Determined Contribution: Working Towards Climate Justice', pp. 3–4.

55. TERI, 'National Energy Map for India: Technology Vision 2030', p. 5.

56. United Nations Economic and Social Council, Official Records, 2001, Supplement No. 9 [Commission on Sustainable Development (2001) Report of the Ninth Session (5 May 2000 and 16–27 April 2001)], available http://www. un.org/documents/ecosoc/docs/2001/e2001-29.pdf, accessed on 25 August 2016.

57. Government of India, 'India's Intended Nationally Determined Contribution: Working Towards Climate Justice', p. 31.

58. V. Thomas, 'Green Safeguards Yield Higher Economic Returns', *The Hindu*, 6 January 2016, p. 11.

59. The article refers to the whole system as incorporating the complete chain from energy production to waste disposal. This paper, however, has not addressed the issue of waste disposal, though it is acknowledged that it is an extremely pertinent issue which requires serious appraisal from an energy justice perspective.

60. Kirsten Jenkins, Darren McCaulay, Raphael Heffron, and Hannes Stephan., 'Energy Justice, a Whole Systems Approach', *Queens Political Review*, 2(II) (February 2014): pp. 74–87, esp. 74, available https://queenspoliticalreview.files.wordpress.com/2014/10/article-5-energy-justice-a-whole-systmes-approach-p74-87.pdf, accessed on 25 August 2016.

61. Jenkins, et al., 'Energy Justice, a Whole Systems Approach', p. 75.

62. C. Clark, 'Environmental Justice and Energy Production: Coal-Fired Power Plants in Illinois', p. 6.

63. People of Color Environmental Leadership Summit, 'Principles of Environment Justice' (1991), available www.ejnet.org/ej/principles.html, accessed on 25 August 2016.

64. L. Guruswamy, 'Energy Justice and Sustainable Development', *Colorado Journal of International Law and Policy*, 21(II) (Spring 2010), pp. 231–76, esp. 260, available http://www.colorado.edu/law/sites/default/files/Vol.21.2.pdf, accessed on 25 August 2016.

65. R.J. Heffron, Darren McCauley, and Benjamin Sovacool, 'Resolving Society's Energy Trilemma through the Energy Justice Metric', *Energy Policy*, 87 (September 2015), pp. 168–76, 169, doi: 10.1016/j.enpol.2015.08.033.

66. M. Wolsink, 'Fair Distribution of Power Generating Capacity: Justice in Microgrids Utilizing the Common Pool of Renewable Energy', in K. Bickerstaff, G Walker, and H Bulkeley (eds), *Energy in a Changing Climate: Social Equity and Low-carbon Energy* (London, New York: Zed Books, 2013), p. 119.

67. Wolsink, 'Fair Distribution of Power Generating Capacity: Justice in Microgrids', pp. 117–19.

68. Planning Commission, 'Integrated Energy Policy', p. 21.

69. Social Statistics Division, Central Statistics Office, Ministry of Statistics and Program Implementation (MOSPI), Government of India, 'Statistics Related to Climate Change—India', 2015, available http://mospi.nic.in/sites/default/files/publication_reports/climateChangeStat2015.pdf, accessed on 14 January 2016.

70. MOSPI, 'Statistics Related to Climate Change—India', p. 84.

71. MOSPI, 'Statistics Related to Climate Change—India', p. 171.

72. MOSPI, 'Statistics Related to Climate Change—India', p. 274.

73. MOSPI, 'Statistics Related to Climate Change—India', p. 275.

74. MOSPI, 'Statistics Related to Climate Change—India', p. 271.

75. MOSPI, 'Statistics Related to Climate Change—India', p. 270.

76. MOSPI, 'Statistics Related to Climate Change—India', p. 268.

77. P.S. Jha, 'Going Solar', *Indian Express* (25 November 2015), p. 11.

78. K. Chandrashekharan, 'India to add 2000MW Solar Power Capacity in Jan-Mar', *Economic Times* (15 January 2016), p. 17.

79. Department of Economic Affairs, Ministry of Finance, Government of India, 'Mid Year Economic Analysis 2015–16' (November 2015), p. 37, available http://www.finmin.nic.in/reports/MYR201516English.pdf, accessed on 13 January 2016.

80. 'Industry Skeptical about Proposed 5 Ultra Mega Power Plants', *Business Standard* (10 March 2015), available http://www.business-standard.com/article/news-ians/industry-sceptical-about-proposed-5-ultra-mega-power-plants-115031000667_1.html, accessed on 25 August 2016.

81. World Coal Association, 'The Case for Coal: India's Energy Trilemma' (2015), p. 3, available https://www.worldcoal.org/sites/default/files/WCA%20 report%20-%20India's%20Energy%20Trilemma.pdf, accessed on 25 August 2016.

82. A. Sasi, 'Green Push Needs a Balancing Act', *Indian Express* (14 October 2015), p. 21.

83. Ministry of New and Renewable Energy, Government of India, 'Draft National Renewable Energy Act, 2015' (2015), available http://mnre.gov.in/file-manager/UserFiles/draft-rea-2015.pdf, accessed on 25 August 2016.

84. Ministry of Power, Government of India, 'Rajiv Gandhi Grameen Vidyuktikaran Yojana', available http://powermin.nic.in/sites/default/files/uploads/Guidelines_for_Village_Electrification_DDG_under_RGGVY_0.pdf, accessed on 13 January 2016.

85. Ministry of Power, Government of India, 'Amendments to RGGVY Guidelines', amended on 1 April 2013, available http://www.ddugjy.gov.in/mis/portal/DDG/Amendment-to-DDG-Guidelines-dated-01.04.2013.pdf, accessed on 15 January 2016.

86. Ministry of Power, Government of India, 'Deen Dayal Upadhyay Gram Jyoti Yojana', available http://www.ddugjy.gov.in/mis/portal/memo/DDUGJY_Guidelines.pdf, accessed on 13 January 2016.

87. More information, though not updated, about the various schemes are available from the Ministry's website at http://mnre.gov.in/schemes/decentralized-systems/, accessed on 15 January 2016.

88. Dr R. Singh, 'Editorial', *Infra Live*, 2(X) (February 2016), p. 3.

89. See for instance, the public notice issued by Ministry of Power in November 2015 inviting comments from public on New Electricity Policy. The first issue sought for review is 'Integration of Intermittent type of renewable energy source to the Grid'.

90. Press Information Bureau, Government of India, 'UDAY (Ujwal DISCOM Assurance Yojana) for financial turnaround of Power Distribution

Companies' (5 November 2015), available http://pib.nic.in/newsite/PrintRelease.aspx?relid=130261, accessed on 14 January 2016.

91. Government of India, 'India's Intended Nationally Determined Contribution: Working Towards Climate Justice', p. 27.

92. S. Das, 'IREDA to Start Selling Rs 1,716 crore Tax-free Bonds', *ET Bureau* (4 January 2016), available http://articles.economictimes.indiatimes.com/2016-01-04/news/69511901_1_taxfree-bonds-tax-free-bonds-ireda, accessed on 25 August 2016.

93. Government of India, 'India's Intended Nationally Determined Contribution: Working Towards Climate Justice', p. 27.

94. S. Jha and Y. Bhargava, 'Don't Change the Goalpost Every Day, Automakers Tell Government', *The Hindu* (15 January 2016), p. 16.

95. V. Mehta, Executive Chairman, Brookings India, 'End of the Oil Age', *Indian Express* (4 January 2016), p. 13.

96. S. Srivastava, 'Pilot Project: E-vehicles may Soon become the Official Ride of the Government', *Indian Express* (11 January 2016), p. 11.

97. The Ministry of Petroleum and Natural Gas launched the 'Give It Up' campaign to encourage people to voluntarily give up LPG subsidies so that it can be directed towards the truly needy people. More can be read about the scheme on www.giveitup.in, accessed on 25 August 2016.

98. NITI Aayog, Government of India, 'Report on India's Renewable Electricity Roadmap 2030: Towards Accelerated Renewable Electricity Deployment', Executive Summary (February 2015), available http://niti.gov.in/writereaddata/files/document_publication/Report_on_India's_RE_Roadmap_2030-full_report-web.pdf, accessed on 25 August 2016.

99. 'New Power Tariff Policy to Boost Investment in Clean Energy', *Firstpost* (21 January 2016), available http://www.firstpost.com/business/centres-nod-to-new-power-tariff-policy-to-boost-investment-clean-energy-2592150.html.

100. M.A. Delucchi and M.Z. Jacobson, 'Providing All Global Energy with Wind, Water and Solar Power Part II: Reliability, System and Transmission Costs and Policies', *Energy Policy*, 39 (2011), pp. 1170–90, doi:10.1016/j.enpol.2010.11.045.

101. More on this can be read from the official website of the Ministry of New and Renewable Energy at www.mnre.gov.in/schemes/offgrid/waste-to-energy/, accessed on 25 August 2016.

102. Presidential Documents, 'Federal Actions to Address Environmental Justice in Minority Populations and Low-Income Populations', *Executive Order 12898*, 59(XXXII) (11 February 1994), available http://portal.hud.gov/hudportal/HUD?src=/program_offices/fair_housing_equal_opp/FHLaws/EXO12898, accessed on 17 January 2016.

About the Editor and Contributors

Editor

Usha Tandon is professor and head at Campus Law Centre, University of Delhi, India. Her areas of specialization and interest include environmental law, human rights of women, and population law. She has successfully completed several national and international research projects including her recent contribution as a national expert for the World Bank's Report, 2017 'Enabling the Business of Agriculture: Access to Water'. She has presented her research work at various international conferences in USA, France, UK, Netherlands, Germany, South Korea, Singapore, China, Malaysia, Nepal, among others. She has authored and edited five books and published around fifty research articles in journals of repute in India and abroad.

Contributors

Abdul Haseeb Ansari is professor at the International Islamic University, Malaysia (IIUM). He is also deputy dean (postgraduate and research) Ahmad Ibrahim Faculty of Laws and a member of the Council of Professors of the university. He has special interest in environmental law, revenue law, international trade law, and comparative jurisprudence and has widely contributed to the knowledge by publishing over 100 articles and six books. He has membership of some of the international institutions of high repute and bagged many awards for his research work including outstanding paper award of the Emerald Literati Award, UK and maximum citation award from the IIUM.

Upendra Baxi is an emeritus professor of law, University of Delhi, India and University of Warwick, England. He was the vice chancellor, University of South Gujrat (1982–5) and the University of Delhi (1990–4). His areas of special expertise include, comparative constitutionalism, social theory of human rights, human rights responsibilities in corporate governance and business conduct, and materiality of globalization. He has taught various courses at Universities of Sydney, Duke University, the American University, the New York University Law School Global Law Program, and the University of Toronto. He is the recipient of Padma Shri, the fourth highest civilian award in India. His most recent work is the fourth edition of *The Future of Human Rights* and *Very Short Introduction to Human Rights* (Oxford University Press; forthcoming).

Neelakshi Bhadauria is a law graduate from the Jindal Global Law School, OP Jindal Global University, Sonipat, India. She works as an associate at the law firm Cyril Amarchand Mangaldas.

Klaus Bosselmann is professor of environmental law and founding director of the New Zealand Centre for Environmental Law at the University of Auckland, New Zealand. He provided consultancy for the OECD, UN, EU, and the governments of Germany and New Zealand, and was a delegate at the Earth Summits in Rio de Janeiro (1992 and 2012). He is an executive member of several professional organizations and chair of the IUCN World Commission on Environmental Law Ethics Specialist Group. He has authored or edited 30 books and over 100 articles and contributing chapters. For his pioneering work on ecological law he received numerous awards including the Inaugural Senior Scholarship Prize of the IUCN Academy of Environmental Law, the global body of environmental law scholars.

V. Chandralekha teaches law at SDM Law College Mangalore, Karnataka, India. Currently she is pursuing her PhD from the University of Mysore in the area of 'Energy Law and Policy in India—A Human Rights Approach'. She has participated and presented papers at international- and national-level seminars and co-authored an article about Real Estate Laws.

Pranay Chitale is a final year student at Jindal Global Law School, Sonipat, India. He has worked on several articles with Professor Armin Rosencranz.

Neeraj Kumar Gupta is pursuing his PhD at Faculty of Law, University of Delhi, India as a Junior Research Fellow on the topic—'Eco-tax for Sustainable Development in India; Prospects and Challenges'. His areas of interest include environmental law, constitutional law, jurisprudence, and intellectual property rights. He has participated in various conferences and seminars.

C.M. Jariwala is professor and dean (academics), National Law University, Lucknow, India. He is a member of several academic bodies under the Government of Uttar Pradesh, Government of India, and University Grants Commission, India. He specializes in constitutional law and environmental law and has published two books and 72 articles in national and international journals mainly dealing with constitutional law and environmental law. He had been head and dean, Law School, Banaras Hindu University, India and also professor and dean (academics), National Law University, Bhopal, India.

Akash Kumar is associate advocate at TRA Law. He has obtained his LLM (Business Laws) from National Law School of India University, Bengaluru and BBA, LLB (IPR Hons.) from National Law University, Odisha, India.

Surender Kumar is professor at the Department of Economics, Delhi School of Economics and Director (Acting), Agricultural Economics Research Centre, University of Delhi, India. He had been a visiting fellow at the University of Illinois at Urbana-Champaign, USA and Senior JSPS Fellow at the Yokohama National University Yokohama, Japan. Kumar was one of the lead authors for IPCC AR5, Working Group II. He is also a Government of India nominee at the 19th Council (2015–19) of the Institute of Cost Accountants of India. He has been working with a concentrated focus on environmental and resource economics and climate change and energy economics. He also extends his ambit of research to productivity and efficiency measurement and applied econometrics. Kumar has authored five books and about 50 papers in reputed journals.

Pushpa Kumar Lakshmanan teaches environmental law and policy at Nalanda University, India. As a Fulbright scholar, he pursued post-doctoral research at Harvard Law School, Harvard University, USA. He is a research professor (hon), at World Institute of Scientific Exploration, Baltimore, USA. He is an interdisciplinary scholar specializing in international environmental law, primarily on the convention on biological diversity, climate change, global environmental governance, and sustainable development. His current focus lies in comparative study of environmental and biodiversity laws of South Asian and ASEAN nations. He is the founding convenor of Green Fulbrighters Forum, India.

Amar Roopanand Mahadew teaches environmental law, public international law, and human rights law at Department of Law, Faculty of Law and Management, University of Mauritius. He holds research interest in the areas of human rights, environmental issues, and other contemporary legal issues. He holds an LLM in Human Rights and Democratization from the Centre for Human Rights, University of Pretoria, South Africa. He is pursuing his PhD at the University of Western Cape, Cape Town, South Africa.

Erimma Gloria Orie teaches environmental law, oil and gas law, and commercial law at the National Open University of Nigeria. She is the head of department, Private and Property Law of the university and also represents the Faculty of Law on the Board of PG School of the university. Her PhD research work is on climate change. She has published several peer-reviewed articles. She is a qualified Mediator and a member of several professional bodies like the Nigerian Bar Association, Maritime Law Association, Chartered Institute of Arbitrators of Nigeria, and the World Commission on Environmental Law (IUCN).

Nikita Pattajoshi teaches at Tamil Nadu National Law School, Tiruchirapalli. Her areas of interest include intellectual property rights, environmental law, and constitutional law.

Armin Rosencranz, AB Princeton and J.D., PhD Stanford, is professor of law at Jindal Global University, Sonipat, India. He taught for many years as a consulting professor at Stanford, where he is a former trustee. As a lawyer and political scientist, he has taught a variety of environmental policy courses at Stanford, and has received a number of teaching awards. Until 1996, Professor Rosencranz headed Pacific Environment, an international NGO that he founded in 1987. He has been co-editor of *Climate Change Science and Policy* (2010) and taught a course on energy and climate at Stanford. He is co-author of 'Environmental Law and Policy in India' (Oxford University Press; third edition forthcoming).

Shivanandy Shetty is a partner at ERM India, Gurugram, India, and his previous stint involved the position of director at the SGS India, Gurugram. He has been awarded PhD by the University of Delhi on the topic, 'Impact of Voluntary Environment Programs on Environmental and Economic Performance of Firms in India'. He has published papers in reputed journals like *Environmental Economics and Policy Studies*, *Economic and Political Weekly*, and so on. He has over 26 years of experience in the field of Environment and Sustainability and has worked on several projects in India and across the world.

Pratibha Tandon is research associate at National Law University, Delhi (NLUD), India. Her areas of interest include corporate laws, international law, environmental law, criminal law, and human rights. She has written on diverse fields and got her articles published in reputed journals such as *Journal of Constitutional and Parliamentary Studies* and *National Journal of Comparative Law*.

Sanjay Upadhyay is advocate, Supreme Court of India, and managing partner of the India's first environmental law firm, Enviro Legal Defence Firm. He is also an India Visiting Fellow at the Boalt Hall School of Law, University of California, Berkeley and a legal intern to the Earth Justice Legal Defence Fund, San Francisco. He started his professional career at the WWF, India. He has served as an environmental and development law expert to international, multilateral, national, and state institutions including World Bank, IUCN, FAO, UNDP, ILO, and so on. He has been part of drafting committees of several forest, wildlife, and biodiversity related legislations both at the national and state levels.

Maansi Verma works as a legislative and policy researcher and assists members of parliament in their legislative duties. She graduated in law from Faculty of Law, University of Delhi, and has been a Young India Fellow (2014–15) and a LAMP Fellow (2015–16). She has keen interest in exploring the possibilities of participatory law and policy making with the aid of technology and has founded *Maadhyam*, a digital medium connecting policy stakeholders with policy makers.

Rajnish Wadehra studied India's energy and development policy choices as an honorary visiting fellow at Observer Research Foundation for a year in 2014–15. He currently writes on India's transition from a planned to a market economy with a focus on energy, while simultaneously pursuing his masters degree (full-time) in public policy at Jindal School of Government and Public Policy, Haryana, India.

M. Afzal Wani is professor of law and dean, University School of Law and Legal Studies, GGS Indraprastha University, Delhi, India, with more than 30 years of experience in teaching and research. He has authored many books and research papers on the subjects of his interest which include peace, development, conventional and constitutionalism, rights of women, protection of children, rights of senior citizens, consumer protection, and Islamic jurisprudence. He was member (par-time) of 19th Law Commission of India (2009–12) and is a member—Delhi Legal Services Authority, for the last ten years. With expertise in comparative islamic and modern jurisprudence, he coordinated programmes under the Rule of Law Project of USAID. He also coordinated UGC sponsored refresher courses for law teachers at Indian Law Institute, Delhi, India.

Index